Multinationals, Clusters and Innovation

Multinationals, Clusters and Innovation

Does Public Policy Matter?

Edited by

Ana Teresa Tavares

and

Aurora Teixeira

palgrave macmillan

First published 2006 by
PALGRAVE MACMILLAN
Houndmills, Basingstoke, Hampshire RG21 6XS and
175 Fifth Avenue, New York, N.Y. 10010
Companies and representatives throughout the world

PALGRAVE MACMILLAN is the global academic imprint of the Palgrave Macmillan division of St. Martin's Press, LLC and of Palgrave Macmillan Ltd. Macmillan® is a registered trademark in the United States, United Kingdom and other countries. Palgrave is a registered trademark in the European Union and other countries.

ISBN-13: 978–1–4039–9725–8 hardback
ISBN-10: 1–4039–9725–X hardback

This book is printed on paper suitable for recycling and made from fully managed and sustained forest sources.

A catalogue record for this book is available from the British Library.

Library of Congress Cataloging-in-Publication Data

Multinationals, clusters, and innovation : does public policy matter / edited by Ana Teresa Tavares and Aurora Teixeira.
p.cm.
Includes bibliographical references and index.
ISBN 1–4039–9725–X (cloth)
1. International business enterprises – Government policy. 2. Industrial clusters – Government policy. 3. Technological innovations – Government policy. I. Tavares, Ana Teresa. II. Teixeira, Aurora.

HD2755.5.M8443 2006
338.8′8—dc22 2006043217

10 9 8 7 6 5 4 3 2 1
15 14 13 12 11 10 09 08 07 06

Printed and bound in Great Britain by
Antony Rowe Ltd, Chippenham and Eastbourne

To Frederick
To João, little João and Sara

Contents

List of Figures

List of Tables

Notes on Contributors

Ulf Andersson is Professor of International Business at the Department of Business Studies, Uppsala University. His primary research interests concern strategic management of multinational corporations, knowledge development and sharing in multinational corporations, subsidiary development and subsidiary influence. He has published in *Strategic Management Journal, Organization Studies, International Business Review, Management International Review* and several other international journals. He has published book chapters in a number of anthologies, in conference proceedings, and has also acted as editor.

Christian Bellak is Associate Professor at the Department of Economics, Vienna University of Economics. His research interests are industrial policy, multinational enterprises and foreign direct investment, particularly with regard to development and to taxation. He has published several articles on these topics in international journals and in edited books.

Teresa Morais da Costa has a B.Sc. in Engineering and Industrial Management and a M.Sc. in Engineering Policy and Management of Technology from Instituto Superior Técnico, Technical University of Lisbon. Her M.Sc. dissertation analysed the impact of foreign direct investment in the Portuguese automotive industry, focusing on the Autoeuropa case. She has also an MBA from ISCTE in Lisbon. Since 2003 she has been a project specialist at Intéli (a think tank based in Lisbon) working in the area of foreign direct investment.

Bernhard Dachs is a Researcher at the Department of Technology Policy, ARC systems research. His areas of expertise are the economics of innovation and technological change, in particular with regard to the internationalization of science and technology, and the analysis of national and international technology policy. He has been involved in a number of consultancy projects for Austrian as well as international organizations.

Jože P. Damijan is Assistant Professor at the University of Ljubljana, and Research Fellow at the Institute for Economic Research, Ljubljana. He is also affiliated to LICOS Centre for countries in transition at the KU Leuven, Belgium. He did his Ph.D. studies at the CA Universitat Kiel, Germany, and received a Ph.D. degree from the University of Ljubljana, Faculty of Economics. His main topics of research consist of international economics and trade policy, economic geography, foreign direct investment and factors enhancing firm productivity growth. He has published recently in *World*

Economy, Weltwirtschaftliches Archiv, World Development, Economic Systems, Economic and Social Review, Eastern European Economics, Economic and Business Review, Prague Economic Papers and contributed to many monographs. He is currently a Deputy President of the Council of Economic Advisers to the Prime Minister of Slovenia.

Bernd Ebersberger is currently Senior Researcher in the department of Innovation Systems and Policy at Fraunhofer ISI in Karlsruhe, Germany. Previously he worked for the University of Augsburg and for the Technical Research Center of Finland (VTT) Espoo, Finland. His main areas of activity and publication include all aspects of the economics of innovation; in particular: comparative research on corporate and public innovation strategies, internationalization of industrial R&D and public research, evaluation of research and technology policy schemes; new schemes in innovation policy, dynamics of national innovation systems.

Manuel P. Ferreira is Assistant Professor at the Escola Superior de Tecnologia e Gestão, Instituto Politécnico de Leiria, Portugal. He obtained his Ph.D. in 2005 at the David Eccles School of Business (University of Utah), in strategy and international management. Among his research interests are examining entrepreneurial and strategic issues in industry clusters, as well as the inter-firm competitive dynamics and knowledge flows among clustered firms. His research has also focused on studying capabilities-based strategies of MNEs, cross-border acquisitions and network effects in transition economies. In addition to several presentations at leading international business and management conferences, he has published in various Portuguese and international journals.

Paulo Negreiros Figueiredo is Professor of technological learning and industrial innovation in the Brazilian School of Public and Business Administration (EBAPE), Getulio Vargas Foundation (FGV). He holds a Ph.D. in technology and innovation management from SPRU – Science and Technology Policy Research, University of Sussex, UK. In 1999 he created – and still heads – the Research Program on Technological Learning and Industrial Innovation Management in Brazil at EBAPE FGV. He is the author of *Technological Learning and Competitive Performance*, and his research has been published in various academic journals including *Research Policy, Technovation, International Journal of Technology Management, Industrial and Corporate Change* and *Oxford Development Studies*.

Sourafel Girma is a Reader in Industrial Economics at the University of Nottingham Business School. He is an applied microeconometrician with research interests that include the econometric evaluation of public policies, the effects of globalization on the performance of firms and the analysis of mergers and acquisitions. He has published widely in economic journals and consulted policy makers on several occasions.

Axèle Giroud is Senior Lecturer in International Business at Bradford University School of Management. Her previous work was in France in the University of Paris 1, Panthéon-Sorbonne. Axèle specializes in Asia and multinational firms' activities in the region, and she has conducted four major research projects in this part of the world. She has taken part in several international seminars and has published articles and book chapters as well as completing several research reports for major international organizations. Recently she has published in *Transnationals, Technology and Economic Development*, and *Multinationals and Asia: Organizational and Institutional Relationships*.

Anne Marie Gleeson graduated from the University of Limerick in 1996 with a B.B.S., majoring in Accounting and Economics. In 2000 she obtained an M.Econ.Sc. from University College Dublin. She has been lecturing in economics and quantitative studies at Waterford Institute of Technology since 1998 and is currently on funded research leave at Trinity College, Dublin, where she is undertaking a Ph.D. in Economics. Her areas of interest include FDI and SME linkages, human capital and international trade issues.

Yundan Gong is a doctoral student at the University of Nottingham Business School. She is currently researching the effects of globalization on firm level convergence in China. Before starting her postgraduate studies, Yundan worked in the Chinese banking sector.

Holger Görg is a Reader in International Economics in the School of Economics at the University of Nottingham and a Research Fellow at the Leverhulme Centre for Research on Globalization and Economic Policy (GEP) and IZA Bonn. He joined Nottingham in 2000 as a post-doctoral researcher in GEP. Before joining the School he worked as a Lecturer in Economics at the University of Ulster at Jordanstown and University College Cork. He completed a Ph.D. in Economics in 1999 at Trinity College, Dublin where he worked on the impact of foreign direct investment on the Irish economy. His research interests are in empirical international trade and industrial organization focusing in particular on the activities of multinational companies, foreign direct investment and international outsourcing.

William Hesterly is the Zeke Dumke Professor of Management at the David Eccles School of Business, University of Utah. His research on organizational economics, vertical integration, organizational forms and entrepreneurial networks has appeared in several top journals, such as the *Academy of Management Review, Organization Science, Strategic Management Journal, Journal of Management, Business History* and *Journal of Economic Behavior and Organization*. He is the co-author, with Jay Barney, of *Strategic Management and Competitive Advantage*. Currently, he is studying the sources of value creation in firms and what determines who captures the value from a firm's competitive advantage. Professor Hesterly serves on the editorial board of

Strategic Organization and has previously served on the boards of *Organization Science* and the *Journal of Management*. He has also been consultant to various Fortune 500 firms in the electronic, office equipment, paper, telecommunications, energy, aerospace, and medical equipment industries, and to smaller firms in several other industries.

Simona Iammarino is a Senior Lecturer at the Science and Technology Policy Research (SPRU), University of Sussex, UK, and Lecturer (on leave) at the University of Rome 'La Sapienza'. Her main research interests lie in the following areas: multinational corporations and the location of technological activities; regional systems of innovation and socio-economic differentials in Europe; geographical and industrial clusters; and theory, empirics and policy implications. Her affiliations include, among others, the Centro Interdipartimentale di Economia Internazionale (CIDEI), University of Rome 'La Sapienza', the Institute of International Affairs (IAI), Rome, the Centre for International Business and Management (CIBAM), Judge Institute of Management Studies, Cambridge, and the Centre for Institutional Performance, University of Reading (UK).

Andreja Jaklič is Assistant Professor at the Faculty of Social Sciences, University of Ljubljana. Her teaching areas are international economic relations, international economics and economics. Her main research topics include internationalization and foreign direct investment, especially outward foreign direct investment, firm growth and multinational companies in Central and Eastern Europe and transition economies. In recent projects her research has broadened to include information and communication technology and innovation. She has co-authored the monograph (with M. Svetličič) *Enhanced Transition through Outward Internationalization: Outward FDI by Slovenian Firms*, as well as contributing to other monographs. Her articles are published in *Transnational Corporations, Eastern European Economics, Economics and Business Review, Services Industries Journal* and *Journal of East European Management Studies*.

Markus Leibrecht is Assistant Professor at the Department of Economics, Vienna University of Economics. He was Staff Economist of the Econometric Modelling Group, Economic Analysis Division of the Austrian National Bank and Research and Teaching Assistant at the Department of Public Economics, Vienna University of Economics. His research interests are tax policy, foreign direct investment and applied econometrics. He has published articles on these topics in both journals and edited books.

Katrin Männik has an M.A. in Economics and is a Ph.D. student involved in research and teaching at the Faculty of Economics and Business Administration, University of Tartu, and Research and Study Fellow at SPRU (University of Sussex) during 2003–05. Her specific field of research concerns the international knowledge development and transfer via foreign direct investments, subsidiary autonomy, dynamic capabilities and innovation policies in catching-up economies. Her research publications include articles in working paper series, journals and in edited books. Previously, she has also been a policy expert in technology and innovation policy planning in Estonia.

Marina Papanastassiou is Professor at the Copenhagen Business School. She gained her Ph.D. from the University of Reading. Marina has recently published in *R&D Management, Applied Economics* and *Management International Review*. She was the president of the European International Business Academy in 2002.

Robert Pearce is Reader in International Business at Reading University Business School. His recent research has focused on roles of subsidiaries and of R&D in multinational enterprises. A current interest is to relate this understanding of multinationals' strategy to the economic issue raised by globalization. He has published widely in international journals. His most recent books are *Multinationals, Technology and National Competitiveness* (with Marina Papanastassiou) and *Multinationals and Transition* (with Julia Manea).

Torben Pedersen, Ph.D., is Professor of International Business at the Copenhagen Business School's Center for Strategic Management and Globalization. He has published over 50 articles and books concerning the managerial and strategic aspects of multinational companies. His research has appeared in publications such as the *Strategic Management Journal*, the *Journal of International Business Studies* and the *International Review of Law and Economics*. He recently published a managerial oriented book entitled *Managing Global Offshoring Strategies: A Case Approach*. Together with Professor Nicolai J. Foss, he received the AIB Best Paper Award in 2003.

Magnus Persson is a doctoral student at the Department of Business Studies, Uppsala University. His research interests concern the effectiveness and efficiency of knowledge transfer in MNCs. He has published chapters in anthologies and participated in several international conferences with competitive papers. He was one of the finalists of the Haymes Prize at Academy of International Business in 2004.

Ewen Peters is a Strategy Consultant and Visiting Senior Research Fellow with SIBU at the University of Strathclyde in Glasgow. His current research interests lie in innovation and the sources of competitive advantage for international businesses and local economies in the twenty-first century knowledge economy. He has published several contributions in book chapters and articles in international journals.

Matija Rojec is Associate Professor at Faculty of Social Sciences, University of Ljubljana and Adviser to the Government at the Institute of Macroeconomic Analysis and Development. His areas of expertise are foreign direct investment (especially focusing on Central and Eastern Europe), transition economics (mainly regarding company restructuring), European integration (especially the enlargement process) and enterprise restructuring. He has published articles in journals such as *World Development, Transnational Corporations, Management International Review, Eastern European Economics, Post Communist Economies, Economic Systems, Industry and Innovation, International Relations and Development, Journal of East–West Business, Prague Economic Papers,* among others. He is also a member of the editorial boards of several academic journals and has contributed to and edited monographs published by vanous publishers.

Frances Ruane is Professor of Economics at Trinity College, Dublin. Apart from Trinity College, she has worked at the Irish Industrial Development Authority and Central Bank, and at Queen's University, Kingston, Ontario. Her postgraduate studies were undertaken at the University of Oxford (Nuffield College). She has published widely on theoretical and empirical aspects of foreign direct investment. She has also written extensively on policy issues relevant to Irish and European economic development. She is currently a member of the editorial boards of the *Journal of International and Economic Policy* and the *International Review of Economics and Finance.*

Francesca Sanna-Randaccio is Professor of Economics at the Department of Systems and Computer Sciences of the University of Rome 'La Sapienza'. She studied at the University of Rome, Johns Hopkins University (M.A. in International Relations) and Oxford University (M.Litt. in Economics). She has been a visiting scholar at the University of California at Berkeley, Duke University, the London School of Economics and visiting professor at the University of Toulouse and the Free University of Bozen. She has published a book and several articles in the fields of international economics, industrial organization and the economics of innovation. In recent years her research has focused on the interaction between firms' multinational expansion and innovative strategy, R&D internationalization, the impact of FDI on host and home countries and the effect of national and multilateral FDI policies.

Maria Savona is a post-doctoral fellow at BETA, Bureau d'Economie Théorique et Appliquée (UMR CNRS 7522), Université Louis Pasteur in Strasbourg, France. She holds a Ph.D. in Science and Technology Policy from SPRU, University of Sussex, UK. Her main research interests lie in the following areas: technological innovation and its economic impact in services: theory and measurement issues; structural change and sectoral patterns of economic growth; MNEs and international re-location of production.

Julie Sutherland holds a Bachelor of Economics (1988) from the University of Newcastle, Australia and a Graduate Diploma and Masters in Development Economics (1995) from the Australian National University. She graduated with a Diploma in Statistics (1999) and a Ph.D. in Economics

(2004) from Trinity College, Dublin. Her research interests include the industrial development process in Ireland in the context of the performance of local and multinational enterprises in the Irish manufacturing sector, with an emphasis on the productivity and export ability of firms. She is a Research Associate of the Institute of International Integration Studies, Trinity College and spends a significant amount of time each year in Ireland undertaking empirical research on industrial development.

Ana Teresa Tavares is Assistant Professor of international economics at the University of Porto (CEMPRE, Faculty of Economics). She has, and had, several Visiting Research Fellow/Visiting Professor positions in international universities. Her areas of specialization include multinational subsidiaries' strategies and evolution, the impact of multinationals on host economies, and policies vis-à-vis multinationals. She has published several articles on these topics in journals such as *Regional Studies, International Business Review, Transnational Corporations* and *Journal of Industry, Competition and Trade*, among others. She is also author of several book chapters and is preparing two books on themes related to multinational enterprises. She is Associate Editor (Strategy/International Business) of the *International Journal of Management Reviews*. She has been a consultant to various institutions on regional development and on public policy towards FDI.

Aurora A.C. Teixeira is Assistant Professor in economics of innovation at FEP (Faculty of Economics, University of Porto) and research fellow at CEMPRE. She did her Ph.D studies at SPRU (Science and Technology Policy Research), and received a Ph.D. from the University of Sussex (UK). Her recent research has focused on the role of human capital and R&D stocks in countries' long-term economic growth, and the relevance of international R&D alliances for regional and clustering dynamics. Her most recent works have been published in *Organisational Transformation and Social Change, Portuguese Economic Journal, Portuguese Journal of Social Science*, and *Brazilian Innovation Review*. She is the author of a book on Portuguese long-term economic growth published by the Portuguese Economic and Social Council and is a Council member of the European Association for Evolutionary Political Economy (EAEPE).

Nick von Tunzelmann has been a Reader (1984–97) and Professor (1997–2005) of the Economics of Science and Technology at SPRU (Science and Technology Policy Research), University of Sussex. In 2005 he was appointed R.M. Phillips Professor of Science & Technology Policy. Since 2001 he has also acted as Director of Research at SPRU. Before he went to SPRU he had been a lecturer in Economic History at the University of Cambridge and a Fellow of St John's College, Cambridge (1970–84). His doctorate is from Oxford University (Nuffield College). His current main interests lie in interlinking micro, meso and macro levels of analysis and appropriate policy integration.

Francisco Veloso is Assistant Professor at the Department of Engineering and Public Policy in Carnegie Mellon University. He also has an appointment with the Faculty of Business and Economics of Universidade Católica Portuguesa. His work focuses on how firms and regions acquire and organize technological capabilities. He has published several papers in international peered-review journals and worked with various international organizations in these areas. Francisco Veloso holds a Ph.D. in Technology, Management and Policy from the Massachusetts Institute of Technology, an M.Sc. in Management of Technology from ISEG and a Diploma in Physics Engineering from IST, both schools of the Technical University of Lisbon.

Alexandre Videira is currently an adviser on the Staff of the Secretary of State for Industry and Innovation, Portugal. He was Senior Project Manager at Intéli – Intelligence for Innovation, where he developed several initiatives in the area of Technology and Innovation Management. His research interests are technology and innovation management, organizational learning, industrial policy and innovation policy. He holds a M.Sc. in Engineering Policy and Management of Technology at IST (Instituto Superior Técnico) – Technical University of Lisbon, and a B.Sc. in Chemical Engineering from IST.

Celeste Amorim Varum is Assistant Professor at the University of Aveiro, Director of the Economics Programme. Previously she worked at the University of Minho, ISAG and IESF (Portugal) and at the University of Reading (UK). She taught Microeconomics, Industrial Organization, Innovation and Macroeconomics. She has provided consultancy and occupied administrative functions at public institutions and private firms. She holds a Ph.D. in Economics (Diffusion of Innovations in countries in the Periphery–Core Transition), 2002, and an M.Sc. in Economic Integration and International Business, both from the University of Reading, UK, and an Economics Degree from the University of Évora (Portugal), 1995. Her current research interests are innovation, business services and multinationals. She has published and presented studies on these topics in several international journals and at conferences.

Stephen Young is a Professor in the School of Business and Management at the University of Glasgow. Together with Neil Young he founded SIBU (Strathclyde International Business Unit) in 1983. His interests are in international business and particularly the interface between management, economic development and public policy. He has published articles in leading international journals, and various books and monographs. His most recent books include *Multinationals and Public Policy* and *The New Economic Analysis of Multinationals: An Agenda for Management, Policy and Research* (with S. Guisinger and T. Brewer). He has been a Visiting Professor in several international universities and also served on the boards of a variety of organizations.

Preface

This is a remarkable book for several reasons. In the first place is the range of contributions – from all parts of Europe, from Asia and from Latin America. In a book that deals with globalization and foreign direct investment (FDI), this is of course what the reader would hope for but it is actually still quite rare to find such a widespread coverage.

Secondly, although the geographic and political range of the contributions is very wide, the editors and the contributors have avoided the temptation to make the book simply a quantitative study of scale and trends in foreign direct investment in various parts of the world. They have concentrated instead on qualitative analysis of the nature of this FDI and its relationship with host countries. This again is what most readers would hope to find as there are other sources for the global statistics.

Thirdly, a very valuable feature of the book is its attention to the relationships between the inward investment and host-country institutions, especially of course the already established firms and industries. It is frequently assumed that FDI is simply a one-way flow of new capital, new equipment, skills and new technology to the benefit of the importing host country. Many chapters show that this is by no means always the case. There is a two-way relationship in which the incoming firm also benefits from the knowledge and techniques already available in the host country.

This means that analysis of the networks developed around the incoming firms in the host countries is by no means a simple business and the study of their innovations in particular is not a simple matter. For this reason a particularly welcome feature of the book lies in the fact that most of the authors participate both in business studies research and teaching and in the more specialized innovation studies characteristic of evolutionary economics and technology policy research organizations. The confluence of these two streams of research lends much greater realism to their work.

Finally, it is obvious that this is a very rapidly changing feature of the world economy and that any studies can soon become out of date. Almost all of these chapters deal with the twenty-first-century developments although they often also have a longer historical perspective. For this reason it is to be hoped that this book is soon circulating widely in business schools, in universities, in business and government, as we all have much to learn from it.

Chris Freeman
Emeritus Professor of Science and Technology Policy
University of Sussex

1
Introduction

Aurora Teixeira and Ana Teresa Tavares

There is hardly any doubt that multinational enterprises (MNEs), clusters, and innovation are three major engines of growth and development. Moreover, they are often related, though in various ways, and with different degrees of strength.

Multinational firms are sought after (literally chased) by nearly all countries nowadays. The forces of globalization, the growing unemployment pressure in many regions in the world (especially in developed countries) and the need for fast technological renewal stimulate this authentic 'race' (Oxelheim and Ghauri, 2003) for foreign direct investment (FDI). This state of affairs is also made possible given the lack of a minimum multilateral regulatory framework as regards MNE- and FDI-related policies (Young and Tavares, 2004).

One of the key issues in order to thrive in this demanding and often hostile environment, as well as to take into account the fact that most FDI is sequential, is that attention should shift from merely attracting FDI (the 'quantity' aspect) to enhancing the quality of such inward investment. Both the quantity and the quality dimensions are analysed in this volume. Nonetheless, as most of the literature until quite recently emphasized FDI attraction, neglecting to a considerable extent qualitative issues, a deep look into the factors that make an investment distinctive as regards its quality is a central objective of this book.

Hence, topics such as innovation, linkages and industrial agglomeration are highlighted. These three aspects are undoubtedly extremely important from both a host economy and from a MNE subsidiary viewpoint (and also potentially from the host economy and the headquarters perspective), as innovative firms, developing strong linkages with the local industrial fabric, and benefiting from being situated in vibrant and resourceful clusters are likely to have greater survival probability, and an enhanced position in the group. It is also much more promising from a host economy standpoint, given the potential for spillovers that innovation, linkages and agglomeration enhance (Tavares and Young, 2005).

The literature on subsidiary strategies and evolution (White and Poynter, 1984; Birkinshaw, 2000), and on the MNE as a differentiated network (Hedlund, 1986; Forsgren and Johanson, 1992), highlights exactly the different impact of MNEs' activities according to their role and evolutionary path in the host economy and intra-group (Pearce, 2001). This view is adopted in several of the chapters here.

This volume is divided into five parts (see Table 1.1 for a summary). The first one deals primarily with *Multinationals* and *Innovation*, analysing themes such as the relevance and magnitude of knowledge spillovers between MNEs and domestic firms, and host-country environment and MNEs. Emphasizing the latter, less focused, direction Dachs and Ebersberger examine how MNEs draw on knowledge from their host economies, i.e. the spillovers from host country to affiliate. Using data from the most recent Community Innovation Survey (CIS) relative to Austria, where MNEs maintain a strong presence in science and technology, the authors scrutinize the differences between Foreign Owned (FO) and Domestic Owned (DO) companies as regards their propensity to enter into co-operative arrangements, and their evaluation of various information sources.

The importance of knowledge spillovers is also noted in Chapter 3. Damijan, Jaklič and Rojec investigate internal and external sources of innovation and evaluate their impact on the productivity growth of a large sample of Slovenian firms (in the period 1996–2002). The authors test whether, and to what extent, firms' ability to innovate is induced by their own R&D activity, and by factors external to the firm, and what are the most important channels of external knowledge spillovers that contribute to productivity growth.

In Chapter 4, focusing on a rather under-explored source of FDI spillovers (product innovation) and on an under-researched type of companies (state-owned enterprises – SOEs), Girma, Gong and Görg assess the potential role foreign MNEs can have in the process of enhancing innovative performance. At the forefront of this chapter, the authors provide an overview of the outstanding recent performance of China in terms of inward investment attraction, and of the success of the proactive (economic liberalization) policies Chinese policy-makers adopted in order to attract FDI. Subsequently, and employing alternative econometric strategies to a sample of 30 000 Chinese SOEs for the period 1999 to 2002, Girma et al. assess whether R&D intensity influences the rate and incidence of new product innovation. While this chapter carries the optimistic message that FDI acts as a channel of product innovation externalities to Chinese SOEs, it also underscores the importance of getting the firm-level fundamentals right.

Iammarino, Sanna-Randaccio and Savona (Ch. 5) provide fresh insights on the locational and innovative choices of foreign MNEs relative to those of nationally-owned firms, and the integration between MNE innovative activities and territorial systems in a country – Italy – with a relatively weak

Table 1.1 Overview of the book's main issues

Chapters/authors	Countries	Industries	Number of firms	Period of study	Main issues
Chapter 2 – Dachs and Ebersberger	Austria	All	1287	1998–2000	
Chapter 3 – Damijan, Jaklič and Rojec	Slovenia	All		1996–2002	
Chapter 4 – Girma, Gong and Görg	China	All (State owned firms)	30 000	8313	Multinationals and innovation
Chapter 5 – Iammarino, Sanna-Randaccio and Savona	Italy	All	15 512	1998–2000	
Chapter 6 – Ferreira, Tavares and Hesterly	Portugal	Molds	Cluster*	1940–2004	
Chapter 7 – Gleeson, Ruane and Sutherland	Ireland	Manufacturing	Census data of all plants with ≥3 employees	1985–2001	Clusters and industrial development
Chapter 8 – Peters and Young	Ireland, Scotland	Biotechnology and life-sciences	2	1969–2004; 1997–2004	
Chapter 9 – Pedersen	Austria, Denmark, Finland, Germany, Norway, Sweden, UK	All	2107	2000	Linkages between multinationals and local firms
Chapter 10 – Andersson and Persson	Europe, North America	Manufacturing	97	2000s	
Chapter 11 – Giroud	Malaysia	Electrical and electronics	120	1996; 2001	
Chapter 12 – Figueiredo	Brazil	Electrical and electronics; motorcycle and bicycle	46	1999–2002	Innovation and linkages
Chapter 13 – Costa, Videira and Veloso	Portugal	Automotive	1	1999–2004	
Chapter 14 – Varum	Portugal, Spain	Automotive	4	1990s	
Chapter 15 – Männik and von Tunzelmann	Estonia, Hungary, Poland, Slovakia, Slovenia (CEEs)	All	433	1998–2002	
Chapter 16 – Bellak and Leibrecht		–	235 observations**	1996–2002	Policy issues
Chapter 17 – Pearce and Papanastassiou	–	–	–	–	

Note: * Currently comprises about 200 firms employing 5000 workers.

** Panel data involving 7 home-countries (i.e., Austria, Germany, France, Italy, The Netherlands, United Kingdom and the United States of America) and the CEEC-5 (i.e., Czech Republic, Hungary, Poland, Slovak Republic and Slovenia), from 1996 to 2002.

degree of both active multinationality and attractiveness towards foreign firms. Based on a similar data source (CIS) as Dachs and Ebersberger (Chapter 2), Iammarino et al. try to assess whether the perception of the importance of obstacles to innovation vary among type of firms and regions and whether such obstacles influence firms' locational choices and the potential level of innovative activity. Additionally, the locational patterns of foreign MNEs' innovative activities in Italy is researched and related to their obstacles' perception. Controlling for a series of regressors, among which firm size, sectoral specificities, ownership type, geographical location, and innovativeness, in a sample of 15 512 firms, Iammarino et al. reveal that the evaluation of obstacles is a relevant symptom of a higher awareness of innovative firms of the problems encountered when carrying out innovation activities.

The second part of the book is devoted to the issue of *Clusters* and *Industrial Development*. It encompasses a broad and relevant range of topics both conceptually focused (Chapter 6) and empirically driven (Chapters 7 and 8). Building on the idea that only by observing the genesis of the ties between firms one is better able to understand the formation and evolution of the inter-firm organizational forms that preside in clusters, thus enabling to understand how clusters develop, in Chapter 6 Ferreira, Tavares and Hesterly provide a rich conceptual account of the evolution of industry clusters through spin-offs highlighting the role of flagship firms. The authors posit that some clusters may emerge and develop more endogenously than usually considered in the literature, through the gestation of new firms. By focusing on entrepreneurial issues at the genesis of at least some clusters, Ferreira et al. contribute primarily to research on the evolution of clusters as a function of entrepreneurial activity by entrepreneurs that are insiders to the cluster, and also to the literature on the value of networks to support entrepreneurial dynamism and success. Moreover, they also formulate some insights on the possible role of 'flagship' firms, specifically MNEs in originating clusters, further complementing extant clusters' research. This chapter presents a 'motherhood model' offering a rather positive perspective of firms' spin-offs focused essentially on employees exiting because there is a promising entrepreneurial opportunity, often complementary to the parent's activity. Although ample empirical evidence in different countries and industries exists testifying that this motherhood process is a realistic model explaining the emergence and development of several clusters, in order to illustrate such perspective the authors examined the particular case of the plastic molds cluster of Marinha Grande, Portugal – an international leader in various niches of the industry.

Although the case of Ireland has been producing a considerable amount of studies related to FDI, proactive policies to promote spatial and sectoral clustering, the success of such policies is little analysed or understood. Moreover, the association of spatial and sectoral MNE clustering and the concentration of local (Irish-owned) enterprises have not yet been measured. In Chapter 7, Gleeson, Ruane and Sutherland compare the spatial and sectoral concentration of manufacturing enterprises in the Irish economy in 1985 and 2001 in order to establish the impact of industrial and regional policy objectives on the clustering of enterprises during this period. Using entropy indices, the pattern of spatial changes found suggests that market forces were already driving firms out of more concentrated locations prior to the introduction of policies to promote greater spatial dispersion in the late 1990s. MNEs have become more sectorally specialized over the period, which is not surprising as policy is deliberately selective in attracting MNEs to key high tech manufacturing sectors.

In Chapter 8 Peters and Young debate whether it is advisable or feasible for peripheral areas – namely, Ireland and Scotland – to target the biotechnology sector, where financial demands and risks (but also potential returns) are high. The example of Ireland highlights the challenges for public policy in taking a position on the structure of the biotech sector, and seeking to nurture a flagship firm. The authors emphasize that there are advantages related to this approach that can be integrated closely with a cluster model of development, but the risks of picking winners are high. In Scotland, while the number of players suggests a buoyant, innovative industry, most are small start-ups. The Scottish approach is termed a 'nursery-model', where an environment of early-stage development revolving around university science is encouraged. The high costs and risks of progressing along the value chain suggest high failure rates for such firms and a high propensity for mergers and acquisitions. The evidence from this chapter is that it is indeed advisable for peripheral areas to target the biotech sector, especially where the basic preconditions exist, notably, a strong science-base; but the critical issue is still the feasibility of this strategy. In particular, and according to the authors, policy-makers have not resolved the fundamental challenge of how to promote the successful emergence and development of significant international businesses of scale, i.e. in filling the gaps in the local milieu.

Local milieu development is intimately related to *linkages* among firms, specifically between *Multinationals* and *Local Firms*. This topic is detailed in Part III of the book. Evidence is gathered (Pedersen, Chapter 9), based on a large-sample data set of more than 2100 subsidiaries located in seven different European countries, and shows that in fact local dynamism in the business environment is an important source for the upgrading of subsidiary competences. This chapter undertakes the first statistical large scale test of Birkinshaw and Hood's (1998) seminal paper on subsidiary evolution, which proposed a model suggesting that such evolution is determined by headquarter assignment, dynamism of local business environment and subsidiary initiatives. The effect of these three determining factors on subsidiary development is tested simultaneously.

Following the same line of the previous chapter, in Chapter 10 Andersson and Persson argue that knowledge-creating activities in subsidiaries are linked to their local environment. They stress, however, that the environment provides opportunities for knowledge acquisition and creation, but it does not induce knowledge development per se. These authors examine how linkages (relationships) between the subsidiary and the local environment influence MNE *re*-investments in host-country subsidiaries. Such investigation is likely to contribute to our current understanding of MNE investment behavior: despite evidence showing that re-investments are a major source of resource inflows into nations, research examining MNE re-investments is relatively sparse. Based on a cross-industry, cross-country data set, from 20 business areas for 97 subsidiaries (92 located in Europe, five in North America), face-to-face interviews with managers at both HQ and subsidiary levels, and 514 relationships to customers and suppliers viewed as important, Andersson and Persson show indeed that the local environment represents an important source of knowledge for the subsidiary and, more specifically, that close relationships with network partners may enhance the importance of the subsidiary within the MNE.

Another type of (however complementary) linkages – supply relationships and supply networks – is featured by Giroud (Chapter 11). The central debate in this chapter is about the background and specificities of multinationals' supply relations in the Malaysian electrical and electronics industry, assessing the success of government policies implemented to encourage supply linkages by MNEs. Complementing other FDI policies Giroud suggests that the latest Malaysian Master Plan emphasized the need for inter-firm linkages and cluster development. Following this policy stance soft incentives have been launched to try and encourage MNE-local supplier linkages. More specifically, the Malaysian government tried to foster linkages through its Industrial Linkages Program (ILP), encouraging (through tax incentives) large companies to undertake actions to enhance the quality of local vendors. Small local vendors also receive incentives when they take part in the program. Based on data gathered through a direct survey of 95 foreign affiliates in 1996 and 25 of these firms in 2001–02, Giroud evaluates the dynamics of local purchasing in this period.

Further research on *Innovations* and *Linkages* is presented in Part IV, including some appealing case studies: local firms and MNEs' subsidiaries located in a developing area of a large late-industrializing country, Brazil (Chapter 12); a large 'cluster-generating' MNE subsidiary from the automobile sector located in Portugal (Chapter 13); and buyer–supplier linkages in the automotive sector in Portugal and Spain (Chapter 14).

In Chapter 12, Figueiredo examines the extent to which globalization of innovative technological capabilities has been spreading to a developing area – the Industrial Pole of Manaus (Northern Brazil) – by investigating the extent to which firms' capabilities have progressed from basic operations levels into innovative levels, not only within electro-electronics firms, but also in two-wheel vehicles' firms and key suppliers. Additionally, he assesses the association between technological capabilities and export performance. Figueiredo argues against over-generalizations and assumptions related to the non-globalization of innovative capabilities, the general deterioration of industrial technological capabilities, and low competitiveness as a result of a change into an outward-looking pattern of industrial development.

It is hardly debatable that MNEs can be credited with positive contributions to economic development if they are able to build a robust level of backward linkages. Aiming at a more profound understanding of how domestic value creation by foreign MNEs increases the potential for creating wealth in the region, Costa, Videira and Veloso (Chapter 13) propose an innovative 'combined model', which extends traditional input–output (I–O) value estimations by coupling it with a detailed account of the structure of existing and potential domestic purchases and an engineering-based cost estimation method, called System Cost Modeling (SCM). Using Autoeuropa (AE) (Ford–VW project) as case study, Costa et al. analyse what the impact of this project would have been under different circumstances and explore alternative value generation configurations. The application of the SCM/I–O combined method to the AE investment shows that variations in the level of incorporation of domestic supplies mean dramatic differences in the value generated by a given investment project. The magnitude of these differences endorses the view that certain density levels in the set of backward linkages might be needed to propel the industry to new levels of value creation.

Once more, using the automobile industry as a framework of analysis, Varum (Chapter 14) examines buyer–supplier (B–S) linkages as a key arrangement inducing innovation and skills' development among domestic Portuguese and Spanish firms. Instead of looking at the transfer of technology, she looks at the development of capabilities at the organizational and managerial levels, and skills of workers of domestic firms, issues overlooked in the literature. Varum argues that B–S linkages motivated and forced domestic suppliers to develop their practices and capabilities. Specifically, buyers acted as a catalyst in promoting learning among domestic suppliers by setting and enforcing strict performance standards.

The final part of the book includes three studies mainly focused on policy-related themes. The first two chapters provide evidence on Central and Eastern Europe (CEE) countries whereas the last chapter (Chapter 17) is general in scope, framing multinationals within National Systems of Innovation, and discussing corresponding strategies and policies.

Männik and von Tunzelmann's study (Chapter 15) is concerned with the interaction between autonomy at the national/regional/local level, and autonomy between MNE parent companies and their subsidiaries. It analyses in detail the role of subsidiaries of foreign MNEs in five CEE countries (Estonia, Hungary, Poland, Slovakia and Slovenia) at the 'functional' level.

Their empirical work uses a database drawn from a questionnaire distributed in those five CEE countries covering 433 firms spread over a range of manufacturing sectors. The authors address multi-dimensionality in subsidiary autonomy, encompassing technology, marketing, managerial and financing dimensions of autonomy. The results support the hypothesis of the heterogeneity of autonomy. The authors highlight a non-linear relation between subsidiary autonomy and performance, stressing that excessive dependence on the parent company might impede the development of subsidiary's absorptive capacity while excessive independence might leave the local unit in a circle of 'internationally uncompetitive' knowledge.

During the past few years lowering corporate tax rates has been regarded as a key policy instrument to attract FDI of foreign MNEs in several countries, including the 'CEEC-5' (i.e. Czech Republic, Hungary, Poland, Slovak Republic and Slovenia). Tax rates tend to be an attractive policy tool for policy-makers because they can easily be changed and are thought to affect the behavior of economic agents immediately. In Chapter 16, Bellak and Leibrecht investigate whether there is a significant empirical relationship between corporate taxation and FDI. Their empirical study, which is based on a panel data set, includes the bilateral effective tax rates instead of the statutory tax rate and as controls mainly variables which intend to capture the differences in location (L) advantages of the CEEC-5. The authors focus on FDI from the main home countries (i.e., Austria, Germany, France, Italy, The Netherlands, United Kingdom and the United States of America) to the CEEC-5, because the latter are in the center of the ongoing public debate within the EU about an increase of (harmful) tax competition. The time span considered here ranges from 1996 to 2002. Using this panel of seven home countries, five host countries, and seven years, Bellak and Leibrecht confirm the importance of the tax rate as a FDI determinant.

The need to understand the qualitative contribution of MNEs' participation in the host economy, rather than the mere quantitative magnitude of FDI attraction, is emphasized by Pearce and Papanastassiou in Chapter 17. The authors outline an idealized approach to a Global Innovation Strategy (GIS) by a contemporary MNE, implemented through two stages each of which is articulated around a different type of R&D laboratory. In a first (pre-competitive) stage, the GIS is positioned around a network of Internationally Interdependent Laboratories (IIL) and seeks to contribute to achieving the first aim (or benefit) of globalizing innovation activity. In a second stage, the GIS is implemented through product mandate (PM) subsidiaries and associated Locally Integrated Laboratories (LIL). According to Pearce and Papanastassiou, the persistence of PM/LIL units often depends (externally) on the parent MNE group's ability to generate new innovation potentials, and (internally) on the host country's ability to support those aspects of its NSI that enable the subsidiary to continue to attract elements of the group's creative programs.

After such stimulating and encompassing contributions, we felt we need to set the record straight and ask whether public policy matters. This comes from a preoccupation with the fact that the vast majority of the literature, although extremely interesting and relevant, tends to fall short of suggesting practical and clear avenues for specific policy intervention. The term 'intervention' should not be misunderstood: here it is meant in a benign and constructive way, allowing market forces to flourish, but helping especially where information asymmetry, market failures and liabilities of newness and smallness (of, for instance, local and recently set up firms) prevent better developments. We think that public policy matters (when well planned in a systemic way, selective, targeted and coordinated), and will explain why (in Chapter 18), based on the topics focused on in this volume. Our perspective is positive, believing that there is always something to improve in order to foster growth, development and prosperity.

References

Birkinshaw, J., *Entrepreneurship in the Global Firm* (London: Sage, 2000).

Birkinshaw, J. and Hood, N., 'Multinational subsidiary evolution: capability and charter change in foreign-owned subsidiary companies', *Academy of Management Review*, 23(4) (1998) 773–95.

Forsgren, M. and Johanson, J., *Managing Networks in International Business* (Philadelphia: Gordon and Breach, 1992).

Hedlund, G., 'The hypermodern MNC: a heterarchy?', *Human Resource Management*, 25(1) (1986) 9–35.

Oxelheim, L. and Ghauri, P., *European Union and the Race for Foreign Direct Investment in Europe* (Oxford: Elsevier, 2003).

Pearce, R., 'Multinationals and industrialisation: the bases of "inward investment" policy', *International Journal of the Economics of Business*, 8(1) (2001) 51–73.

Tavares, A.T. and Young, S., 'FDI and multinationals: patterns, impact and policies', *International Journal of the Economics of Business*, 12(1) (2005) 3–16.

White, R. and Poynter, T., 'Strategies for foreign-owned subsidiaries in Canada', *Business Quarterly* (Summer) (1984) 59–69.

Young, S. and Tavares, A.T., 'Multilateral rules on FDI: do we need them? Will we get them? A developing country perspective', *Transnational Corporations*, 13(1) (2004) 1–30.

Part I

Multinationals and Innovation

2
Knowledge Flows between Multinational Enterprises and National Innovation Systems: The Case of Austria[1]

Bernhard Dachs and Bernd Ebersberger

2.1 Introduction

The international business (IB) literature regards knowledge spillovers from foreign affiliates to domestic enterprises as one of the major benefits a host economy derives from the presence of multinational enterprises (MNEs) (Blomström and Kokko, 1998). This chapter deals with the opposite situation, and sets out to measure the extent to which innovative MNEs utilize the knowledge base of their host countries. We believe that spillovers from the host country to MNE affiliates are extremely relevant to science and technology policy. Given the fact that innovation is an interactive process that depends heavily on external sources (Edquist, 2005), spillovers may be one of the most important reasons for MNEs to undertake innovative activities at any particular location.

As a consequence, if innovative MNE affiliates mainly exchange knowledge within the group to which they belong, and have only little or no exchange with the host country, we may conclude that MNEs are not dependent on local knowledge in the innovation process. MNE research, development and innovation activities are therefore extremely mobile, and can be moved elsewhere to benefit from cost advantages without losing spillovers from the host country knowledge base. On the other hand, if we find strong links with domestic enterprises and universities, it may be an indication that the innovative activities of MNEs are less 'footloose' than the critics of globalization assume.

We will examine here the degree to which MNE affiliates are 'embedded' in the knowledge base of their host countries in the case of Austria, which is one of the countries where MNEs maintain a strong presence in science and technology. In 2002 about a third of all gross expenditure for R&D in the Austrian business enterprise sector was funded by sources from abroad (OECD, 2005), and 40 per cent of all Austrian patent activities can be attributed to MNE affiliates (Dachs and Schibany, 2004).

The chapter is structured as follows: Section 2.2 presents some theoretical considerations and empirical results on knowledge spillovers and foreign enterprises. Section 2.3 presents our data source. The methodology and results of the empirical analysis are laid out in Section 2.4. Finally, Section 2.5 draws some conclusions from the results for theory as well as policy.

2.2 Theoretical background and research hypothesis

There are two quite distinct bodies of literature in economics dealing with the internationalization of corporate innovative activities. The first consists of the IB literature focusing on explaining foreign direct investment (FDI) and the existence of MNEs. The second is made up of studies in the neo-Schumpeterian tradition, which regard knowledge diffusion between various actors as a crucial element in explaining innovation and technological change.

Knowledge plays a key role in the microeconomic approach in the IB literature (see, for example, Markusen, 2002). FDI, exports, licensing etc. are seen as different ways of utilizing superior, company-specific assets such as technology, products, brands, or superior management skills. These assets (or 'knowledge capital' in Markusen's terminology) may be the result of innovative activity in the home country. Superior assets give MNEs an advantage over local companies, enabling them to enter foreign markets. According to this view MNEs locate R&D and design facilities abroad in order to adapt existing products and technologies to local needs, tastes and regulations.

When MNEs make use of these assets by setting up affiliates and producing and selling products outside their home countries, parts of their superior assets spill over to domestic enterprises.[2] Spillovers emanating to domestic enterprises from the affiliates of MNEs are assumed to be a major benefit for the host economies from FDI and have been widely recognized as such in IB literature (see Blomström and Kokko, 1998; and Keller, 2004, for a review). According to Blomström and Kokko (2003), these spillovers are the strongest argument for countries to try and attract inward investment.

In contrast to this view a number of studies suggest that MNE affiliates' activities abroad also add to the company's existing stock of knowledge instead of merely utilizing it. This strategy has been labeled 'strategic asset seeking' (Dunning and Narula, 1995). An important part of this strategy for enterprises is not only to absorb spillovers from clients and competitors, but also from universities and other research institutes in the host country. The scientific and technological specialization of the host country may therefore be an important incentive in deciding where to locate a MNE affiliate.

The competences of overseas MNEs' R&D facilities appear to have extended in recent decades as a consequence of this strategy. Apart from adjusting and customizing existing parent-company technologies, foreign affiliates increasingly create new technologies and products, or absorb new technological or market trends by acting as 'surveillance outposts' (Miller, 1994) or 'antennas' (Florida, 1997). A number of enterprises have even established 'technology centers' or 'global centers of excellence' to concentrate the company's research activities in a specific technology or product group. Decentralization, however, has its price, and includes some trade-offs, as described in a model by Sanna-Randaccio and Veugelers (2003). MNEs need to organize the transfer and distribution of knowledge internally throughout their organizations. Such flows of information may be two-way, incurring further cost and losses. Central R&D departments may lose some economies of scale in their activities, whilst decentralized R&D increases the possibility of knowledge 'leaking out' to local competitors. Moreover, decentralizing the innovative activities of MNEs also implies a higher degree of independence, enabling affiliates to establish external relationships with local companies and universities in their host countries (Zanfei, 2000).

The two types of spillovers discussed in the literature are summarized by Kuemmerle (1999), who distinguishes two basic ways in which foreign affiliates innovate and interact with their environments:

- 'Home base exploiting' (HBE), where foreign affiliates mostly exploit existing knowledge to support foreign production by doing minor development work in adjusting existing technologies and products. The knowledge relevant to the affiliate's innovative activities mainly originates from within the multinational group, from the parent company, or from affiliates in other countries. External links to companies, universities or public laboratories in the host country are of only minor importance to the affiliate's innovative achievements. The most important information sources contributing to the affiliate's innovative performance reside inside the MNE.
- 'Home base augmenting' (HBA), where foreign affiliates actively contribute to the stock of knowledge and the range of products of the group as centers of excellence. The knowledge relevant to the affiliate's innovative activities originates not only from within the group, but also from its environment in the host country. External links to companies, universities or public laboratories in the host country are therefore of much greater importance, and foreign and domestic enterprises participate in a more dynamic and vivid exchange than in HBE mode. We may also assume that direct links to various external information sources such

as universities or public research centers are more important than in HBE mode.

Our research question is straightforward: which one of these two modes is predominant? A comparable, or even higher level of knowledge exchange between foreign affiliates and their environments than Austrian enterprises would indicate HBA. If, on the other hand, HBE is more frequent, we will find a lower level of interaction.

Both the theoretical arguments presented above and empirical evidence suggest that we may expect to find evidence in favor of HBE. Previous studies with patent data (Patel and Vega, 1999; le Bas and Sierra, 2002) have shown that the HBE strategy is more common than HBA: in the majority of cases, MNEs locate their affiliates in host countries that are less specialized in the MNE's technological strengths. HBA (MNE and host country both specialized in a certain technology) is encountered rarely.

We will test for differences with respect to two specific transfer mechanisms: (1) *innovation co-operations*, where we differentiate between the location of the collaboration partners, and (2) the *information sources* the company uses for innovation, where it is necessary to differentiate between sources from within the corporate group and those from outside it. Both channels are well described in the literature (Schartinger et al., 2002; Tether, 2002; Caloghirou et al., 2003).

Innovation co-operations are formal agreements on a contractual basis that include not only knowledge exchange but also allow for the risks of innovation projects to be shared, or economies of scale in R&D to be exploited. In contrast to the concept of spillovers, where knowledge is only transferred one way, co-operative arrangements involve mutual learning and an exchange in both directions. Moreover, co-operative activities also help to build trust and create a common 'codebook' that facilitates the exchange of knowledge via 'tacit' channels. Since co-operative arrangements constitute a certain degree of commitment, they require a high degree of independence for the affiliate.

On the other hand, *information sources* relevant to the company's innovation activities are far less formal and encompass various channels of knowledge transfer, such as demonstrations, imitation and reverse engineering, or loose supplier–customer relationships. These may imply a less intense exchange but appear to be closer to the concept of spillovers used in the IB literature (Blomström and Kokko, 1998, p. 3), since we can assume that most information flows are not mutual, and do not include adequate financial compensation.

2.3 Data

This analysis draws on the results reported in the third and most recent wave of the Community Innovation Survey (CIS 3). CIS 3 is a survey based on a common questionnaire administered by EUROSTAT aimed at assessing various aspects of companies' innovative behavior and performance.

Foreign-owned companies (FOCs) are identified in the questionnaire by a question about the location of the head office if the enterprise is part of a group (EUROSTAT, 2004, p. 290). Our sample consists of 618 enterprises belonging to a group, and 669 which do not. Of the group members, 390 were domestically-owned companies (DOCs), and 228 were foreign-owned companies (FOCs). Affiliates of parent companies based in Germany or Switzerland were the largest group of FOCs.

In contrast to most extant empirical studies, where spillovers are measured indirectly by relating productivity changes in a domestic industry to the presence of FOCs, CIS data provides direct indicators of knowledge transfer. We look at the two potential channels for the diffusion of knowledge between foreign affiliates and domestic companies referred to above:

- *Innovation co-operation* (EUROSTAT, 2004, p. 295) is described in the questionnaire as 'active participation in joint innovation and R&D projects with other organizations (either enterprises or non-commercial organizations)'. Simply contracting out work where there is no active co-operation is not defined as co-operation in the survey. Innovation co-operation can be differentiated by partner (partner from the group, external companies, universities, research centers) and geographical scope (domestic partner or foreign partner).
- *Sources of information* (EUROSTAT, 2004, p. 296) used in the company's innovation activities are defined as *internal sources* (within the enterprise or other enterprises of the group), *market sources* (suppliers, customers, competitors, consultants or commercial R&D), *science sources* (universities and public laboratories) or *other sources* (conferences, exhibitions, etc.).

2.4 Empirical analysis

The research question essentially amounts to assessing the impact of foreign ownership on a company's co-operation and sourcing of innovation. However, FOCs and DOCs exhibit some structural differences, which make comparisons difficult.

Markusen (2002), for example, shows that only the most successful companies in a country become MNEs, while others remain domestic and export. In a third case, they operate solely in their home markets. Bellak (2004) finds evidence of a superior performance by FOCs with respect to wages, skills, labor relations, capital intensity, (labor) productivity, company growth and profitability. When we compare FOCs and DOCs, we therefore face a selection bias problem, since we are comparing one population with only the most successful portion of the other population.

To overcome this bias, we need to contrast the *actual* behavior of an FOC with the *counterfactual*, yet unobservable, behavior of the same company in the case of Austrian ownership. This section discusses how we deal with this missing data problem. Let us illustrate the problem by examining the following equation:

$$E(\Delta) = E(\pi_F) - E(\tilde{\pi}_F) \tag{2.1}$$

where π_F denotes the behavioral variable of the FOC, $\tilde{\pi}_F$ is the counterfactual behavior of the FOC in the case of domestic ownership. $E(\Delta)$ is the impact of foreign ownership, as it describes the difference in behavior between the company under foreign ownership compared to its behavior in the counterfactual situation of Austrian ownership. As $\tilde{\pi}_F$ is not observable, it must be estimated. Any estimation of $\tilde{\pi}_F$ should check for a potential selection bias, since foreign ownership and membership of a corporate group cannot be considered a random event. Essentially, this means that companies belonging to a corporate group differ from companies that do not, with respect to various indicators such as size and investment behavior. It also means that companies belonging part to a foreign-owned group are different from companies that are part of Austrian groups.

Table 2.1 shows that FOCs are significantly different from Austrian companies. FOCs are larger and invest a smaller proportion of their turnover than the DOCs in the sample. FOCs also tend to be more frequent in the high technology sector.

Table 2.1 Average characteristics by ownership

	Part of a corporate group			**Foreign ownership**		
Variables	**Yes**	**No**	**Sig.**	**Yes**	**No**	**Sig.**
EMPL	0.283	0.062	***	0.252	0.150	***
*EMPL*²	0.475	0.027	***	0.339	0.221	
INVEST	0.070	0.080		0.049	0.081	**
SEHT	0.129	0.060	***	0.184	0.074	***
SMELT	0.097	0.099		0.075	0.103	
SELT	0.084	0.227	***	0.079	0.176	***
SEKIS	0.325	0.049	***	0.110	0.197	***
SEOS	0.364	0.565	***	0.553	0.450	***
APPRCOND	1.248	1.166	***	1.211	1.204	
SEEXCHR	0.199	0.166	***	0.262	0.164	***
SPEED	0.410	0.465	***	0.466	0.433	**
N	618	669		228	1059	

Notes: (1) Equality of means is tested by a t-test assuming unequal variances. * (**, ***) indicates significance at the 5% (1%, 0.1%) level. (2) See text below for explanation of variables.

Source: own calculations.

Due to the selection bias, the counterfactual behavior $\tilde{\pi}_F$ of an FOC cannot be estimated from the outcome π_A of observations of DOCs as $E(\pi_F) \neq E(\pi_A)$. Rubin (1977) introduces the conditional independence assumption (CIA) to help overcome the selection problem. Briefly, in our context this assumption states that knowledge sourcing behavior is independent of company ownership as long as the companies compared share the same exogenous characteristics **x**. Assuming the validity of this assumption, it follows that $E(\tilde{\pi}_F, \mathbf{x}) = E(\pi_A, \mathbf{x})$. Hence, the counterfactual outcome can be estimated using the observed outcome of DOCs as long as both FOC and DOC share the same exogenous characteristics **x**. The average effect of ownership is

$$E(\Delta) = E(\pi_F, \mathbf{x}) - E(\pi_A, \mathbf{x}) \tag{2.2}$$

To estimate the counterfactual situation for an FOC, one has to balance the sample of DOCs in such a way as to resemble the sample of FOCs with respect to their characteristics **x**. The literature on the construction of matching samples contains several approaches to the construction of the counterfactual group. Supposing **x** contains only one variable, it would be intuitive to look for a DOC that has exactly the same value of **x** as the FOC. However, if the number of matching criteria were large, it would hardly be possible to find any such Austrian case – the curse of dimensionality. To overcome this difficulty, Rosenbaum and Rubin (1983) introduced propensity score matching. The idea is to collapse the number of matching criteria down to a low dimensional scale.

In our exercise, we use three dimensions to match the companies: (1) the probability of their belonging to a corporate group based on exogenous company characteristics captured by a propensity score measure (2) the probability of their being foreign-owned based on exogenous company characteristics, also captured by the propensity score,[3] and (3) the size of the company. The probability of belonging to a corporate group and the probability of being foreign-owned are estimated for the whole sample.

Table 2.2 displays the results of the Heckman probit regression taking account of the fact that, by definition, only companies that are part of a corporate group can be foreign-owned. The propensity scores used as matching criteria are computed from this Heckman probit regression, where we use several company-specific as well as sector-specific exogenous variables: *ORCH* as an indicator of organizational change within a company,[4] *EMPL* and $EMPL^2$ as size variables for the number of employees and the squared number of employees, and *INVEST* as a company's investment expenditure. The sector specific-variables are *SEHT, SEMLT, SEKIS* and *SEOS*, which refer to a company's sectoral affiliation, appropriability conditions (*APPRCOND*), innovation dynamics (*SPEED*) and the export orientation of the sector (*SECEXHSHR*), which refer to the two-digit NACE sectors. After matching the samples of FOCs and DOCs, they are then comparable

Table 2.2 Heckman probit regression of foreign ownership and belonging to a corporate group

Dependent variable	Foreign ownership			Part of a corporate group		
	Coef.	**Std. err.**	**Sig.**	**Coef.**	**Std. err.**	**Sig.**
ORCH	–	–		0.252	0.075	***
EMPL	0.381	0.227	*	2.178	0.239	***
EMPL^2	−0.063	0.046		−0.280	0.042	***
INVEST	−0.131	0.229		0.122	0.173	
SEHT	0.400	0.349		1.042	0.399	***
SEMLT	−0.083	0.221		0.450	0.157	***
SEKIS	0.957	0.272	***	2.046	0.187	***
SEOS	1.018	0.194	***	0.604	0.135	***
APPRCOND	0.242	0.236		0.586	0.195	***
SECEXSHR	1.845	0.276	***	1.138	0.200	***
SPEED	0.333	0.328		−0.763	0.260	***
CONST	−2.833	0.449	***	−1.975	0.332	***
Wald	81.74	***		LR	5.82	**

Note: * (**, ***) indicates significance at the 5%, (1%, 0.1%) level. The likelihood ratio test does not permit the correlation coefficient ρ to be zero at the 1% level.

with respect to probability (distribution) of belonging to a corporate group or being foreign-owned. They are also comparable with respect to size distribution.

In this analysis, we apply kernel-based matching[5] to estimate the counterfactual based on the whole sub-sample of Austrian companies. The counterfactual behavior for an FOC i is generated as a convex combination of the behavior of all DOCs observed.

$$\tilde{\pi}_i = \sum_{j \in I_A} \lambda_{ij}\pi_j \text{ with } \sum_{j \in I_A} \lambda_{ij} = 1 \text{ and } i \in I_F \tag{2.3}$$

where the estimated counterfactual of the observation i is denoted by $\tilde{\pi}_i$. The index set of foreign (Austrian) owned observations is denoted by $I_{F(A)}$.

The weights λ_{ij} of the convex combinations are determined by a kernel regression on the distances in matching criteria space between FOCs i and DOCs.

$$\lambda_{ij} = \frac{K(d_{ij})/h}{\sum_{k \in I_o} K(d_{ik})/h} \quad i \in I_F;\ j \in I_A \tag{2.4}$$

where $K(d_{ij})$ is a kernel function which monotonically decreases by d_{ij}. As d_{ij} is the distance between observation i and observation j, $K(d_{ij})$ downweights observations j that are further away from observation i. Due to the dimensionality $k > 1$ of $\mathbf{x}$, the Mahalanobis metric is used to measure d_{ij}.

This selection of matching criteria ensures that FOCs and DOCs are as likely to belong to a corporate group as they are to be foreign-owned. Furthermore, they are also of comparable size. In the kernel regression, we employ a Gaussian kernel and a bandwidth suggested by a variant of Silverman's rule of thumb (Silverman, 1986; Bergemann et al., 2001).

2.5 Results

Table 2.3 presents the results of our analysis of FOCs versus DOCs. The effect indicates whether foreign ownership affects their propensity to co-operate positively (+) or negatively (−). Rows in which we found no significant differences are marked with (.).

Foreign ownership does not affect the general propensity of a company to co-operate on innovation (Table 2.3, first row). Hence the effects observed further down the table are not caused by a generally higher likelihood of one of the two groups entering into co-operations for innovation. Group

Table 2.3 Effects of foreign ownership on co-operative behavior

		Effect			Effect
Co-operation in general	co	.			
Domestic co-operation	copdom	–	International co-operation	copglo	.
Dom vertical co-operation	coverd	–	Int vertical co-operation	coverg	.
Dom co-operation with customers	cocld	–	Int co-operation with customers	coclg	–
Dom co-operation with suppliers	cosd	–	Int co-operation with suppliers	cosg	.
Dom horizontal co-operation	cohord	.	Int horizontal co-operation	cohorg	.
Dom science co-operation	coscid	–	Int science co-operation	coscig	.
Dom co-operation with universities	cound	–	Int co-operation with universities	coung	.
Dom co-operation with gov. research	cogd	.	Int co-operation with gov. research	cogg	.
Dom co-operation with consultants	coctd	.	Int co-operation with consultants	coctg	.

Note: '.' indicates no significant effect. '+' indicates a positive effect due to public ownership, while '−' denotes a negative effect significant at the 10% level.

Source: own calculations.

ownership cannot therefore be the reason for the differences observed. Ownership is the only company-specific characteristic that differs in the two samples. We can therefore argue that the differences in collaborative behavior are caused by foreign ownership.

Although there were no differences in co-operation behavior in general, we observed distinct patterns of co-operation once we broke the co-operation partners down into Austrian and international co-operation partners.

Our analysis reveals a significantly lower propensity for FOCs to enter into co-operation agreements with *national* partners. Our observation that domestic co-operation, i.e. co-operation with national partners, is less frequent among FOCs is attributable to their reduced propensity to collaborate with vertical partners (clients, suppliers) and with national science institutions, and is not due to differences in horizontal co-operation or to co-operation with consultants. The fact that foreign ownership causes reduced co-operation with national science institutions stems from its impact on companies' likelihood to co-operate with Austrian universities for innovation. The likelihood of collaborating with Austrian governmental or non-profit research institutes appears to be largely unaffected by ownership. The impact of foreign ownership on vertical innovation co-operation is supported by a reduced propensity to collaborate with any type of national vertical partner.

With respect to the knowledge sources used for innovation, our results clearly show that FOCs focus more on internal than on external sources (Table 2.4). This is in line with the theory that MNEs possess superior knowledge which they utilize in foreign markets. Information sources from within a corporate group are used more, and more highly appreciated among FOCs than they are among their Austrian-owned counterparts. Foreign ownership has an adverse effect on the use of knowledge from external sources.

Table 2.4 Effects of foreign ownership on the evaluation of various knowledge sources

Sources of knowledge		Effect
Group internal knowledge sources	sgrpa	+
External sourcing from companies	sext	−
Sourcing from science	ssci	.
Sourcing from fairs, exhibitions	sfair	−
Internal vs. external sourcing	sintern	+

Note: '.' indicates no significant effect. '+' indicates a positive effect due to public ownership, while '−' denotes a negative effect significant at the 10% level.

Source: own calculations.

Table 2.5 Likelihood of foreign and domestic enterprises entering into co-operative arrangements

	FOC	DOC
All co-operations	0.307	0.355
Domestic co-operations	0.228	0.333
International co-operations	0.238	0.249

Source: own calculations.

Nevertheless, sourcing from science institutions is not significantly affected by foreign ownership, meaning that information from universities and research centers is equally appreciated by both FOCs and DOCs.

The observation that FOCs tend to collaborate less with domestic partners and tend to rely on internal rather than external sources for their information is convincing evidence that the Austrian subsidiaries of MNEs pursue a *home base exploiting* strategy. However, contacts between FOCs and their host-country environments are far from non-existent, and the companies do not pursue a pure home base exploiting strategy. We also observed a surprisingly high proportion of foreign affiliates entering into co-operative arrangements. The likelihood of a FOC to enter into any type of co-operation is 0.307, compared with 0.355 in the case of an Austrian affiliate (Table 2.5). The difference between both the groups is even smaller for international co-operations, but larger for domestic ones.

2.6 Conclusions

The results tend to confirm the 'home base exploiting' rather than the 'home base augmenting' view with respect to the intensity with which foreign-owned companies draw on knowledge bases in their home countries. Foreign-owned companies in Austria prefer to rely on MNE internal sources rather than on external knowledge absorbed by either formal co-operation or informal channels. These results are not surprising, since such behavior is predicted in the IB literature. MNEs have some natural disadvantages in foreign markets as compared to domestic companies, which may also include their ability to tap into local knowledge. Moreover, theoretically MNEs are likely to have less incentive to tap into local knowledge anyway, first of all because they may *already* possess knowledge superior to that accessible in the country and, secondly, because the transfer of such knowledge from the affiliate to the parent company is not cost-free.

However, the differences between foreign-owned and domestically-owned companies are small in some cases, and we also found a surprisingly high percentage of foreign affiliates tapping into local knowledge. This is the good news for policy from our exercise. As far as policy is concerned, the results suggest that foreign-owned firms frequently use local expertise. As a result, we conclude that the R&D activities of foreign companies in Austria are less mobile than some may have feared, simply because they rely to a certain extent on the local knowledge and expertise of their host country. Moreover, we can also predict that domestic organizations benefit from the presence of foreign affiliates, since co-operative arrangements usually imply the mutual exchange of knowledge, and intense contacts increase the likelihood of spillovers.

From this perspective, there are no grounds for fearing that foreign-owned companies could keep their know-how secret and/or restrict their affiliates' imports of know-how developed elsewhere, as cited by Sanna-Randaccio and Veugelers (2003, p. 18). By contrast, a lively exchange of knowledge between foreign enterprises and domestic and international partners appears to be taking place.

Governments which fear that ongoing internationalization may endanger the capabilities of their national innovation systems by resulting in the relocation of MNE R&D facilities to locations abroad may be well advised to increase the frequency and quality of their contacts with other organizations in the national innovation system. This may be achieved by providing incentives to enterprises to link up with universities, research centers or other enterprises in the country. In order to facilitate the exchange of knowledge and information the governments of a number of OECD countries have already taken action to strengthen such links, particularly between the public science system and industry (OECD, 2004). One of the side effects of such initiatives may be to embed foreign-owned companies more strongly.

Notes

1 The authors would like to thank Christian Bellak, Werner Hölzl, Hans Lööf and the participants of the workshop on the preparation of this book for their helpful comments, and Karl Messmann (Statistics Austria) for supplying the data.
2 Knowledge or technology spillovers are used in a very general sense in the IB literature and include both, market-mediated and non-mediated knowledge diffusion.
3 For the analysis of case 1 we only use the propensity score for the foreign ownership.
4 Organizational change is only used in the group membership equation of the Heckman probit.
5 Recently matching estimators have been applied and discussed, amongst others, by Heckman et al. (1998a; 1998b). Most recent contributions in the field of innovation studies include Almus and Czarnitzki (2003), Czarnitzki and Fier (2002; 2003) and Czarnitzki et al. (2004).

References

Almus, M. and Czarnitzki, D., 'The effects of public R&D subsidies on firms' innovation activities: the case of Eastern Germany', *Journal of Business and Economic Statistics*, 21(2) (2003) 226–36.

Bellak, C., 'How domestic and foreign firms differ and why does it matter?', *Journal of Economic Surveys*, 18(4) (2004) 483–514.

Bergemann, A., Fitzenberger, B. and Speckesser, S., 'Evaluating the employment effects of public sector sponsored training in East Germany: conditional difference-in-differences and Ashenfelter's dip', (Mannheim: University of Mannheim, 2001).

Blomström, M. and Kokko, A., 'Multinational corporations and spillovers', *Journal of Economic Surveys*, 12(3) (1998) 247–77.

Blomström, M. and Kokko, A., 'The economics of foreign direct investment incentives', *NBER Working Paper Series* No. 9489 (Cambridge, MA: NBER, 2003).

Caloghirou, Y., Ioannides, S. and Vonortas, N.S., 'Research joint ventures', *Journal of Economic Surveys*, 17(4) (2003) 541–70.

Czarnitzki, D. and Fier, A., 'Do innovation subsidies crowd out private investment? Evidence from the German service sector', *Applied Economics Quarterly*, 48(1) (2002) 1–25.

Czarnitzki, D. and Fier, A., 'Publicly funded R&D collaborations and patent outcome in Germany', *ZEW Discussion Paper*, Vol. 03-24 (Mannheim: ZEW, 2003).

Czarnitzki, D., Ebersberger, B. and Fier, A., 'The relationship between R&D collaboration, subsidies and patenting activity: empirical evidence from Finland and Germany', *ZEW Discussion Paper*, Vol. 04-37 (Mannheim: ZEW, 2004).

Dachs, B. and Schibany, A., 'The internationalisation of innovative activities in Austria measured by patent data', *Plattform fteval Newsletter*, 22 (August) (2004).

Dunning, J.H. and Narula, R., 'The R&D activities of foreign firms in the United States', *International Studies of Management & Organization*, 25(1–2) (1995) 39–72.

Edquist, C., 'Systems of innovation: perspectives and challenges', in Fagerberg, J., Mowery, D. and Nelson, R.R. (eds), *The Oxford Handbook of Innovation* (Oxford: Oxford University Press, 2005).

EUROSTAT, *Innovation in Europe. Results for the EU, Iceland and Norway* (Luxembourg: Eurostat, 2004).

Florida, R., 'The globalization of R&D: results of a survey of foreign-affiliated R&D laboratories in the USA', *Research Policy*, 26(1) (1997) 85–103.

Heckman, J., Ichimura, H. and Smith, J., 'Matching as an econometric evaluation estimator', *Review of Economic Studies*, 65(2) (1998a) 261–94.

Heckman, J., Ichimura, H., Smith, J. and Todd, P., 'Characterizing selection bias using experimental data', *Econometrics*, 66 (1998b) 1017–98.

Keller, W., 'International technology diffusion', *Journal of Economic Literature*, 42(3) (2004) 752–82.

Kuemmerle, W., 'Foreign direct investment in industrial research in the pharmaceutical and electronics industries – results from a survey of multinational firms', *Research Policy*, 28(2–3) (1999) 179–93.

Le Bas, C. and Sierra, C., ' "Location versus home country advantages" in R&D activities: some further results on multinationals' locational strategies', *Research Policy*, 31(4) (2002) 589–609.

Markusen, J.R., *Multinational Firms and the Theory of International Trade* (Cambridge, MA: MIT Press, 2002).

Miller, R., 'Global R&D networks and large-scale innovations: The case of the automobile industry', *Research Policy*, 23(1) (1994) 27–46.

OECD, *Science, Technology and Industry Outlook* (Paris: Organization for Economic Cooperation and Development, 2004).

OECD, *Main Science and Technology Indicators* (Vol. 2005/1) (Paris: Organization for Economic Cooperation and Development, 2005).

Patel, P. and Vega, M., 'Patterns of internationalisation of corporate technology: location vs. home country advantages', *Research Policy*, 28(2–3) (1999) 145–55.

Rosenbaum, P.R. and Rubin, D.B., 'The central role of the propensity score in observational studies for causal effects', *Biometrika*, 70 (1983) 41–55.

Rubin, D.B., 'Assignment to treatment group on the basis of covariate', *Journal of Educational Statistics*, 2 (1977) 1–26.

Sanna-Randaccio, F. and Veugelers, R., 'Global innovation strategies of MNEs: implications for host economies', in Cantwell, J. and Molero, J. (eds), *Multinational Enterprises, Innovative Strategies and Systems of Innovation* (Cheltenham: Edward Elgar, 2003).

Schartinger, D., Rammer, C., Fischer, M. and Fröhlich, J., 'Knowledge interactions between universities and industry in Austria: sectoral patterns and determinants', *Research Policy*, 31(3) (2002) 303–28.

Silverman, B.W., *Density Estimation* (London: Chapman and Hall, 1986).

Tether, B.S., 'Who co-operates for innovation, and why: an empirical analysis', *Research Policy*, 31(6) (2002) 947–67.

Zanfei, A., 'Transnational firms and the changing organisation of innovative activities', *Cambridge Journal of Economics*, 24(5) (2000) 515–42.

3

Do External Knowledge Spillovers Induce Firms' Innovations? Evidence from Slovenia

Jože P. Damijan, Andreja Jaklič and Matija Rojec

3.1 Introduction

According to new growth theory, technological progress is endogenous and driven by an intentional investment of resources by profit-seeking firms. Still, innovation activities in firms depend heavily on external sources (Fagerberg, 2005). For most countries foreign sources of technology are of dominant importance for productivity growth (Eaton and Kortum, 1999; Keller, 2002). Therefore, economic analysis of innovation recognizes international knowledge flows (through FDI, trade, licensing and international technological collaborations) as important determinants of the development and diffusion of innovations. Here, the notion of technology and knowledge spillovers is central. It is based on theories of endogenous technical change of the early 1990s (Romer, 1990; Grossman and Helpman, 1991; Aghion and Howitt, 1998), claiming that the return to technological investments is partly private and partly public (Keller, 2004). Because of the non-rival character of technology, an innovation that is produced by one firm may also be used by another firm, without incurring very much additional cost (Smolny, 2000). These are technology or knowledge spillovers.

The central objective of this chapter is to test whether, and to what extent, firms' ability to innovate is induced by firms' own R&D activity, and by factors external to the firm, and what are the most important channels of external knowledge spillovers that contribute to productivity growth. These can be in the form of direct technology transfer (through FDI, trade, licensing, importing, and so on), learning effects (innovation spillovers and learning-by-exporting), as well as in the form of public R&D subsidies. So far, most of these channels have been studied separately. Here, we analyse the impact and determine the relative importance of knowledge transfer through inward FDI and trade versus that of R&D subsidies and firms' own R&D for innovation activity of firms within an integrated dynamic model.

Most of the existing empirical studies assess the impact of own R&D and knowledge spillovers on innovation activity in an indirect way, by estimating either the rate of return to firms' own R&D expenditures or the impact of external knowledge spillovers on firms' productivity growth. We estimate the impact of firms' internal R&D capital and external R&D spillovers on firms' productivity growth in a two step procedure. In the first step we evaluate the impact of firms' own R&D capital and external knowledge spillovers on firms' innovation activity. In the second step we then estimate the efficiency of firms' innovation activity; that is, we estimate the impact of innovation activity on firms' productivity growth (regardless of the source of innovation activity), while controlling for other sources of productivity growth.

Using firm level data on innovation activity combined with firm financial data for a large sample of Slovenian firms in the period 1996–2002, we find several interesting results. First, firms' own R&D expenditures, as well as external knowledge spillovers, such as R&D subsidies, foreign ownership and intra-sector innovation spillovers, do enhance firms' ability to innovate. Second, innovations do contribute substantially to firms' total factor productivity (TFP) growth. And third, foreign ownership has a double impact on firms' TFP growth – it first enhances firms' ability to innovate, and then it additionally contributes to firms' TFP growth via superior organization techniques and other channels of knowledge diffusion.

The chapter is structured as follows. Section 3.2 reviews literature on R&D as a determinant of firm's innovation capacity and of the absorptive capacity for external knowledge spillovers. Channels of external knowledge spillovers are reviewed in the following section. Section 3.4 provides empirical evidence; after brief descriptive analysis, the determinants of innovation activity are identified, followed by the estimations of the effect of the innovation activity on firms' productivity growth. Section 3.5 concludes and derives policy implications.

3.2 Own R&D as a determinant of firm's innovation activity and of firm's capacity to absorb external knowledge spillovers

Own R&D is the crucial determinant of firms' innovation activity and of firms' capacity to absorb external knowledge. Thus, R&D can be thought of as having two complementary effects on a firm's innovation activity and productivity growth. First, R&D directly expands a firm's technology level by leading to new innovations, which is called the innovation effect. On the other hand, it increases a firm's absorptive capacity – the ability to identify, assimilate and exploit outside knowledge, which is usually called the learning or the absorption effect (Cohen and Levinthal, 1989). These effects are both included in our model.

Theoretical foundations for the innovation effect are supplied by the literature on endogenous innovation and growth (for instance, Romer, 1990; Grossman and Helpman, 1991; Aghion and Howitt, 1998). Cameron et al. (2005) quote a body of empirical work in favor of a positive influence of R&D on productivity growth. Important references include Griliches (1980), Mansfield and Romeo (1980), Griliches and Lichtenberg (1984), Hall and Mairesse (1995), and Griffith et al. (2004). The R&D capital model has been the ruling research paradigm to investigate the relationship between firms' innovation and productivity growth. This approach adds some measure of knowledge capital, computed from the data on R&D, to the list of inputs entering the production function. The knowledge capital does not depend only on a firm's own research effort, but also on the pool of general knowledge a firm has access to; that is, a firm may learn from innovations of other firms (Ornaghi, 2004). This is how technological externalities or spillovers enter into the model.

A firm's capacity to absorb external knowledge spillovers, through FDI and other types of technology transfer, has been broadly analysed and emphasized in the literature. In general, empirical evidence demonstrates that FDI can contribute to overall domestic productivity growth only when the technology gap between domestic and foreign firms is not too large and when a sufficient absorptive capacity is available in domestic firms (Kokko, 1994; Kokko et al., 1996; Borensztein et al., 1998; Kinoshita, 2000; Damijan et al., 2003a). Other determinants of absorptive capacity include technological intensity of an industry (Keller and Yeaple, 2003), human capital (Borensztein et al., 1998; Hoppe, 2005), company size (Ornaghi, 2004), business climate (Moran, 2001; Keller, 2004), agglomeration of foreign subsidiaries in a host country (Sgard, 2001), and so forth.

3.3 Channels of external knowledge spillovers

The literature identifies three principal channels of international technology transfer. The first is licensing agreements, which of late provide a less important source as the latest and most valuable technologies are not available on license (UNCTAD, 2000). The second is FDI that provides probably the most important and the cheapest channel of direct technology transfer as well as of indirect knowledge spillovers. The third is international trade, in particular, imports of intermediate products and capital equipment as well as through learning-by-exporting. There is an amount of literature on individual channels of international technology transfer but the papers that compare various channels in a way that generates evidence about the relative importance of each of them are few (see, for instance, Keller and Yeaple, 2003; Alvarez and Robertson, 2004; Keller, 2004).

3.3.1 Direct FDI effects (foreign vs domestic ownership)

In dealing with FDI as a source of foreign technology and productivity growth, one should distinguish between direct effects of FDI and FDI spillovers. Direct effects relate to the impact of foreign ownership on the technology transfer to foreign subsidiaries, and on the productivity of the latter; they relate to the issue of why are foreign subsidiaries (or MNEs in general) more efficient than domestic companies (or non-MNEs in general). Thus, in measuring the contribution of FDI to the technological upgrading of a host country, one should first take into account the technological endowment of foreign subsidiaries, which can be expected to be superior to that of local producers (Sgard, 2001).

There is ample empirical evidence on positive direct technology transfer from a MNE to its foreign affiliates in terms of higher productivity levels and growth. Studies using firm-level panel data include Haddad and Harrison (1993), Aitken and Harrison (1999), Blomström and Sjöholm (1999), Girma et al. (2001), Blalock (2001), Barry et al. (2002), Damijan et al. (2003b), among others. FDI as a source of foreign technology and productivity growth has been particularly important for firms in transition economies because of the urgent need to restructure quickly. FDI may also be the cheapest means for technology transfer, as the recipient firm normally does not have to finance the acquisition of new technology. It also tends to transfer newer technology more quickly than licensing agreements and international trade (Mansfield and Romeo, 1980).

3.3.2 FDI spillovers

The issue of FDI spillovers is the most extensively analysed channel of external knowledge spillovers in the literature. Knowledge spillovers from FDI take place when the entry or presence of foreign subsidiaries, which have typically better technologies and organizational skills than domestic firms, increases knowledge of domestic firms and MNEs do not fully internalize the value of these benefits (Javorcik, 2003). The presence of a foreign subsidiary can thus increase the technological learning in the host economy indirectly through knowledge spillovers to domestic firms.

Blomström and Kokko (1998) identify at least four ways how technology might be diffused from foreign subsidiaries to other firms in the economy: (i) demonstration-imitation effect, (ii) competition effect, (iii) foreign linkage effect and (iv) training effect. Demonstration effect occurs if domestic firms learn superior production technologies from arm's length relationships with foreign subsidiaries. Competition effect is when competition from foreign subsidiaries forces domestic rivals to update production technologies and techniques to become more productive (Lim, 2001; Griffith et al., 2004). Foreign linkage effect goes through engaging of domestic suppliers for foreign subsidiaries (Markusen and Venables, 1999; Griffith et al., 2004) and by foreign subsidiaries giving access to new specialized intermediate inputs also for domestic firms (Rodríguez-Clare, 1996), or because domestic firms use local intermediate goods suppliers whose productivity has been raised through the know-how supplied by foreign subsidiaries (Keller and Yeaple, 2003). Training effect is present if there are movements of highly skilled staff from MNEs to domestic firms; these employees may take with them knowledge that may be usefully applied in domestic firms (Lim, 2001; Keller and Yeaple, 2003; Griffith et al., 2004). Not all spillovers are positive, as FDI can generate negative externalities when foreign firms with superior technology force domestic firms to exit, since they shift away demand from them – this is the crowding-out or business-stealing effect (Haddad and Harrison, 1993; Aitken and Harrison, 1999; Djankov and Hoekman, 2000).

FDI spillovers' literature further distinguishes between technology spillovers through FDI that occur between firms that are vertically integrated with the MNE (vertical, inter-industry spillovers to domestic firms in upstream and downstream industries), or in direct competition with it (horizontal, intra-industry spillovers). Since MNEs have an incentive to prevent information leakages that would enhance the performance of their local competitors, but at the same time may want to transfer knowledge to their local suppliers, spillovers from FDI are more likely to be vertical than horizontal in nature (Javorcik, 2003). The authors who explicitly bring the notion of vertical and horizontal spillovers in the literature are Blalock (2001), Damijan et al. (2003b) and Javorcik (2003), all providing evidence of positive FDI spillovers through backward linkages.

The substantial body of literature on FDI spillovers, which has been developed over the last 30 years, has produced mixed empirical results. Empirical analyses have found positive, neutral, as well as negative FDI spillover effects. The evidence suggests that there can be FDI spillovers, but they do not occur everywhere to the same degree (Keller, 2004); there is also no strong consensus on the associated magnitudes of FDI spillovers (Blomström et al., 2000), nor on the causality (Lim, 2001). In contrast to earlier literature (Haddad and Harrison, 1993; Aitken and Harrison, 1999; Djankov and Hoekman, 2000; Konings, 2001), recent firm level based studies speak in favor of positive and, in some cases also, economically large spillovers associated with FDI (Girma et al., 2001; Damijan et al., 2003b; Keller and Yeaple, 2003; Javorcik, 2003).

3.3.3 Imports and learning-by-exporting

International trade works as a channel for technology transfer either through imports of intermediate products and capital equipment, or through learning-by-exporting into industrial countries. Several authors have recently examined the issue of technological externalities associated with trade. A first set of papers has looked for international R&D spillovers driven by imports. Overall evidence supports the notion that importing is associated with technology spillovers, but we do not know how strong is the diffusion through embodied technology in intermediate goods versus other technology diffusion associated with imports (Coe and Helpman, 1995; Eaton and Kortum, 2001; Keller and Yeaple, 2003; Keller, 2004).

Comparing with imports, there is much less evidence for knowledge spillovers via learning-by-exporting. Conventional wisdom is that learning-by-exporting effects are non-existent and this is consistent with current evidence. According to Keller (2004), learning-by-exporting effects have been found in the case study literature, whereas authors of econometric studies take a much more skeptical view.

3.3.4 R&D subsidies

R&D subsidies are another external source of innovation inducement. In contrast to other types of external knowledge spillovers, they do not provide knowledge as such. What they do provide is money for additional own R&D or for more comprehensive and better use of external sources of technology. In our context the crucial issue of R&D subsidies is whether there are any positive spillovers from public to private R&D expenditures. Are R&D subsidies stimulating or displacing company-financed R&D (David et al., 1999)? The standard rationale for government support of R&D is rooted in the belief that some form of market failure exists that leads the private sector to underinvest in R&D (Nelson, 1959; Arrow, 1962). Underinvestment in R&D occurs because the output of R&D is characterized by its public good nature, which implies that benefits are not fully appropriable by the investor but generate spillovers that might be captured by competitors (Romer, 1990; Griliches, 1998). Economic incentives therefore do not generally lead firms to undertake the first best level of R&D spending. Publicly supported R&D thus supposes to augment or complement private R&D expenditures. Yet David et al. (1999), who survey the body of available econometric evidence on the subject, find ambivalent results. The survey does not offer a definite empirical conclusion regarding the sign and magnitude of the relationship between public and private R&D. The empirical evidence suggests that there is some substitution between private and government funded R&D.

3.4 Determinants of firms' innovation and the impact of innovation on firms' productivity growth in Slovenia

3.4.1 Innovation activity of Slovenian firms

In this section we make use of the official innovation surveys (based on OECD methodology) in order to reveal the determinants of the innovation activity by Slovenian firms. Innovation surveys are being conducted by the Slovenian Statistical office every second year. Up to now there have been four such extensive innovation surveys carried out – in 1996, 1998, 2000 and 2002. The surveys are being carried out among a wide sample of firms with no conditions put on actual R&D activity by these firms. The number of firms covered by innovation survey has been constantly increasing during the 1996–2002 period (stratified random sampling). Hence, these surveys allow for a broad picture of determinants of the innovation activity and its impact on performance of Slovenian firms.

Descriptive statistics of innovation activity by Slovenian firms

In this sub-section we show some descriptive statistics of firms' innovation activity. Innovation activity has been analysed with regard to the type of ownership, a firm's size as well as technological intensity of sectors. Table 3.1 reveals that the innovation activity, which captures both the innovation of products and services as well as innovation of processes,[1] is pretty low in Slovenia. Only about 20 per cent of Slovenian firms are innovative, that is, have claimed to have conducted at least one innovation in the respective two-year period. What is striking is the negative trend of innovation activity of Slovenian firms, showing that the share of innovative Slovenian firms is shrinking from 1998 to 2002.[2] The share is shrinking despite the fact that total innovation expenditure is increasing. This is predominantly due to the low innovation activity of indigenous firms (only 17 per cent of firms with domestic owners are innovative). Among foreign-owned firms (firms with 10 per cent or higher foreign equity share) the share of innovative firms is twice

Table 3.1 R&D expenditures and innovation activity of Slovenian firms by type of ownership, 1996–2002 (in per cent)

	N	**R&D/Sales (Innovative firms)**	**R&D/Sales (Non-innovative firms)**	**Fraction of innovative firms**
All firms				
1996	1454	1.5	0.026	21.7
1998	1777	1.6	0.003	23.0
2000	2518	6.0	0.021	21.2
2002	2564	6.5	0.015	20.6
Domestic				
1996	1148	1.4	0.027	18.6
1998	1371	1.5	0.003	19.5
2000	1923	7.1	0.023	17.5
2002	1935	6.4	0.004	17.3
Foreign				
1996	306	1.8	0.023	33.3
1998	406	1.9	0.003	34.7
2000	595	4.1	0.012	32.9
2002	629	6.6	0.055	30.5

Source: Statistical Office of the Republic of Slovenia; own calculations.

as high as in domestic firms. This points toward more competitive and innovation friendly environment in foreign-owned firms.

The breakdown according to firms' size shows that the share of innovative firms is five to six times higher among large firms (more than 250 employees) and three to four times higher among the medium-sized ones (50–250 employees) than among small firms (less than 50 employees). Again, Table 3.2 reveals significant differences among indigenous and foreign-owned firms. Foreign-owned firms, especially if they are of medium or large size, are more likely to be innovative than firms with domestic owners. More precisely, 30–35 per cent of foreign-owned medium-sized firms are on average innovative, while this ratio with domestic firms is only about 25 per cent. With large firms this difference is narrower as 60 per cent of foreign-owned firms relative to 55 per cent of domestic firms are innovative.

Let us now look into differences in innovation activity among firms in different technology intensity groups.[3] Table 3.3 demonstrates that, in Slovenia, the most innovative are firms in the medium-high technology sectors, such as electrical appliances, automotive production, machinery and chemical production. But there again, foreign-owned firms exhibit up to 20 percentage points higher figures of innovation activity. High-technology sector firms also exhibit above average innovation activity, but substantially lower than firms in medium-high technology sectors (25 per cent relative to 35 per cent, respectively). For foreign-owned firms these differences in

Table 3.2 R&D expenditures and innovation activity of Slovenian firms by size and ownership type, 1996–2002 (in per cent)

	N		**R&D/Sales (Innovative firms)**		**R&D/Sales (Non-innovative firms)**		**Fraction of innovative firms**	
	Domestic	**Foreign**	**Domestic**	**Foreign**	**Domestic**	**Foreign**	**Domestic**	**Foreign**
Small								
1996	578	67	1.6	2.2	0.011	0.000	8.8	13.4
1998	790	121	1.0	2.2	0.000	0.000	10.5	11.6
2000	1358	265	9.4	5.4	0.021	0.000	11.4	14.7
2002	1424	281	9.0	16.1	0.000	0.016	12.4	11.7
Medium								
1996	438	146	1.4	1.9	0.017	0.011	22.6	27.4
1998	447	183	2.1	1.8	0.008	0.000	25.5	35.5
2000	445	215	5.5	4.5	0.030	0.005	26.3	40.9
2002	406	222	4.1	4.9	0.019	0.144	24.9	36.9
Large								
1996	132	93	1.2	1.8	0.198	0.087	48.5	57.0
1998	126	102	1.0	1.9	0.003	0.022	56.3	60.8
2000	120	115	4.7	2.9	0.025	0.092	54.2	60.0
2002	105	126	2.6	4.3	0.010	0.000	54.3	61.1

Source: Statistical Office of the Republic of Slovenia; own calculations.

Table 3.3 R&D expenditures and innovation activity of Slovenian firms by technology-defined sectors and ownership type, 1996–2002 (in per cent)

	N		R&D/Sales (Innovative firms)		R&D/Sales (Non-innovative firms)		Fraction of innovative firms	
	Domestic	Foreign	Domestic	Foreign	Domestic	Foreign	Domestic	Foreign
Low tech								
1996	314	98	0.7	0.6	0.026	0.003	17.8	31.6
1998	333	110	0.8	0.9	0.004	0.000	20.1	39.1
2000	423	138	4.2	3.1	0.004	0.002	15.6	39.1
2002	413	147	3.5	4.8	0.004	0.015	14.8	40.1
Medium-low tech								
1996	451	96	0.7	0.5	0.005	0.015	12.0	18.8
1998	548	149	0.8	1.0	0.001	0.000	11.1	23.5
2000	867	256	5.4	3.7	0.007	0.020	11.0	20.7
2002	923	266	5.6	4.5	0.005	0.000	10.7	18.8
Medium-high tech								
1996	154	61	2.3	2.6	0.011	0.062	31.2	50.8
1998	203	71	2.0	2.3	0.000	0.025	35.0	49.3
2000	245	103	5.4	4.1	0.000	0.012	30.6	47.6
2002	243	101	4.1	3.4	0.000	0.101	34.2	39.6
High tech								
1996	229	51	2.0	3.6	0.087	0.047	24.5	43.1
1998	287	76	2.2	4.0	0.007	0.000	24.0	36.8
2000	339	90	9.6	5.9	0.117	0.000	25.4	42.2
2002	329	107	11.3	7.2	0.002	0.240	26.1	35.5

Source: Statistical Office of the Republic of Slovenia; own calculations.

innovation activity across sectors are less prominent since, with the exception of the medium-low technology intensity sectors foreign-owned firms seem to be equally inclined to innovation activity at a rate of about 40–50 per cent.

What is especially striking in Tables 3.1–3.3 is that higher innovation activity by foreign-owned firms is not necessarily backed by their higher own R&D expenditures (relative to total sales). The fact is that in the last two innovation surveys (2000, 2002) foreign-owned firms show proportionally less R&D expenditures compared to indigenous firms. Hence, their higher ability to innovate must be driven by other factors, such as constant transfer of technology and other knowledge spillovers from their parent companies. The next sub-section delves further into this issue.

Determinants of firms' innovation in Slovenia

In this sub-section we explore the question of what drives the innovation activity of Slovenian firms. An integrated approach to the analysis of the impact of external knowledge spillovers on firms' innovation activity (conceptualized in

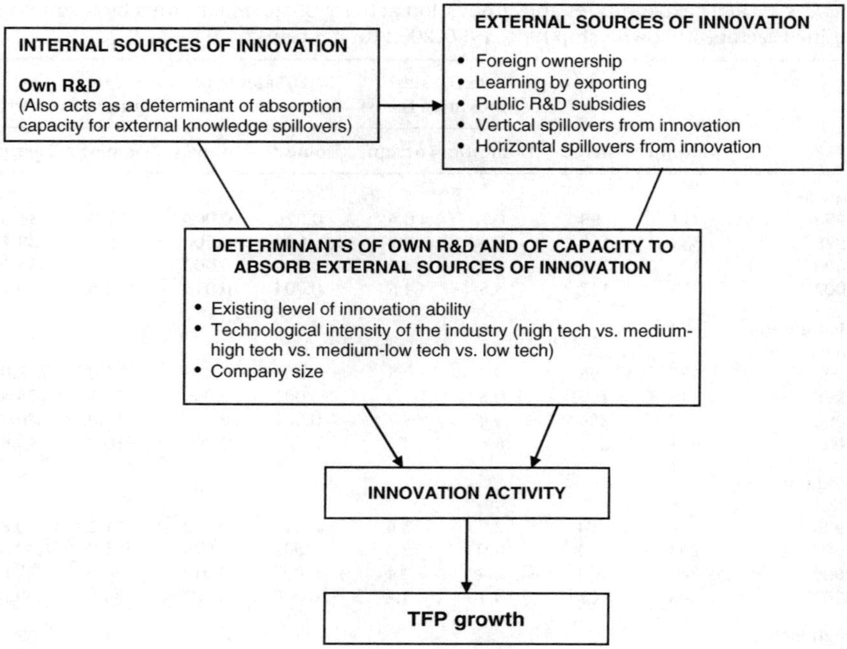

Figure 3.1 Interlinkages of sources and determinants of firms' innovation activity used in the model*

Note: *Based on the findings of econometric literature and depending on data availability.

Figure 3.1) suggests the most important determinants of firms' innovation activity. The list of those available to empirical testing includes:

- Following the endogenous innovation and growth theory, firms' own R&D is crucial for their innovation activity, and also for their capacity to absorb external knowledge spillovers. The higher their own R&D expenditures, the higher the firms' innovation activity, and the higher the absorption of external knowledge.
- External knowledge spillovers are becoming increasingly important for firms' overall innovation activity. However, the relevance of different channels of external knowledge spillovers varies considerably. Direct FDI effect, embodied in foreign technology and organizational knowledge transferred by foreign parent companies to their subsidiaries, might be by far the most important channel of external knowledge spillovers. As regards knowledge spillovers via trade, the learning-by-exporting effect is not obvious.[4]

- The spillover effect from public R&D subsidies in the form of additional funds to the private R&D expenditures of subsidy recipients is not conclusive. One may expect results in the range between weak complementarity and weak substitution.
- Firms' (and countries') absorptive capacity is very important for the magnitude of external knowledge spillovers. External knowledge spillovers will increase with own R&D expenditures, foreign ownership, technological intensity of the industry (distinction among high tech, medium-high tech, medium-low tech and low tech) and size of the firm.[5]

Table 3.4 shows the importance of the above explained determinants of innovation activity by Slovenian firms. It demonstrates that innovation activity of firms is persistent over time and that firms that have been innovative two years ago are more likely to innovate again. The table also shows innovative firms are likely to be larger in terms of employment, do invest much more into R&D and also attract higher proportion of subsidies, either public or foreign.[6] At the same time, innovative firms are also more inclined to exports and are more likely foreign owned. Surprisingly, innovative firms do not seem to be more productive in terms of value added per employee (measured to the individual sector average).

In order to reveal the importance of these individual factors on firms' innovation activity we estimate the probability $INOV_{it}[0, 1]$ of a firm i in period t to innovate:

$$\Pr(INOV_{it} = 1|\mathbf{M}_{it}) = G(\omega\mathbf{M}_{it}) \tag{3.1}$$

where $\mathbf{M}_{it}$ is a matrix of operational characteristics of firms. We assume that errors are IID distributed and have an independent extreme-value distribution. The dependent variable $INOV_{it}$ is equal to 1 if a firm has made any innovation of products (services) or production processes in period t, and 0 otherwise. The control variables contained in $\mathbf{M}_{it}$ are those listed in Table 3.4, that is, a dummy for past innovation activity (lagged one period, that is, two years), firm size (number of employees), firm relative productivity (firm value added per employee relative to the average productivity of particular sector), share of R&D expenditures in total sales, export propensity and dummy for foreign ownership as well as three variables for the importance of R&D subsidies (total R&D subsidies, public R&D subsidies and R&D subsidies received from abroad, all calculated as a ratio to total firm's R&D expenditures). In the model we also include horizontal and vertical spillovers from innovation activity of other firms. Horizontal spillovers are being measured by the number of innovations done in the same sector. Vertical spillovers are calculated as the number of innovations conducted in a related sector multiplied by the respective input–output coefficient, where the latter reflects the strength of input–output relationships between sectors.

Table 3.4 Determinants of firms' innovation in Slovenia, 1996–2002 (in per cent)

	N	INOV_t-2	rVA/ Employee	Employment	R&D/Sales	R&D/VA	Total subsidies/ R&D	Public subsidies/ R&D	Foreign subsidies/ R&D	Exports/ Sales	IFDI (foreign owned)
Innovative firms											
1996	316	–	1.26	346.7	1.55	5.39	5.39	3.12	0.27	43.9	0.388
1998	409	0.643	0.84	312.9	1.62	5.96	4.07	2.42	0.85	43.1	0.397
2000	533	0.554	1.11	278.5	6.02	19.22	4.33	3.42	0.59	38.1	0.368
2002	527	0.694	1.09	283.6	6.47	18.42	4.98	3.14	1.08	43.7	0.364
Non-innovative firms											
1996	1138	–	1.19	122.8	0.026	0.101	0.180	0.066	0.054	25.7	0.254
1998	1368	0.095	1.11	96.5	0.003	0.006	0.004	0.004	0.000	27.3	0.237
2000	1985	0.122	1.01	68.5	0.021	0.047	0.013	0.013	0.000	21.6	0.201
2002	2037	0.113	0.99	67.5	0.015	0.038	0.016	0.000	0.001	22.8	0.215

Source: Statistical Office of the Republic of Slovenia; own calculations.

In other words, the more interlinked are two sectors through bilateral supply and demand links, and the higher the innovation activity in both sectors, the larger is the scope for positive vertical knowledge spillovers between both sectors. The model also takes into account the technology intensity of the sectors in which firms are operating. It is expected that firms operating in technologically more sophisticated sectors will have a higher probability to innovate in order to remain competitive or to build their technological advantage over competitors. Due to a short and non-balanced panel, we do not include the time dummies.

We estimate a probit model using the bi-annual data for a set of manufacturing as well as non-manufacturing firms in Slovenia in the period 1996–2002. Results for two separate probit estimations are given in Table 3.5. Both estimations show that firms' present innovation activity is heavily dependent on previous innovation activity. More specifically, there is a 82 per cent probability that a firm will innovate either a product or process if it was innovative in the previous period. Firm size positively affects a firm's ability to innovate, most likely due to the scale effect; that is, a large scale of

Table 3.5 Firms' probability to innovate• in Slovenia, 1996–2002 (results of a probit model)

	Model 1		Model 2	
	Coef.	z-stat	Coef.	z-stat
$INOV_{t-2}$	0.821***	11.5	0.822***	11.5
Size	0.495***	10.0	0.497***	10.0
rVA/Emp	0.003	0.4	0.003	0.4
R&D/Sales	117.259***	25.2	118.173***	25.2
Total sub./R&D	7.217***	5.1		
Public sub./R&D			8.497***	4.3
Foreign sub./R&D			17.678*	1.7
IFDI	0.119*	1.7	0.117*	1.7
EX/Sales	0.112	1.1	0.103	1.0
HS_INOV	0.008***	3.3	0.009***	3.4
VS_INOV	−0.003	−0.4	−0.002	−0.4
ML tech	−0.043	−0.4	−0.056	−0.5
MH tech	−0.035	−0.3	−0.045	−0.4
H tech	−0.133	−1.0	−0.162	−1.2
Const.	−2.602***	−18.7	−2.603***	−18.7
Number of obs	4167		4167	
LR chi2(12)	2888.5		2897.5	
Prob > chi2	0.00		0.00	
Pseudo R2	0.616		0.618	

Notes: Dep.var.: $INOV_t$

• Product and process innovation are treated equally.

*, ** and *** denote significance of coefficients at the 10%, 5% and 1%, respectively.

sales allows the raising of enough funds for substantial R&D expenditures. This is confirmed by the highly significant and positive sign of the firm's own R&D expenditures. While literature is inconclusive regarding the importance of R&D subsidies, our results show that both public R&D subsidies, as well as R&D subsidies received from abroad (both measured as a share of firm's total R&D subsidies), significantly help Slovenian firms to increase their ability to innovate.

Foreign ownership does stimulate firms to innovate while exporting is not shown to have a significant impact on firms' innovation activity. Horizontal knowledge spillovers seem to drive firms' innovation activity, while vertical knowledge spillovers are shown not to be important. This can be interpreted in the sense that a highly competitive environment in terms of high innovation activity of competitors pushes individual firms to engage in R&D and innovation activity. On the other hand, technological linkages to other sectors seem to be rather weak.

Interestingly and highly striking are the facts that a firm's individual productivity and technology intensity of sectors in which it operates do not dominate their innovation activity. One would expect that firms would build their productivity growth upon persistent innovation activity. Accordingly, it is somehow strange that firms engaged in medium-high and high technology intensive sectors are not more likely to be innovative than their counterparts in less technologically sophisticated sectors. Especially since the share of innovation expenditure in sales was also considerably higher in the high technology class.[7]

3.4.2 Impact of innovation on firms' productivity growth in Slovenia

This sub-section is aimed at exploring the efficiency of innovations for firms TFP growth. In empirical work we are following the great body of literature on contribution of R&D to firms' TFP growth. Typically, a growth accounting approach in the form of a standard Cobb–Douglas production function is used in this type of analysis (Griliches, 1998). We start from the following production function:

$$Y_{it} = Ae^{\lambda t} K_{it}^{\alpha} L_{it}^{\beta} R_{it}^{\gamma} e^{\varepsilon_{it}}, \tag{3.2}$$

where Y_{it} is value added in firm i at time t, and K, L, and R represent the capital stock, employment and research capital used in production, respectively. A is a constant and λ represents the rate of disembodied technical change; e is the error term capturing all firm specific disturbances as well as measurement errors, and so on. The production function is homogenous of degree r in K, L and R, such that $g = \alpha + \beta + \gamma \neq 1$, which implies that Y may have non-constant returns to scale. α, β and γ are the elasticities of production with respect to capital, labor and R&D capital. Our main focus is placed toward the estimated elasticity γ, which reflects the marginal productivity or rate of return of output to R&D capital. By log-linearizing one can easily rewrite (3.2) in the form of first differences:

$$\Delta y_{it} = \lambda + \alpha \Delta k_{it} + \beta \Delta l_{it} + \gamma \Delta r_{it} + \Delta \varepsilon_{it} \tag{3.3}$$

Note that after controlling for standard inputs (labor and capital) the estimate of γ returns the contribution of R&D capital to TFP growth. We assume that R&D capital contains a set of factors that enhance innovation activity and are either internal or external to the firm. Hence, one can write R as a function of firm's internal R&D capital $\mathbf{F}_{it}$ and of various spillover effects $\mathbf{Z}_{it}$:

$$R_{it} = f^{i}(\mathbf{F}_{it}, \mathbf{Z}_{it}), \tag{3.4}$$

where $\mathbf{F}_{it}$ contains firms' own R&D expenditures, measured as a share of R&D expenditures relative to firms' total sales. $\mathbf{Z}_{it}$ captures all spillover effects that enhance firms' ability to innovate, such as foreign ownership, learning-by-exporting (exports to sales ratio), public R&D subsidies received either from national or international sources, as well as innovation spillovers received from other firms in the same sector or from other sectors.

Note that in a panel data framework equation (3.2) is typically subject to firm specific time invariant disturbances, which one can take control of by using one of the standard panel data econometric techniques (within or between estimators). Alternatively, one can get rid of firm specific effects by estimating the equation as in (3.3), where by first-differencing the time invariant firm specific effects are simply wiped out. Another problem with the time-series cross-section specification of (3.2) is a potential endogeneity between the inputs and the output, which may lead to biased estimation of input coefficients. However, in such a short and unbalanced panel dataset with mostly two to three observations per firm there is little one can do about it. Correcting for this endogeneity both by using the Olley–Pakes method or general method of moments (GMM) requires longer time series of input and output data in order to be efficiently used as lagged instruments for a firm's present performance.

In the first specification, we follow other empirical studies and estimate (3.3) by including only R&D expenditures (relative to sales) as a measure of R&D capital. This estimate gives us the upper bound of possible return of output to R&D capital. Indeed, as shown in Table 3.6 (see Model 1) the estimated elasticity of R&D capital with respect to output growth for Slovenian firms in the period 1996–2002 is about 0.24. This estimate is within the bounds of returns – between 0.04 and 0.56 – found by other empirical studies with similar model specification (Griliches, 1998).

Table 3.6 Impact of R&D and innovation on firm TFP growth of Slovenian firms, 1996–2002

	Model 1		Model 2		Model 3	
	Coef.	t-stat	Coef.	t-stat	Coef.	t-stat
ΔCapital	0.029***	4.5	0.025*	3.4	0.021***	3.0
ΔLabor	0.446***	13.4	0.446***	13.2	0.451***	13.4
ΔR&D/Sales	0.238*	1.9				
INOV			0.069*	1.8		
P[INOV]					0.083**	2.2
IFDI			0.062	1.8	0.051*	1.8
INOV * IFDI			−0.051	−0.8		
EX/Sales			0.052	1.3		
HS_INOV			0.001	1.5		
VS_INOV			0.002	1.4		
ML tech			−0.055	−1.2		
MH tech			0.036	0.7		
H tech			0.054	0.5		
Const.	−0.205***	−3.0	−0.302***	−3.6	−0.185***	−2.6
Time dummies	No		Yes		Yes	
Number of obs	3144		3073		3073	
F-test	72.81		21.63		45.65	
Adj R-sq.	0.064		0.069		0.068	

Notes: Dep.var.: ΔVA
*, ** and *** denote significance of coefficients at the 10%, 5% and 1%, respectively.

However, in our second specification (see Model 2 in Table 3.6) we go one step further by estimating the impact of innovations, which is the effective result of R&D, on firm TFP growth. This is our preferred estimation returning the estimate of the rate of return to innovation of 0.069. It demonstrates that in an average Slovenian firm innovation results in TFP growth by 6.9 per cent. In addition, foreign ownership enhances firms' TFP growth by an additional 6.2 per cent, but our results also show that innovations have the same impact on TFP growth both in foreign-owned and indigenous firms (see interaction term INOV*IFDI). Hence, foreign ownership has a double impact on firms' TFP growth. It first enhances firms' ability to innovate (already demonstrated in the previous sub-section), but then it also contributes additionally to firms' TFP growth via superior organization techniques, and so on.

Other external spillover variables included in our specification of Model 2, such as export propensity and vertical innovation spillovers, do not seem to have any further impact on firms' TFP growth. As demonstrated in the previous sub-section, it is very likely that these external knowledge spillovers only induce firms' ability to innovate but do not affect firms' TFP growth per se. We check for this by including the predicted value of innovation that we have estimated in the probit model of 'innovation production' (we take predicted values of Model 1 in Table 3.5). By including predicted values of innovation we implicitly controlled for endogeneity between R&D expenditures and fixed firms' characteristics (such as size, productivity, and so forth). The results of including this predicted innovation variable (see Model 3 in Table 3.6) include a slightly higher estimate of the return to innovation (estimate of γ increases to 0.83). But again, foreign ownership is shown to contribute additionally 5.1 per cent to firm TFP growth.

According to the above findings, we can draw three important conclusions for Slovenian firms. First, firms' own R&D expenditures as well as external knowledge spillovers, such as national and international public R&D subsidies, foreign ownership and intra-sector innovation spillovers, do enhance firms' ability to innovate. Second, innovations as a result of firms' R&D do contribute substantially to firm's TFP growth. And third, foreign ownership has a double impact on firms' TFP growth – it enhances firms' ability to innovate, but then it also contributes additionally to firms' TFP growth via superior organization techniques, and so on.

3.5 Conclusions

In spite of a rapidly increasing number of studies related to innovation, numerous questions related to the process of innovation are still open. Empirical studies mostly explore multiple determinants and effects of innovation activity. Along with the increasing number and complexity of significant determinants of innovation activity and channels of knowledge diffusion identified, exploring their relative importance and simultaneousness effects remains an important research challenge.

Evidence from Slovenia is based on simultaneous estimation of the importance of internal and external sources of innovation and evaluation of their impact on productivity growth. Own R&D expenditures and previous innovation activity (used as variables of internal sources) are consistently confirmed as significant determinants of innovation activity; however, they are much more efficient when accompanied by diffusion of knowledge from outside sources. External knowledge spillovers, either domestic or international are thus found to be an important innovation incentive. R&D subsidies, both domestic and from international sources, and intra-sectoral innovation spillovers complement internal sources and significantly increase the ability of Slovenian firms to innovate. Inward FDI significantly increases firms' ability to innovate as well. Compared to domestic firms, foreign-owned firms even show lower average level of R&D expenditures, suggesting that innovation activity must be driven by other factors such as knowledge and technology spillovers. Exporting, on the other hand, has not been found to be an important channel for knowledge diffusion, or an innovation inducement. Productivity and a sector's technological intensity also are not found to influence innovation activity significantly.

The importance of the external factors suggests that firms, though being productive, technologically intensive and innovative in the past, and in spite of their own R&D activity, are less likely to be self-sufficient in their current and future innovation activity. As R&D activity is frequently a result of non-cooperative strategies, and the character of technology and innovation is non-rival, spillovers are particularly important. External innovation incentives resulting from R&D subsidies, foreign investment and competitive environment (horizontal innovation spillovers) should thus be taken into account as important complementary sources. Exploiting external spillovers also complements the major effect of R&D sources, which reflects in considerable increase of TFP. For Slovenia, the estimated rate of return to R&D capital by using a growth accounting approach amounts to 0.24 and range within the boundaries found by other studies with similar model estimations. Foreign ownership thus has, similarly as R&D, a double impact on firms' TFP growth – it enhances firms' ability to absorb knowledge and innovate, but then it also contributes additionally to firms' TFP growth via superior organization techniques and other channels of knowledge diffusion.

Policy implications, based on the above conclusions, are clear if albeit rather general. To stimulate innovation activity (and TFP growth) the right way for a government is to increase the fraction of innovative firms; that is, to stimulate non-innovative firms to launch innovation activity, to increase R&D subsidies, to attract inward FDI and, in general, to create and upgrade the competitive environment in each sector.

Notes

1 We do not discriminate between innovation of products (services) and innovation of processes. The analysis of determinants of both types of innovation activity shows almost identical results, which justifies the decision to treat them together in one single variable.

2 Case studies and discussions with managers suggest that firms (especially the smaller ones) sometimes tend not to declare themselves any more as innovative firms since the status of the innovative firm exposes them to additional administrative (monitoring, reporting) and financial burden, without gaining any additional advantage.

3 Individual sectors are classified into four technology intensity groups (low technology, medium-low technology, medium-high technology and high technology) according to OECD methodology.

4 Those from imports of capital equipment and intermediate inputs are likely to be positive (but were unfortunately unavailable in this research), while that is much less certain, if at all, for learning-by-exporting.

5 In the literature, the stock of human capital is treated as one of the main determinants of a firm's absorption capacity (see, for instance, Borensztein et al., 1998; Cameron et al., 2005; Hoppe, 2005). The available data do not allow us to test this proposition in our model.

6 However, R&D subsidies on average do not represent a significant share of R&D expenditure. According to the survey innovation expenditure was mostly covered by firms' own funds.

7 8.5 per cent for high, 2.5 per cent for medium-high, 2.7 per cent for medium-low and 1.4 per cent for low technology sectors for total sample.

References

Aghion, P. and Howitt, P., *Endogenous Growth Theory* (Cambridge, MA: MIT Press, 1998).

Aitken, B.J. and Harrison, A.E., 'Do domestic firms benefit from direct foreign investments? Evidence from Venezuela', *American Economic Review*, 89 (1999) 605–18.

Alvarez, R. and Robertson, R., 'Exposure to foreign markets and plant level innovation: evidence from Chile and Mexico', *Journal of International Trade and Economic Development*, 13 (2004), No. 1, 57–87.

Arrow, K., 'Economic welfare and the allocation of resources to invention', in Nelson, R.R. (ed.), *The Rate and Direction of Inventive Activity* (Princeton: Princeton University Press, 1962), pp. 609–25.

Barry, F., Görg, H. and Strobl, E., 'Productivity spillovers and labour-market crowding-out: interactions between foreign and domestic firms in Irish manufacturing', *mimeo* (Dublin: University College Dublin, 2002).

Blalock, G., 'Technology from foreign direct investment: strategic transfer through supply chains', *mimeo* (Berkeley: University of California, 2001).

Blomström, M. and Kokko, A., 'Multinational corporations and spillovers', *Journal of Economic Surveys*, 12 (1998) 247–77.

Blomström, M. and Sjöholm, F., 'Technology transfer and spillovers: does local participation with multinationals matter?', *European Economic Review*, 43 (1999) 915–23.

Blomström, M., Globerman, S. and Kokko, A., 'The determinants of host country spillovers from foreign direct investment', *CEPR Discussion Paper 2350* (London: Centre for Economic Policy Research, 2000).

Borensztein, E., de Gregorio, J. and Lee, J.W., 'How does foreign direct investment affect economic growth?', *Journal of International Economics*, 45 (1998) 115–35.

Cameron, G., Proudman, J. and Redding, S., 'Technological convergence, R&D, trade and productivity growth', *European Economic Review*, 49 (2005) 775–807.

Coe, D. and Helpman, E., 'International R&D Spillovers', *European Economic Review*, 39 (1995) 859–87.

Cohen, W. and Levinthal, D., 'Innovation and learning: the two faces of R&D', *Economic Journal*, 99 (1989) 569–96.

Damijan, J.P., Knell, M., Majcen, B. and Rojec, M., 'The role of FDI, R&D accumulation and trade in transferring technology to transition countries: evidence from firm panel data for eight transition countries', *Economic Systems*, 27 (2003a) 189–204.

Damijan, J.P., Knell, M., Majcen, B. and Rojec, M., 'Technology transfer through FDI in top-10 transition countries: How important are direct effects, horizontal and vertical spillovers', *William Davidson Working Paper No. 549* (Ann Arbor: William Davidson Institute, University of Michigan, 2003b).

David, P.A., Hall, B.H. and Toole, A.A., 'Is public R&D a complement or a substitute from private R&D? A review of the econometric evidence', *NBER Working Paper Series*, no. 7373 (Cambridge, MA: National Bureau of Economic Research, 1999).

Djankov, S. and Hoekman, B., 'Foreign investment and productivity growth in Czech enterprises', *World Bank Economic Review*, 14 (2000), No. 1, 49–64.

Eaton, J. and Kortum, S., 'International patenting and technology diffusion: theory and measurement', *International Economic Review*, 40 (1999) 537–70.

Eaton, J. and Kortum, S., 'Trade in capital goods', *European Economic Review*, 45 (2001) 1195–235.

Fagerberg, J., 'Innovation: a guide to the literature', in Fagerberg, J., Mowery, D.C., and Nelson, R.R. (eds), *The Oxford Handbook of Innovation* (Oxford: Oxford University Press, 2005).

Girma, S., Greenaway, D. and Wakelin, K., 'Who benefits from foreign direct investment in the UK?', *Scottish Journal of Political Economy*, 48 (2001) 119–33.

Griffith, R., Redding, S. and Simpson, H., 'Foreign Ownership and Productivity: New Evidence from the Service Sector and the R&D Lab', *CEP Discussion Paper*, no. 649 (London: Centre for Economic Performance, London School of Economics and Political Science, 2004).

Griliches, Z., *R&D and Productivity* (Chicago: University of Chicago Press, 1998).

Griliches, Z., 'Returns to R&D expenditures in the private sector', in Kendrick, K. and Vaccara, B. (eds), *New Developments in Productivity Measurement* (Chicago, IL: Chicago University Press, 1980).

Griliches, Z. and Lichtenberg, F., 'R&D and productivity growth at the industry level: is there still a relationship?', in Griliches, Z. (ed.), *R&D, Patents and Productivity* (Chicago, IL: NBER and Chicago University Press, 1984).

Grossman, G. and Helpman, E., *Innovation and Growth in the World Economy* (Cambridge, MA: MIT Press, 1991).

Haddad, M. and Harrison, A., 'Are there positive spillovers from direct foreign investments? Evidence from panel data for Morocco', *Journal of Development Economics*, 42 (1993) 51–74.

Hall, B. and Mairesse, J., 'Exploring the relationship between R&D and productivity in French manufacturing firms', *Journal of Econometrics*, 65 (1995) 263–94.

Hoppe, M., 'Technology Transfer through Trade', *Nota di Lavoro* 19.2005 (Milano: Fondazione Eni Enrico Mattei, 2005).

Javorcik, B.S., 'Does foreign direct investment increase the productivity of domestic firms? In search of spillovers through backward linkages', *William Davidson Working Paper* no. 548, March (Ann Arbor: William Davidson Institute, University of Michigan, 2003).

Keller, W., 'Geographical localization of international technology diffusion', *American Economic Review*, 92 (2002) 120–42.

Keller, W., 'International technology diffusion', *mimeo* (Austin: University of Texas, 2004).

Keller, W. and Yeaple, S.R., 'Multinational enterprises, international trade, and productivity growth: firm level evidence from the United States', *NBER Working Paper* no. 9504, February (Cambridge, MA: National Bureau of Economic Research, 2003).

Kinoshita, Y., *R&D and Technology Spillovers via FDI: Innovation and Absorptive Capacity* (Prague: CERGE-EI, 2000), mimeo.

Kokko, A., 'Technology, market characteristics and spillovers', *Journal of Development Economics*, 43 (1994) 279–93.

Kokko, A., Tansini, R. and Zejan, M., 'Local technological capability and spillovers from FDI in the Uruguayan manufacturing sector', *Journal of Development Studies*, 34 (1996) 602–11.

Konings, J., 'The effects of foreign direct investment on domestic firms: evidence from firm level panel data in emerging economies', *CEPR Discussion Paper* no. 2586 (London: Centre for Economic Policy Research, 2001).

Lim, E.G., 'Determinants of, and the relation between foreign direct investment and growth: a summary of the recent literature', *IMF Working Paper* WP/01/175 (Washington, DC: International Monetary Fund, 2001).

Mansfield, E. and Romeo, M., 'Technology transfer to overseas subsidiaries by U.S. based firms', *Quarterly Journal of Economics*, 95 (1980) 737–50.

Markusen, J. and Venables, A., 'Foreign direct investment as a catalyst for industrial development', *European Economic Review*, 43 (1999) 335–56.

Moran, T., *Parental Supervision: The New Paradigm for Foreign Direct Investment and Development* (Washington, DC: Institute for International Economics, 2001).

Nelson, R., 'The simple economics of basic scientific research', *Journal of Political Economy*, 67 (1959) 297–306.

Ornaghi, C., 'From innovation to productivity', unpublished Ph.D. Dissertation (Madrid: Universidad Carlos III de Madrid, Departamento de Economia, 2004).

Rodríguez-Clare, A., 'Multinationals, linkages, and economic development', *American Economic Review*, 86 (1996) 852–73.

Romer, P., 'Endogenous technological change', *Journal of Political Economy*, 98 (1990) S71–S102.

Sgard, J., 'Direct foreign investments and productivity growth in Hungarian firms, 1992–1999', *Document de travail* no. 01.19, Décembre (Paris: CEPII – Centre d'edutes prospectives et d'informations internationales, 2001).

Smolny, W., *Endogenous Innovations and Knowledge Spillovers* (Heidelberg: Physica-Verlag, 2000).

UNCTAD, *World Investment Report 2000* (New York and Geneva: United Nations, 2000).

4

Product Innovation as a Source of FDI Spillovers? The Case of Chinese State-owned Enterprises

Sourafel Girma, Yundan Gong and Holger Görg[1]

4.1 Introduction

A quarter of a century of relentless market reform has witnessed the transformation of China from an almost autarchic economy to one that is deeply embedded within the global production network. For example, China is a leading destination of foreign direct investment (FDI) in the world[2] and has been described as the 'export processing zone of the world' (Lin, 2002).

Chinese state-owned enterprises (SOEs) – the once dominant players in the domestic industrial landscape – had to adjust in the face of this globalization (e.g., Jefferson and Rawski, 1994; Naughton, 1995). In the pre-reform era, the state was responsible for all strategic decisions facing individual firms (e.g. wage setting and pricing), as well as the redistribution of total industrial profits between firms according to some centrally determined objectives. This meant that SOEs had little incentive to engage in efficiency-seeking transactions.

Starting from the late 1970s the state has been loosening its grip on SOEs, but at the same time making them more and more accountable for their performance. This, coupled with growing international and internal competition, has resulted in a large-scale restructuring of the state sector, mainly via mergers and acquisitions, exit of unprofitable firms and privatization. It is by now generally accepted that the efficiency gain from this reallocation of resources has contributed positively to China's recent phenomenal growth experience, although the performance level of a large percentage of SOEs still leaves a lot to be desired (e.g., Lin et al., 1998).

Of course, restructuring and resource reallocation are not the only factors driving long-term growth. In particular, the accumulation of new knowledge through innovation has also long been regarded as one of the key drivers of economic growth. As observed by Jefferson et al. (forthcoming), among others, if China is to sustain its current growth path it must enhance its capacity to innovate.

An important question in this regard, of course, is how the country can enhance this capability. We concern ourselves here with the role of inward FDI as a means of enhancing innovative performance. We utilize a comprehensive panel data set of some 30 000 enterprises over the period 1999–2002 to address the following main research questions: (i) Does the entry of multinational firms (MNEs) have any impact on the product innovation (defined as the production of new goods or goods involving new technologies) of SOEs in China? (ii) Does the motivation of FDI (i.e. domestic market seeking versus export oriented) matter for innovation spillovers? (iii) Is the extent of spillovers contingent on the nationality of the foreign investor? (iv) Are spillovers through inter-industry linkages more important than intra-industry spillovers? and (v) Are innovation spillovers geographically confined or do they also occur outside the region MNEs locate in?

The rest of the chapter is organized as follows. The major trend of FDI in China is reviewed briefly in Section 4.2. Section 4.3 discusses the theoretical and empirical literature. The data set is described in Section 4.4. Section 4.5 presents the empirical model. Section 4.6 discusses the results from the econometric estimates and Section 4.7 draws some policy implications. Finally Section 4.8 concludes.

4.2 An overview of FDI in China

This section provides a brief overview of the trend of FDI flow into China over the past two and half decades.[3] When the Chinese government initiated economic reforms in the late 1970s, FDI was only allowed in four designated Special Economic Zones (SEZs)[4] and foreign investors were required to have local partners. However, in 1986, the government started to implement more proactive FDI attraction policies. Wholly foreign-owned enterprises were allowed for the first time, and export-oriented joint ventures and those employing advanced technology were encouraged through the provision of tax benefits.

As Figure 4.1 shows, the various policies undertaken to attract FDI have been spectacularly successful. From nearly 0 in 1979 the annual flow of FDI into China reached US$ 53.51 billion in 2003, and China now ranks as the first FDI destination in the world. The dramatic increase in FDI after 1992 has been mainly attributed to a wave of new policies of further economic liberalization. Foreign investors were offered better opportunities to sell their products in the domestic market and allowed to invest in hitherto restricted sectors such as retail trade and finance. But it is worth noting that the devaluation of the Renminbi also played an important part in the surge of FDI during this period.

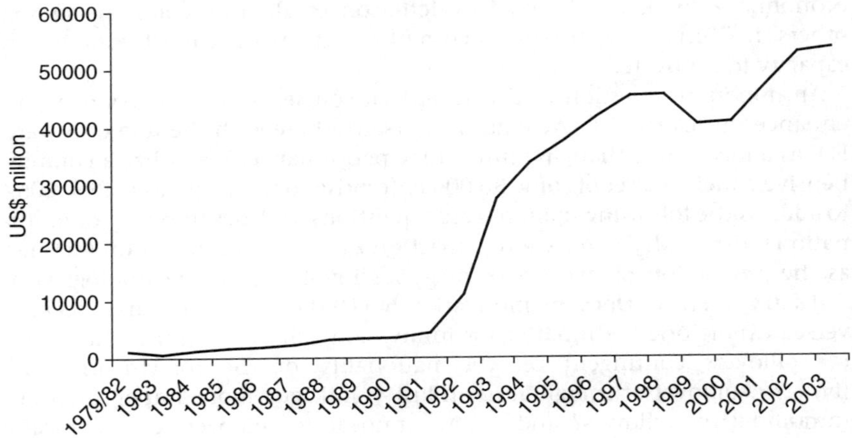

Figure 4.1 FDI flows into China, 1979–2003
Source: China Statistical Yearbook, various issues.

Multinational activity in China is characterized by a very uneven regional distribution. During the period 1987–2000, about 87 per cent of cumulative FDI was located in the coastal regions (Wei, 2003). This is a reflection of the initial policy that restricted FDI to coastal regions and the proximity of those regions to Hong Kong and Taiwan, the main sources of foreign investment. Although western and central regions have started gradually to attract more and more foreign investors, the skewed distribution of FDI in favor of the eastern coastal regions has raised serious concerns that FDI might exacerbate existing regional disparities.

Given the huge potential of the Chinese market it is surprising that only few countries are major FDI sources. Between 1979 and 1991, Hong Kong accounted for nearly two-thirds of total FDI. Most of the investment from Hong Kong is export-oriented and tends to concentrate in labor intensive sectors. During the same time period, the share of Japanese and US FDI was 14 per cent and 10 per cent respectively. By 1998 the average share of FDI from the US had declined to 8 per cent and investment from the European Union countries accounted for less than 7 per cent of total FDI. In general, FDI from OECD countries tends to be directed to more capital-intensive sectors and is predominantly motivated by the desire to access the huge domestic market.

4.3 FDI and product innovation

Why would one expect the increased influx of FDI to have had any effect on product innovation carried out by state-owned enterprises? In order to justify this research question we relate to the well-established literature on productivity spillovers from FDI. Productivity spillovers are said to take place if domestic firms improve their productivity following the influx or increasing presence of foreign multinationals either in the same or in related industries.

The theoretical argument for why one may expect productivity spillovers from foreign MNEs is straightforward. MNEs are expected to have access to some form of firm specific asset (FSA), such as a superior production technique, know-how, or management strategy, which has at least some of the characteristics of a public good and enables the firm to locate profitably abroad (Caves, 1996). These FSAs can be transferred at low or zero cost between subsidiaries of the same firm. The possibility for positive spillovers arises because MNEs may find it difficult to protect a leakage of this FSA to other firms in the host country. The public good characteristics imply that once the FSA is out on the external market it can be used by other firms as well, owing to it being at least to some extent non-rival and non-excludable.

The inability of multinationals to protect their firm-specific asset – giving rise to spillovers – can be due to a number of different channels. First, there may be demonstration effects, whereby domestic firms learn the best practice from MNEs through, for example, imitation or replication of the goods produced by the latter. Secondly, workers who have some knowledge of the FSA may leave the MNE and either move to a domestic firm or set up their own enterprise. Thirdly, domestic firms may learn through customer–supplier relationships with MNEs. For example, multinationals may provide assistance to their domestic suppliers in order to enable them to produce high quality inputs for the MNE. This channel has been stressed as particularly important for spillovers by, e.g., Javorcik (2004). Competition is also frequently argued to lead to spillovers. However, in that case there is no transfer of the firm specific asset; rather, domestic firms on their own accord improve their performance in order to be able to compete successfully with multinationals.[5]

Whether the domestic firm's improved productivity is due to a transfer of the FSA or through competition the empirical literature usually does not consider how the firm actually goes about improving productivity.[6] Rather, these analyses usually relate productivity in domestic firms or industries to a measure of the presence of foreign-owned firms in the same or in vertically related industries. The former measure is included to capture horizontal (or intra-industry) spillovers, while the latter is included to capture vertical spillovers through customer–supplier relationships between domestic and foreign-owned firms.

Here we argue that one important intermediate step through which firms can improve productivity is to implement a better technology through product innovation. If domestic firms learn a better technology from MNEs, or aim at reducing inefficiencies due to competition from multinationals, one way through which they may implement this better technology is through innovation activity. Hence, we argue that the presence of FDI may lead to spillovers on innovative activity of domestic firms, which in turn enables these to increase their ability to innovate and, hence, increase their productivity subsequently as a result of the increase in innovative activity.

That innovative activity is an important determinant of productivity is well established. Coe and Helpman (1995), for example, provided evidence that R&D activity is positively related to productivity growth in a cross-country comparison. Using micro level data, Wang and Tsai (2003) also find a positive productivity effect of R&D in their analysis of Taiwanese manufacturing firms. Furthermore, Hu et al. (2005) also establish that R&D activity positively affects productivity using firm level data for China. Our analysis builds on these findings and asks whether firms' innovative activity is positively affected by the presence of FDI in the same and in related industries.

4.4 Database description and variable construction

We draw on the Annual Report of Industrial Enterprise Statistics compiled by the State Statistical Bureau of China, covering firms accounting for at least 80 per cent of their respective industrial output. The data set includes information on firm ownership structure, industry affiliation, geographic location, establishment year, employment, gross output, sales, product innovation (defined as the production of goods involving new technologies), value added, net fixed assets, exports, R&D and employee training expenditures.[7] Multinational enterprises originating from Hong Kong, Macao, or Taiwan are labeled Ethnic Chinese, whereas other MNEs are called foreign multinationals. The data available to us cover the period 1999 to 2002.

As pointed out in the previous section, domestic firms can benefit from horizontal or vertical spillovers from FDI. Accordingly, we first calculate a variable to capture horizontal spillovers, i.e., effects on domestic firms from MNEs operating in the same industry. The degree of horizontal FDI in industry j for region r at time t, say $HRFDI_{jrt}$ is constructed as the proportion of output accounted for by multinational companies. A distance-weighted measure of foreign presence outside the region but within the same sector is also computed following the literature on neighborhood agglomeration (Adsera, 2000). For a firm in region r and industry j, this is defined as

$$OUTFDI_{jrt} = \sum_{\forall k \neq r} \frac{HRFDI_{jkt}}{d_{kr}^2} \tag{4.1}$$

where d_{kr} is the distance between the largest cities in regions k and r. We distinguish foreign presence in the same region and outside the region as empirical work has shown that, generally, there appears to be a geographical dimension to spillovers: they appear to be stronger if domestic firms are located closely to multinationals (e.g., Girma and Wakelin, 2002).

As has recently been shown by Javorcik (2004) and Girma et al. (2004), spillovers from FDI in related industries may be even more important than horizontal spillovers. Hence, we calculate two measures of foreign presence in backwardly and forwardly linked industries. To gauge the extent of backward linkages (spillovers received by domestic firms in upstream sectors), the backward measure of FDI for region *r* and industry *j* at time *t* is computed as:

$$BRFDI_{jrt} = \sum_{\forall k \neq j} \alpha_{kj} HRFDI_{krt} \tag{4.2}$$

where α_{kj} is the proportion of sector *j*'s output supplied to industry *k*. It is assumed that the greater the proportion of output supplied to an industry with foreign multinational presence, the greater the degree of linkages between foreign and local firms. We refer to this FDI as *downstream FDI*.

The index of regional FDI in upstream (forwardly linked) sectors is calculated in a similar fashion as

$$FRFDI_{jrt} = \sum_{\forall k \neq j} \beta_{kj} HRFDI_{krt} \tag{4.3}$$

where β_{kj} represents the proportion of sector *k*'s output supplied to industry *j*. This measure of FDI, which we label *upstream FDI*, captures the extent of forward linkages local firms in downstream sectors have with MNEs in upstream sectors. Based on *FRFDI* and *BRFDI*, distance-weighted measures of inter-industry FDI outside the region are computed along the lines described in Equation (4.1).

Thus we hypothesize six potential sources of FDI spillovers: horizontal, backward and forward FDI, at both regional and extra-regional levels. Each of these FDI indices is further divided according to the geographic origin of the investment (Ethnic Chinese[8] versus Foreign multinationals) or the market orientation of FDI (domestic versus export markets). The latter distinction has been shown to be important by Girma et al. (2004), who show that generally positive spillovers appear to be stronger from export oriented FDI, while domestic market oriented FDI may be more likely to lead to negative competition effects. As in Girma et al. (2004), the indices for domestic market oriented FDI are computed on the basis of the output of foreign firms sold on the domestic market, whereas export oriented FDI is calculated using the output of the same firms that is exported.

Region-specific sectoral FDI is computed for the 171 three-digit industries for which data is available. The information to construct the backward and forward linkage indices is obtained from the 1997 Input–Output Table of China published by the State Statistical Bureau. The Input–Output Table includes 124 sectors, 84 of which are in manufacturing. To construct distance-weighted indices of FDI outside the region we use data on the road distance between the capitals of the provinces obtained from the Ministry of Transportation and Communication.

For 1999 and 2002, Table 4.1 gives the shares of state-owned and foreign-owned firms in the database across the main regions of China. It is apparent that foreign-owned firms (defined as those with at least 25 per cent share of

Table 4.1 Output share of state-owned (SOEs) and foreign-owned (FOEs) enterprises by province

	Share of SOEs		Share of FOE	
Province	**1999**	**2002**	**1999**	**2002**
Coastal Region				
11 Beijing	0.452	0.384	0.441	0.414
12 Tianjing*	0.236	0.223	0.511	0.528
21 Liaonign*	0.597	0.553	0.198	0.230
31 Shanghai*	0.302	0.232	0.546	0.613
32 Jiangshu*	0.268	0.186	0.265	0.305
33 Zhejiang*	0.191	0.106	0.190	0.200
35 Fujian*	0.207	0.143	0.659	0.657
37 Shandong*	0.356	0.289	0.163	0.163
44 Guangdong*	0.150	0.109	0.626	0.626
46 Hainan*	0.644	0.553	0.232	0.236
Central Region				
13 Hebei *	0.463	0.370	0.124	0.128
14 Shanxi	0.627	0.505	0.062	0.074
15 Neimenggu	0.718	0.609	0.119	0.107
22 Jilin	0.679	0.573	0.200	0.262
23 Heilongjiang	0.683	0.641	0.120	0.108
34 Anhui	0.587	0.506	0.095	0.167
36 Jiangxi	0.708	0.617	0.125	0.133
41 Henan	0.466	0.429	0.073	0.072
42 Hubei	0.566	0.532	0.114	0.133
43 Hunan	0.674	0.571	0.061	0.088
45 Guangxi*	0.662	0.496	0.123	0.206
Western Region				
50 Chongqin	0.632	0.495	0.139	0.156
51 Sichuan	0.580	0.492	0.079	0.089
52 Guizhou	0.801	0.669	0.031	0.043
53 Yunnan	0.771	0.754	0.062	0.068
54 Tibet	0.711	0.874	0.007	0.004
61 Shanxi	0.688	0.699	0.145	0.113
62 Ganshu	0.776	0.662	0.049	0.049
63 Qinghai	0.835	0.705	0.038	0.044
64 Ningxia	0.679	0.584	0.112	0.114
65 Xinjiang	0.816	0.749	0.049	0.028

Notes:
(i) Authors calculations based on the database used in this chapter.
(ii) The numbers preceding the region names refer to the codes used by the State Statistical Bureau of China.
(iii) * provinces with Special Economic Zones and open cities.

Table 4.2 Summary statistics of the variables used in the econometric estimation

	Mean	Std. Dev.
Variable		
Product innovation indicator	0.111	0.314
New product sales/total sales (overall)	0.033	0.130
New product sales/total sales (innovators only)	0.290	0.273
Export sales/ total sales	0.055	0.096
Labor training expenditure/wage	0.006	0.032
R&D expenditure/total sales	0.002	0.023
Age	27.143	17.683
Three-digit industry market share	0.001	0.007
Regional FDI indices		
Sectoral FDI		
Foreign Multinationals	0.1068	0.1565
Ethnic Chinese	0.0785	0.1352
Export oriented	0.0458	0.1103
Domestic markets oriented	0.1395	0.1694
Upstream FDI		
Foreign Multinationals	0.0009	0.0082
Ethnic Chinese	0.0005	0.0051
Export oriented	0.0006	0.0077
Domestic markets oriented	0.0008	0.0061
Downstream FDI		
Foreign Multinationals	0.0005	0.0046
Ethnic Chinese	0.0004	0.0030
Export oriented	0.0002	0.0020
Domestic market oriented	0.0007	0.0054
Indices of FDI outside region		
Sectoral FDI		
Foreign Multinationals	0.1447	1.1957
Ethnic Chinese	0.0677	0.4368
Export oriented	0.0548	0.7123
Domestic markets oriented	0.1577	0.9702
Upstream FDI		
Foreign Multinationals	0.0017	0.0205
Ethnic Chinese	0.0004	0.0046
Export oriented	0.0002	0.0031
Domestic markets oriented	0.0018	0.0210
Downstream FDI		
Foreign Multinationals	0.0005	0.0070
Ethnic Chinese	0.0002	0.0024
Export oriented	0.0002	0.0042
Domestic market oriented	0.0005	0.0056

foreign capital) have significant presence in most regions. But FDI is unevenly distributed across regions, with coastal provinces exhibiting a much higher presence of foreign-owned enterprises. Since the focus of this paper is on SOEs, we restrict our analysis to some 30 000 SOEs for which adequate information is available. Table 4.2 provides some summary statistics of the variables of used in the analysis. The average incidence of innovation across the sample period is 11 per cent, which is pretty low. However, among innovating SOEs, the average innovation rate (i.e. new product sales/total sales) is quite high at 29 per cent.

4.5 The empirical model

Two alternative econometric strategies are employed: an endogenous probit model to analyse the incidence of product innovation, and an endogenous Tobit specification to estimate the determinants of the share of new products in total production.

In the probit model the SOE (indexed by *i*) either innovates at time *t* ($y_{it} = 1$) or it does not ($y_{it} = 0$). The probability of product innovation is modeled as:

$$\Pr ob(y_{it}) = \phi\ (\beta'\mathrm{X} + \gamma'\mathrm{Z}) \qquad (4.4)$$

where ϕ is the normal distribution function, **X** is a vector of firm level determinants of innovation, **Z** is a vector of FDI indices described in the previous section, and the full set of industry and time dummies. The vector **X** consists of exporting intensity, R&D intensity, age, the ratio of employee training expenditure to total wage bills and SOE's market share within the three-digit industry. The choice of these firm level covariates is guided by theoretical considerations as well as existing empirical evidence (e.g. Crepon et al. (1998) and Jefferson et al. (forthcoming)).

The Tobit model can be formulated in terms of a latent variable model as follows:

$$\begin{aligned} S^*_{it} &= \beta'\mathrm{X}_{it} + \gamma'\mathrm{Z}_{it} + \varepsilon_{it} \\ S_{it} &= 0 \qquad\qquad if\, S^*_{it} \leq 0 \\ S_{it} &= S^*_{it} \qquad\quad if\, S^*_{it} > 0 \end{aligned} \qquad (4.5)$$

The same set of covariates used in the probit model is also employed in the Tobit specification, but the dependent variable now is the share of new products in total output, *S*.

In both specifications R&D intensity is potentially endogenous since it is a major input in the product innovation process, and the choice of this input is arguably correlated with factors that determine the firm's decision to innovate. As a result we estimate the models using the instrumental variables technique for Probit and Tobit models, which is due to Rivers and

Vuong (1988) and Smith and Blundell (1986) respectively. In this chapter we instrument R&D intensity with lagged profit rates and total liability/assets, which are both indicators of the financial constraints facing the firms. Here our working assumption is that, conditional on the exogenous regressors of the model, past level of financial constraint affects the incidence of new product innovation through R&D intensity alone.

The estimation of Probit/Tobit models with endogenous regressor (in our case R&D intensity) involves two steps: (i) running a linear regression of R&D intensity on the instrumental variables candidates and all other exogenous regressors, and (ii) estimating the Probit/Tobit models by including the residual term from step (i) in the list of covariates. The residual term is a correction term for the endogeneity problem, and a statistically significant coefficient on this correction term can be taken as evidence in favor of the hypothesis that R&D intensity is indeed endogenous.

4.6 Discussion of the results

In this section the main findings of the chapter are discussed. As can be seen from Table 4.3, the endogeneity-correction term is statistically significant in all specifications, vindicating our approach to employ instrumental variable techniques.

Our econometric estimates indicate that across all models R&D intensity exerts a positive and significant influence on the rate and incidence of new product innovation. This is reassuring given that R&D intensity is a major input in the product innovation process. We also find that SOEs that invest in the training of their employees have a higher propensity to innovate. This result underlines the importance of building human capital in order to engage in innovation.

We document robust evidence that firms that enjoy higher market shares in their industry are more likely to engage in product innovation activity. This is consistent with the view that larger firms have the marketing and technological capacity as well as the willingness to take the risk to introduce new products.

In contrast to Jefferson et al. (forthcoming), we report that older firms are more likely to introduce new products than their younger counterparts. It may be that older SOEs are realizing that their survival depends on the constant upgrading of their productive capabilities and changing their existing way of doing things.

Interestingly enough, we did not find any relationship between product innovation and exporting intensity. This would appear to suggest that among Chinese SOEs exporting is not necessarily an indicator of superior performance (cf. Kraay, 1999). Indeed, Eckaus (2004) establishes a strong correlation between subsidies received by loss-making SOEs and their export performance.

Table 4.3 FDI and product innovation: marginal effects from endogenous Probit and Tobit models

	Tobit	Probit		Tobit	Probit
Export intensity	0.062	0.011		0.054	0.010
	(0.047)	(0.011)		(0.047)	(0.011)
Labor training	0.219	0.068		0.189	0.061
	(0.111)**	(0.031)**		(0.109)*	(0.030)**
R&D intensity	18.242	4.372		21.575	5.091
	(5.861)***	(1.508)***		(5.513)***	(1.417)***
Age	0.006	0.002		0.006	0.002
	(0.000)***	(0.000)***		(0.000)***	(0.000)***
Market share	4.198	2.736		4.087	2.714
	(0.395)***	(0.320)***		(0.390)***	(0.319)***
Endogeneity correction term	−17.661	−4.238		−20.993	−4.956
	(5.864)***	(1.510)***		(5.516)***	(1.419)***
Regional FDI indices					
Sectoral FDI					
Foreign Multinationals	0.008	0.002	Export oriented	−0.002	−0.000
	(0.003)***	(0.001)***		(0.005)	(0.001)
Ethnic Chinese	0.003	−0.002	Domestic oriented	0.013	0.003
	(0.012)	(0.003)		(0.003)***	(0.001)***
Upstream FDI					
Foreign Multinationals	−0.353	−0.110	Export oriented	2.329	0.430
	(0.348)	(0.093)		(1.190)*	(0.357)
Ethnic Chinese	1.970	0.382	Domestic oriented	−0.283	−0.093
	(0.797)**	(0.224)*		(0.341)	(0.089)
Downstream FDI					
Foreign Multinationals	2.037	0.629	Export oriented	−0.019	0.123
	(0.473)***	(0.130)***		(1.038)	(0.222)
Ethnic Chinese	−1.121	−0.376	Domestic oriented	2.325	0.647
	(1.738)	(0.440)		(0.663)***	(0.155)***
Extra-regional FDI indices					
Sectoral FDI					
Foreign Multinationals	0.009	0.002	Export oriented	−0.001	−0
	(0.002)***	(0.001)***		(0.005)	(0.001)
Ethnic Chinese	−0.003	−0.003	Domestic oriented	0.014	0.004
	(0.012)	(0.003)		(0.003)***	(0.001)***
Upstream FDI					
Foreign Multinationals	−0.234	−0.068	Export oriented	1.229	0.204
	(0.320)	(0.078)		(0.969)	(0.256)
Ethnic Chinese	2.009	0.395	Domestic oriented	0.002	−0.018
	(0.771)***	(0.205)*		(0.294)	(0.067)
Downstream FDI					
Foreign Multinationals	2.007	0.589	Export oriented	0.592	0.159
	(0.446)***	(0.120)***		(0.841)	(0.191)
Ethnic Chinese	−1.461	−0.584	Domestic oriented	1.960	0.541
	(1.673)	(0.451)		(0.624)***	(0.144)***
Observations	61389	61389		61389	61389

Notes: (i) Standard errors in parentheses; (ii) *significant at 5%; **significant at 1%; (iii) all specifications include regional, time and industry dummies.

Turning our attention to the relationship between FDI and product innovation, it is evident from Table 4.3 that the nature and extent of product innovation spillovers from multinational firms vary according to the type of FDI under consideration. Foreign multinationals and domestic market-oriented in both sectoral and downstream FDI appear to spur the innovative activity of the average SOEs. By contrast, Ethnic Chinese FDI, which tends to be less technologically intensive and less export-oriented, does not have any robust and discernible effect on the product innovation of SOEs in their sector. This lends support to the view that more export-oriented multinationals are less likely to be embedded within the host economy. Based on aggregate provincial data Cheung and Lin (2003) conclude that the export intensity of FDI firms does not have any influence on domestic R&D activity. Like Cheung and Lin (2003), we conjecture that the absence of innovation externalities from export-oriented MNEs might be because these firms are motivated by the desire to use China's cheaper labor to produce labor-intensive products for exports.

Existing evidence (e.g. Girma and Gong, 2005) suggests that *productivity* spillovers from FDI in China are region-specific, consistent with the widely held belief that there is a very limited economic integration and labor mobility between Chinese provinces. A striking result from our econometric models is the existence of product innovation spillovers from FDI outside the region. It seems new products launched in one region can easily reach another region, and innovation spillovers (probably through reverse engineering) are not geographically limited.

4.7 Policy discussion

The econometric findings that (i) FDI originating from OECD countries and domestic market seeking FDI are significant channels of innovation spillovers, and (ii) in-house R&D and employee skill upgrading are important determinants of innovation activity, suggest a two-pronged policy challenge: (i) concentrate on attracting high-tech domestic market oriented FDI, and (ii) at the same time ensure that SOEs have the right organizational and incentive structures to develop adequate technical capacity.

However, the scope for ensuring the flow of the 'right kind' of FDI can be rather limited, as placing performance requirements on foreign invested firms is against the rules of the WTO, which China joined in 2001. Thus, it appears that getting the firm level fundamentals right is the most potent policy option conducive to fostering SOEs' innovative capacity.

Current government policy of enterprise reform is pursuing the idea of liberalizing (e.g. through privatization) medium and small SOEs, while retaining larger ones and enhancing their efficiency (e.g., Huang et al., 1999). However, several scholars (e.g., Lin et al., 1999) argue that SOE reform in China will only succeed if SOEs are allowed to compete with other enterprises fairly through the elimination of policy-induced burdens, especially the responsibilities of SOEs as providers of medical care, housing and pension for their workers. However, according to Qian (2003), the key problem plaguing SOEs is the Communist Party's control over the appointment of SOEs managers. Qian (2003) argues that since the selection and evaluation of managers is on information through bureaucratic rather than market channels, the best available people are unlikely to be appointed. This view is also echoed by Chang and Wong (2004), who find that the decision-making power of local party committees relative to managers is associated negatively with firm performance. The importance of an efficient managerial incentive structure for promoting in-firm R&D and motivating managers to constantly seek new products and process innovations is also underlined in an extensive UNCTAD report (Gabriele, 2001).

The policy option proposed by other economists is more radical in that privatization is argued to be the only answer to the efficiency problem plaguing Chinese SOEs (e.g., Fan, 1999; Huang et al., 1999). Here the implicit belief is that privatization would provide the right incentive scheme for incumbent managers to make the long-term investment in building up learning and innovative capabilities. However, as the Russian experience demonstrates, privatization without the necessary institutional framework (e.g. a well-defined property rights market) is not guaranteed to lead the efficient utilization of erstwhile public assets (Jefferson, 1998).

4.8 Concluding remarks

This chapter focuses on an under-explored source of FDI spillovers, namely, product innovation, and uncovers robust evidence that domestic market oriented FDI in China facilitates the product innovation process of SOEs. This suggests that Chinese policy-makers should strive harder to attract foreign investors that are likely to have stronger linkages with the domestic economy, the benefits of being the 'export processing zone of the world' notwithstanding.

In spite of their diminishing share in total industrial production, Chinese SOEs still play a vital part in the economy. The welfare of tens of millions of urban workers, the efficiency of the domestic banking sector and the generation of adequate state revenues all depend to a large extent on the success of SOEs; and we have found that SOEs with *internal* R&D activity and investment in human capital development are successful innovators. Thus, while this chapter carries the optimistic message that FDI acts as a channel of product innovation externalities to Chinese SOEs, it also underlines the importance of getting the firm-level fundamentals right.

Notes

1 Holger Görg gratefully acknowledges financial support from the Leverhulme Trust (Grant No. F114/BF).

2 'Trends and Recent Developments in Foreign Direct Investment'. OECD Directorate for Financial, Fiscal and Enterprise Affairs, June 2004.
3 Some of the material in this section draws on Chen (1996, 1997), Lemoine (2000) and Wei (2003).
4 The SEZs consisted of three in Guangdong Province: Shenzhen (across the sea from Hong Kong), Zhuhai (across the border from Macau), Shantou (on the Guangdong coast facing Taiwan), and also Xiamen in Fujian Province (on the other side of the Straits of Taiwan).
5 Competition may also lead to negative spillovers, if domestic firms lose market share to MNEs. These spillover channels have been described extensively in the recent literature, see, for example, Blomström and Kokko (1998) and Görg and Greenaway (2004).
6 An exception is Görg and Strobl (2005) who examine in detail productivity spillovers through the movement of workers from multinationals to domestic firms in Ghana.
7 Nominal values are deflated using industry-specific ex-factory price indices obtained from the China Statistical Yearbook 2004.
8 The distinction between Ethnic Chinese and Foreign FDI can also help assess whether common culture and language matter for technology transfer.

References

Adsera, A., 'Sectoral spillovers and the price of land: a cost analysis', *Regional Studies and Urban Economics*, 30 (2000) 565–85.

Blomström, M. and Kokko, A., 'Multinational corporations and spillovers', *Journal of Economic Surveys*, 12 (1998) 247–77.

Caves, R.E., *Multinational Enterprises and Economic Analysis*, 2nd edn (Cambridge: Cambridge University Press, 1996).

Chang, E. and Wong, S., 'Political control and performance in China's listed firms', *Journal of Comparative Economics*, 32 (2004) 617–36.

Chen, C.H., 'Regional determinants of foreign direct investment in mainland China', *Journal of Economic Studies*, 23(2) (1996) 18–30.

Chen, C., 'Provincial characteristics and foreign direct investment location decision within China', *Chinese Economy Research Unit Working Paper*, No. 97/16 (Adelaide: University of Adelaide, 1997).

Cheung, K. and Lin, P., 'Spillover effects of FDI on innovation in China: evidence from the provincial data', *China Economic Review*, xx (2003) 1–20.

Coe, D.T. and Helpman, E., 'International R&D spillovers', *European Economic Review*, 39 (1995) 859–87.

Crepon, B., Duguet, E. and Mairesse, F., 'Research, innovation, and productivity: an econometric analysis at the firm level', *NBER Working Paper* 6696 (Washington D.C.: NBER, 1998).

Eckaus, R.S., 'China's exports, subsidies to state owned enterprises and the WTO', *MIT Department of Economics Working Paper*, 04-35 (Cambridge, MA: MIT, 2004).

Fan, X., 'How spillovers from FDI differ between China's state and collective firms', *MOCT-MOST* 9(1) (1999) 35–48.

Gabriele, A., 'Science and technology policies, industrial reform and technical progress in China', *UNCTAD Discussion Paper*, No. 155 (Geneva: UNCTAD, 2001).

Girma, S. and Gong, Y., 'FDI, linkages and the efficiency of state-owned enterprises in China', *Mimeo* (Nottingham: University of Nottingham, 2005).

Girma, S. and Wakelin, K., 'Regional underdevelopment: is FDI the solution? A semiparametric analysis', *CEPR Discussion Paper*, 2995 (London: CEPR, 2002).

Girma, S., Görg, H. and Pisu, M., 'Exporting, linkages and productivity spillovers from foreign direct investment', *GEP Research Paper*, 04/30 (Nottingham: University of Nottingham, 2004).

Görg, H. and Greenaway, D., 'Much ado about nothing? Do domestic firms really benefit from foreign direct investment?', *World Bank Research Observer*, 19 (2004) 171–97.

Huang, Y., Woo, W.T. and Duncan, R., 'Understanding the decline of China's state sector', *MOCT-MOST*, 9(1) (1999) 1–15.

Hu, A.G.Z. Jefferson, G.H. and Jinchang, Q., 'R&D and technology transfer: firm level evidence from Chinese industry', *Review of Economics and Statistics*, 87(4)(2005) 780–6.

Javorcik, B.S., 'Does foreign direct investment increase the productivity of domestic firms? In search of spillovers through backward linkages', *American Economic Review*, (2004) 605–27.

Jefferson, G.H., 'China's state enterprises: Public goods, externalities, and Coase', *American Economic Review*, 88(2) (1998) 428–32.

Jefferson, G.H. and Rawski, T.G., 'Enterprise Reform in Chinese Industry', *Journal of Economic Perspectives*, 8(2) (1994) 47–70.

Jefferson, G.H., Huamao, B., Xiaojing, G. and Xiaoyun, Y., 'R&D performance in Chinese industry', *Economics of Innovation and New Technology* (forthcoming).

Kraay, A., 'Exports and economic performance: evidence from a panel of Chinese enterprises', *Revue d'Economie du Développement*, (1999) 183–207.

Lemoine, F., 'FDI and the opening up of China's economy', *CEPII Working Paper*, No. 00-11, June (Paris: CEPII, 2000).

Lin, J.Y. Cai, F. and Li, Z., 'Fair competition and China's state-owned enterprises reform', *MOCT-NOST*, 9 (1999) 61–74.

Lin, J.Y., Fang, C. and Zhou, L., 'Policy burdens, and State-owned enterprise reform', *American Economic Review Papers and Proceedings*, 88(2) (1998) 422–7.

Lin, Y., 'China's accession to WTO: exaggerated fears?', *UNCTAD Discussion Paper*, No. 165 (Geneva: UNCTAD, 2002).

National Bureau of Statistics of China, *China Statistical Yearbook* (Beijing: National Bureau of Statistics of China, 2004).

Naughton, B., *Growing Out of Plan: Chinese Economic Reforms* (New York: Cambridge University Press, 1995).

OECD, *Trends and Recent Developments in Foreign Direct Investment* (Paris: OECD Directorate for Financial, Fiscal and Enterprise Affairs, 2004).

Qian, Y., 'How reform worked in China', in Rodrik, D. (ed.), *In Search of Prosperity: Analytic Narratives on Economic Growth* (Princeton, N.J.: Princeton University Press, 2003) pp. 297–333.

Rivers, D. and Vuong, Q.H., 'Limited information estimators and exogeneity tests for simultaneous probit models', *Journal of Econometrics*, 39 (1988) 347–66.

Smith, R.J. and Blundell, R.W., 'An exogeneity test for a simultaneous equation Tobit model with an application to labor supply', *Econometrica*, 54(4) (1986) 679–85.

Wang, J. and Tsai, K., 'Productivity growth and R&D expenditure in Taiwan's manufacturing firms', *NBER Working Paper*, 9724 (Washington, D.C.: NBER, 2003).

Wei, Y., 'Foreign Direct Investment in China', *Lancaster University Management School Working Paper*, 2003/002 (Lancaster: Lancaster University, 2003).

5

Obstacles to Innovation and Multinational Firms in the Italian Regions: Firm-level Evidence from the Third Community Innovation Survey[1]

Simona Iammarino, Francesca Sanna-Randaccio and Maria Savona

5.1 Introduction

The intense debate on the globalization of innovation has drawn increasing attention to multinational enterprises (MNEs) as creators of innovation across national boundaries (e.g. Cantwell, 1989, 1995; Niosi, 1999; Petit and Sanna-Randaccio, 2000). The development of cross-border corporate integration and intra-border inter-company sectoral integration makes it increasingly important to examine where and how innovative activities by MNEs are internationally dispersed and regionally concentrated (Zanfei, 2000; Cantwell and Piscitello, 2002; Cantwell and Iammarino, 2003).

Notwithstanding the lively ongoing debate, however, too little information is available on the locational and innovative choices of foreign MNEs relative to those of nationally-owned firms, and about the (beneficial or detrimental) integration between MNE innovative activities and territorial systems. This chapter tries to provide some fresh insights on this topic, all the more crucial in a country with a relatively weak degree of both active multinationality and attractiveness towards foreign firms such as Italy. In particular, two main sets of questions are addressed here:

(1) What are the locational patterns of foreign MNEs' innovative activities in Italy? To what extent are innovative behaviors both firm- (i.e. foreign versus domestic, MNEs versus single firms) and region-specific?
(2) Does the perception of the importance of obstacles to innovation vary among type of firms and regions? Do such obstacles influence firms' locational choices and the potential level of innovative activity?

We study how the probability of perceiving as important the obstacles to innovation is affected by different firm-, sector- and region-specific features. Among the firm's characteristics we take into account: size; type of ownership, i.e. whether the firm belongs to a foreign-owned group, a nationally-owned group, or is a single domestic firm; firm's innovativeness. We then consider sectoral specificities and include the macro-area in which the firm is located (north-west; north-east; center; south). The empirical analysis is carried out on Italian firm-level data provided by the third Community Innovation Survey (CIS 3), with reference to the period 1998–2000.

The remainder of the chapter is structured as follows. The next section first of all summarizes the conceptual background relevant to the crucial issue of the interaction between global and local actors in innovative processes. Then it briefly recalls the (few) empirical contributions dealing with the nature and relevance of the obstacles and factors slowing down innovation activities. Section 5.3 provides a description of the CIS 3 firm-level sample and a broad picture of firms' innovative activities in the Italian regions. Descriptive evidence on the perception of the obstacles to innovation across regions and type of firm is also reported. Section 5.4 describes the specification of the model used to explore the factors affecting the probability of perceiving the obstacles as important, and discusses the results of the econometric testing. Finally, Section 5.5 summarizes the empirical evidence and draws relevant implications for public policy.

5.2 Literature background: concepts and some empirical evidence

5.2.1 Innovation and the interaction between 'global' and 'local'

There are two clear observations pointing towards the link between geographical location and industrial performance. The first is that economic and innovative activities show a strong tendency to agglomerate in certain locations, giving rise to patterns of national and regional specialization. The second is that the performance and the growth of firms depend to a large extent on the conditions of the environment in which they operate, and particularly on those in the immediate proximity (Malmberg et al., 1996).

The evolutionary view of firms' growth and technological change allows to identify the major transformations brought about by globalization, which, in the case of the MNE, consist of an increasing cross-border interdependence and integration of all kinds of operations, including those aimed at creating new knowledge and technology. Consequently, in the last decades, there has been a shift in attention away from the MNE as a mere vehicle of technology transfer towards the crucial role it plays as a creator of innovation and technological knowledge (among others, Chesnais, 1988; Cantwell, 1989, 1992; Pearce, 1989; Birkinshaw, 1996). Dunning and Wymbs (1999) have demonstrated that the degree of multinationality is significantly associated with the perception that firms increase their global technological advantage from foreign sources (see also Ietto-Gillies, 2001). They pursue this aim by establishing integrated networks of affiliates, as a means of building a sustainable competitive advantage based much more on capabilities and dynamic improvements than on static efficiency criteria (Dunning and Narula, 1995; Zanfei, 2000; Frost, 2001; Veugelers and Cassiman, 2004).

On the other hand, the importance of contextual factors and systemic interactions in the process of generation and diffusion of innovation has been long recognized as a key determinant of the technological and economic performance of firms, countries and regions. The significance of the 'regional/local dimension' of innovation systems has emerged as the logical consequence of the interactive model (Kline and Rosenberg, 1986), which puts the emphasis on the relations with knowledge sources external to the firm. Such relations – at inter-firm level, between firms and science infrastructure, between the business sector and the institutional environment, etc. – are strongly influenced by spatial proximity mechanisms that favor processes of polarization and cumulativeness (see, for example, Lundvall, 1988; von Hippel, 1989).

Indeed, as emphasized by Dicken, ' "global" and "local" are not fixed scales; rather, they represent the extreme points of a dialectical continuum of complex mutual interactions' (Dicken, 1994, p. 103). As a consequence, structure and behavior of the two 'extreme points' need to be considered within the context of their increasing interdependence, including both endogenous determinants and exogenous variables relevant to the analysis (Cantwell and Iammarino, 2003). In other words, the evaluation of the effects of FDI in innovative activities has to take into account the fact that such effects depend both upon the technological strategy and profile of the firm and upon the characteristics of the host environment (Pearce and Papanastassiou, 1999; Cantwell and Piscitello, 2000; Cantwell and Iammarino, 2001).

The capacity of regions to withstand the (macro) processes of technological change and globalization is determined by the comparative advantage upon which they can rely. Being able to build new competencies quickly involves the ability to establish links at all levels, from the 'global' to the 'local': the extent to which a region attracts innovative resources from outside – spurring its external integration – depends first and foremost upon its extant absorptive capacity (for example, Cooke, 2001; Asheim and Isaksen, 2002; Simmie, 2003; Iammarino, 2005). On the other hand, innovation diffusion is related to the ability of a region to absorb innovation. A crucial element in the model of local accumulation of knowledge involves the attraction of outside resources, which may set off strong cumulative processes (Castellani and Zanfei, 2002). The inflow of knowledge is driven both by actors from the outside attracted into the region, and by local actors that try and tap into outside knowledge.

In this chapter we chose to focus on firms' and regional differences in the perception of obstacles to innovation as the decision of (both domestic and foreign) firms to locate in particular regions and to engage in innovative activities might be affected, *ceteris paribus*, by their evaluation of the difficulties encountered when innovating. Our conjecture is that, other things being equal, the perception of obstacles depends on the type of firm, and firms tend to face different types of problems depending on the socio-economic and institutional contexts in which they are located. This might in turn translate into vicious circles for least attractive areas, further disparities in the concentration of economic activities and in the degree of attractiveness, and lower potential for catching up and growth for the most vulnerable regions in Italy, a country historically characterized by strong territorial imbalances. Should the evidence support this conjecture, it would entail crucial implications in terms of regional and innovation policy and open up room for public intervention.

5.2.2 Obstacles to innovation: evidence from innovation surveys

The empirical literature drawing upon the evidence provided by the Community Innovation Survey (CIS) and exploring the nature and characteristics of technological innovation across firms and sectors is now large and consolidated (for the Italian CIS see, among others, Archibugi et al., 1991; Evangelista et al., 1997). However, among the innovation survey-based empirical literature, very few contributions have analysed the role of obstacles, the extent to which they actually hamper or slow down innovation and the factors affecting their perception, at least as (qualitatively) assessed by the firms themselves. These contributions (Arundel, 1997; Mohnen and Rosa, 2000; Mohnen and Röller, 2001; Baldwin and Lin, 2002; Galia and Legros, 2004; Tourigny and Le, 2004) are based on Canadian and French Innovation Surveys data. Most of them mainly focus on differences in firms' characteristics that may affect the perception of the obstacles, as well as on the extent of complementarities among single obstacles, the latter being claimed to be crucial in drawing policy indications.

The empirical evidence provided by these contributions is surprisingly unanimous in showing that the more a firm is involved in R&D and innovative activities, the more it is likely to attach greater importance to the obstacles faced to innovate. For instance, Baldwin and Lin (2002), following Arundel (1997), first examine whether the perception of obstacles does discriminate between innovators and non-innovators – adopters of advanced technologies *vis à vis* non-adopters in the case analysed by Baldwin and Lin – and then estimate whether such a perception makes a difference in the intensity of innovation among the sub-population of innovators. They find that a larger proportion of innovators compared with non-innovators, evaluate the obstacles as more relevant in affecting their innovative activities. Furthermore, the study shows that the perception of obstacles as more relevant is associated to those firms' characteristics usually conducive to both high innovation intensity – i.e. high-tech sectors, greater firm size and age – and R&D investments *tout court*.

Mohnen and Rosa (2000) carry out the same empirical analysis in the case of Canadian services over the period 1996–98, though confining their test to innovative firms only, and using the R&D intensity as a proxy for innovation intensity. Galia and Legros (2004) base their analysis on CIS 2 data for French manufacturing firms, in order to identify complementarities among obstacles, so to be able to draw policy implications regarding *sets* of obstacles rather than focusing on the single ones. Also these contributions point out a positive association between the propensity/intensity of innovation activities and the likelihood of perceiving as very relevant the obstacles to innovative activities. Hence, the surveyed firms seem to be more prone to attach importance to the problems encountered when they innovate, regardless of their industrial sector, their geographical location (at the national level) and the time-span considered – so to leave room for a good deal of generalizability of these results.

The empirical stylized fact of a positive link between innovation propensity/intensity and the likeliness of evaluating as crucial the barriers to innovation calls for a plausible interpretation. Most of the reported studies seem to indicate that the obstacles to innovation, at least as measured in innovation surveys like the CIS, should not be interpreted as factors preventing innovation or technology adoption. Rather, they should be more generally considered as an indicator of how successfully the firm is to overcome them. In this way, the empirical literature tends to discard the original meaning attached to the obstacles in the CIS – i.e. a factor hampering or slowing down innovation – and to consider firms' assessment as a measure of their ability to overcome them. However, in our view such an interpretation entails a radical turndown *vis-à-vis* the original design of the CIS questionnaire and, as a consequence, of the main purpose of CIS designers. The rationale behind the inclusion of the section on obstacles is to identify potential areas of intervention and draw sound policy implications to remove barriers to innovation. Hence, the starting point of assessing the importance of the obstacles to innovation should adhere to the original meaning as it was conceived by the CIS questionnaire designers. This latter should not be overlooked – or ignored – *a posteriori*, according to the empirical findings.

In this regard, we believe that a more in-depth empirical support should be provided to check the actual generalizability of the (positive) relationship between innovativeness and assessment of obstacle relevance, as well as all the specific factors affecting the perception of obstacles – i.e. firm-, sector- and context-specific variables. On the basis of the insights from the literature, there is ground to expect that MNEs should show a higher innovation propensity than domestic firms,[2] and that, according also to the empirical literature on obstacles, the latter should discriminate among the locational choices of foreign versus domestic firms, as they are to a large extent the expression of local institutional settings. The exploration of regional and firm regularities (foreign MNEs *vis-à-vis* Italian MNEs and single firms) in the perception of the obstacles' relevance may allow us to identify the factors affecting firms' locational and innovative choices, as well as a potential domain for regional and innovation policy.

5.3 The structure of the Italian CIS 3 sample

The CIS is based on a European (EUROSTAT) standardized questionnaire, to which each of the National Statistical Institutes must conform. The Italian CIS 3 questionnaire, in line with the EUROSTAT, contains a section devoted to the factors hampering or slowing down innovative activities, which all respondent firms are required to answer.[3] The types of obstacles are grouped according to whether they are of economic/financial nature; related to the organizational (internal) structure of the firm; and other obstacles.[4] All respondent firms are asked to rate the importance of each of the obstacles as affecting their (innovation) activity, on a 4-point Likert scale, which goes from 0 (not relevant) to 3 (very important). The micro-data used in the empirical analysis come from the Italian CIS 3 provided by the National Institute of Statistics (ISTAT), and covers the period 1998–2000. The sample is composed of 15 512 firms, stratified by industry and size, representative of a universe of 164 599 firms.[5]

Table 5.1 provides a general picture of the structure of the CIS sample. The table reports the total number of sample firms, in absolute values and as a percentage of the general total respectively by:

(i) type of firm (firm belonging to a foreign group, to an Italian group, or single domestic);[6]
(ii) location (firm located in the north-west, north-east, center and South);[7]
(iii) sector (19 sectors, both manufacturing and services).

Table 5.1 reports also the number of innovative firms by each category and the relative percentage out of the total number of firms by category. The distribution of firms by type of ownership shows that a large proportion (77 per cent) of respondents does not belong to groups. About 23 per cent of the respondent firms belong to a group, and only less than 6 per cent of the total has stated to belong to a foreign group, reflecting the relatively marginal dimension of the foreign presence in Italy. Yet, in line with the results of theoretical models and with the bulk of empirical evidence, in the Italian case the percentage of innovators among foreign MNEs (57.5 per cent) is almost double that of single domestic firms (31 per cent),[8] and higher than that of Italian MNEs (50 per cent).

Table 5.1 Structure of the sample and percentage of innovative firms

Variables	Number of firms	Firms (per cent of total)	Number of innovative firms	Innovative firms (per cent)
Type of firms				
Firm belonging to a domestic group	2595	16.7	1301	50.1
Firm belonging to a foreign group	905	5.8	520	57.5
Single domestic firm	12012	77.4	3683	30.7
Total sample	*15512*	*100.0*	*5504*	*35.5*
Location of firms				
Firm located in the north west of Italy	4852	31.3	1939	40.0
Firm located in the north east of Italy	4503	29.0	1804	40.0
Firm located in the center of Italy	2979	19.2	980	32.9
Firm located in the south-islands of Italy	3178	20.5	781	24.6
Total sample	*15512*	*100.0*	*5504*	*35.5*
Sectors				
Extraction	232	1.5	48	20.7
Food, beverage and tobacco	627	4.0	229	36.5
Textile, clothing and leather	1186	7.7	278	23.4
Wood, paper, printing and publishing	1502	9.7	508	33.8
Coke, oil, nuclear, chemicals	617	4.0	351	56.9
Plastic and non metal products	1071	6.9	451	42.1
Metals	1061	6.8	440	41.5
Machinery and equipment	697	4.5	433	62.1
Electrical machinery, electronics and optical	1124	7.3	618	55.0
Transport goods	525	3.4	221	42.1
Other manufacturing	624	4.0	194	31.1
Energy, gas and water	212	1.4	58	27.4
Trade	1722	11.1	408	23.7
Hotels and restaurants	529	3.4	89	16.8
Transport services and communication	1321	8.5	254	19.2
Financial services	770	5.0	409	53.1
Real estate	187	1.2	29	15.5
Computer, R&D, KIBS*	740	4.8	353	47.7
Other business services	765	4.9	133	17.4
Total sample	*15512*	*100.0*	*5504*	*35.5*

Note: * KIBS include engineering and technical consultancy.

As is well known, the Italian economic system is typically characterized by strong geographical polarization of wealth and imbalances of both production and innovative activities, which are among the sharpest in the European regions. The results of previous CIS-based empirical studies on Italian technological innovation at the regional level had shown that the territorial distribution of innovation turns out to be highly concentrated in just a few regions (among others, Iammarino et al., 1998; Evangelista et al., 2001, 2002).[9]

CIS 3 data on the distribution of respondent firms by type across the macro-regions generally confirm such imbalances. Firms belonging to a foreign group are strongly concentrated in the north-west (almost 60 per cent of the total foreign presence in the country). The whole north accounts for almost 80 per cent of foreign MNEs. Conversely, the choice to locate in southern regions is definitely marginal for foreign MNEs, whereas Italian groups' territorial distribution is slightly more balanced (though, the north as a whole actually hosts around 65 per cent of Italian-owned MNEs). The southern part of the country recovers in terms of single domestic firms, whose geographical location is by far the most evenly distributed across the four geographical areas here considered. The Independence Chi-square test conducted on the distribution of firms by type across the macro-regions is significant at the 1 per cent level, showing that foreign groups locate in the north-west of Italy significantly more than expected on the basis of a perfectly random distribution. Similarly, the test allows claiming that foreign groups tend to locate in the other Italian macro-regions significantly less than expected.

Table 5.2 reports the percentages of innovative firms both by type and macro-region. Such percentages relate to the weighted sample (which explains the differences with the values reported in Table 5.1, related to the sample). The evidence produced in Table 5.2 confirms both the 'regional innovative divide' of Italy – with central and, especially, southern regions showing substantially lower innovation propensity compared with the north, irrespective of the type of firm – as well as the 'innovation gap' between foreign and domestic firms, irrespective of their location. It is notable that this 'innovation gap' is particularly wide in the south: the share of innovative firms in the north of Italy (about 35 per cent in both north-west and north-east) is in fact higher than those in the center (29 per cent) and, especially, in the south (20 per cent).

One of the aims of this chapter is to analyse whether the Italian regional innovation divide is reflected, *ceteris paribus* (that is to say, taking into

Table 5.2 Distribution of innovative firms by type and macro-region, weighted sample

Macro-regions	Innovative firms in Italian group (per cent)	Innovative firms in foreign group (per cent)	Innovative firms not belonging to an industrial group (per cent)	Total by macro-region (per cent)
North-west	44.9	53.2	31.5	33.7
North-east	48.4	59.0	32.5	34.4
Center	44.1	49.4	26.8	29.0
South	33.2	46.0	19.1	20.3
Total by type	*44.3*	*53.5*	*28.8*	*30.9*

account size and sectoral effects), by the territorial distribution of foreign MNEs. The description of the sample seems to suggest that this might be the case, as the regional distribution of firms belonging to foreign groups is significantly higher in the north (particularly in the north-west), where, moreover, the percentage of innovators within the category of foreign MNEs is also particularly high. Yet, such descriptive evidence does not allow us to infer any conclusion, not only in terms of causal relationships, but even less so in terms of structural associations among these empirical facts.

Our second conjecture was that regional and firms' specificities in the perception of obstacles to innovation might play a role in the locational and innovative choices of firms in Italy. The sectoral and regional distribution of the share of sampled firms that has perceived as important or very important (values 2 and 3 of the Likert scale) each of the obstacles to innovation shows some interesting evidence. First, the obstacles more frequently indicated as important by firms are of economic/financial nature. The incidence of problems related to high innovation costs, excessive financial risks and lack of financial resources appears to be higher (respectively 30 per cent, 23 per cent and 20 per cent) than that of the obstacles related to firms' internal organization or to institutional rigidities. The lack of skilled personnel is also indicated as a relevant obstacle (for 20 per cent of the firms), as expected given that the scarcity of human capital is unanimously considered not only a crucial variable in explaining firm innovativeness, but often also a distinctive feature between foreign and domestic firms. The least problematic factors seem to concern the collection of information to innovate: lack of information on technology or markets are indicated as being very important by only 13 per cent of firms. Secondly, as far as the sectoral specificities are concerned, a systematic difference occurs between the perception of obstacles in manufacturing and service activities. In particular, firms in services show a lower attitude to perceive as relevant the majority of the obstacles indicated in the questionnaire. This is the case for all the financial-related barriers and for those linked to the lack of skilled personnel and of information on technology and markets. There is instead no remarkable difference, at least on a merely descriptive level, between services and manufacturing when they face problems related to internal organization flexibility, regulatory system or lack of customer response to innovative products and services. Thirdly, the sectors that seem to experience more serious difficulties are machinery and equipment and electrical machinery, electronics and optical. Similarly, sectors with a higher evaluation of the importance of obstacles among services are Computer, R&D and Knowledge Intensive Business Services (KIBS).

At a first glance, these descriptive results on the perceived importance of obstacles are definitely in line with the main findings of the empirical contributions reviewed in Section 5.2.2, according to which a higher evaluation of obstacles is more frequent in firms belonging to the most innovative sectors, or to those with higher R&D and technology adoption. This would simultaneously explain not only the cases of machinery, electronics and knowledge-intensive services, but also, by and large, the systematic differences occurring between manufacturing and service sectors.

On the other hand, turning to the perception of obstacles by macro-region, some peculiar features are found for the whole sample. Rather surprisingly, for most types of obstacles, the respondents located in the north-east are those with the highest perception of their importance. Only the lack of financial resources and regulation rigidities are perceived as more relevant in the south than in other parts of the country. Conversely, with no exception, firms in the north-west and in central regions are those that recognize the difficulties to innovation as less important. In the southern regions, the perception of obstacles is higher than the average particularly in the case of lack of financial resources and regulation rigidities.

This evidence called for a more in-depth exploration of the data, particularly to check whether a systematic difference occurs between (foreign-owned vs. nationally-owned) firms and among regions, as well as between innovators and non-innovators, in the perception of obstacles to innovation.

5.4 Econometric analysis

5.4.1 The model

We estimated the probability of the event 'firm evaluating the obstacle(s) as important or very important' occurring as a function of a series of regressors, among which firm size, sectoral specificities, ownership type, geographical location, innovativeness (that is, whether the firm has introduced an innovation or not). The dependent variables relate to the perceptions of the obstacles to innovation as indicated by firms (section 12.3 of the Italian CIS questionnaire) according to a 4-point Likert scale, which goes from 0 (not relevant) to 3 (highly important).[10] It is important to bear in mind that this variable is qualitative in nature and represents the *evaluation* provided by the respondents to the *perceived* factors hampering innovation activity. The formulation of section 12.3 of the questionnaire[11] does not indicate a direct causal effect between the perception of the obstacle and the choice of introducing or not an innovation.

It should also be recalled that the CIS questionnaire itself classifies the nine obstacles according to their nature (i.e. economic-financial; organizational-internal and others). This feature is important in affecting the choice of the model specification and the estimation method, as firms might tend to assess similarly obstacles belonging to the same category. In other words, the model specification and the estimation method should account (and control) for the fact that the obstacles' rating are correlated due to both the formulation of the questionnaire and the nature itself of the variables considered. Yet, we are interested in assessing the association of the chosen regressors to *each* single obstacle, in the belief that each one has an informative potential per se, controlling for the possible presence of an unobserved structure which correlates obstacles among themselves.[12]

Hence, the nature of the dependent variable and the structure of the questionnaire imposed two econometric choices. First, we estimated the probability that a firm evaluates the obstacle(s) as important using a Probit model estimated by Maximum Likelihood Estimation (MLE). Second, we employed a MLE of a Multivariate Probit Model (MPM) on the nine obstacles. Specifically, the MPM allows the error terms to be freely correlated across equations in a similar fashion as in seemingly unrelated least square regressions (the so-called SUR models). The use of MPM in this work, therefore, allows us to account (and control) for the fact that the nine obstacles' ratings are correlated with one another (Greene, 2000; Cappellari and Jenkins, 2003).

The general specification of the MPM is as follows:

$$y^{*}_{ij} = a_j + b'_j x_{ij} + u_{ij}, \tag{5.1}$$

here
$y_{ij} = 1$, if $y^* = \{2,3\}$ and 0 otherwise
with $i = 1, \ldots \ldots \ldots .n$ (observations)
and $j = 1, \ldots \ldots \ldots .9$ (obstacles, i.e. equations)
The equations' disturbances u_{ij} have a multivariate normal distribution with mean vector 0 and variance-covariance matrix **V**, where **V** has the leading diagonal elements equal to 1 and correlation $\rho_{jk} = \rho_{kj}\ \forall\ j,k \in [1;9]$ as off-diagonals elements. The MLE of MPM was carried out following Cappellari and Jenkins' (2003) set up of the **mvprobit** program in STATA.[13] Appendix A.5.1 displays the list of dependent and independent variables included in the estimations.

The set of regressors included in the estimation procedure relates to:

(i) firm specific characteristics;
(ii) regional location;
(iii) industrial sector.

The first set (i) of regressors includes all firms' specific characteristics, and in particular a proxy for size (log value of the number of employees in 1998); three dummies identifying the type of firms, namely whether the firm belongs to a foreign group, an Italian group or whether the firm is a single (Italian) one. Further, we include a dummy (innovativeness) for firms that have introduced at least a product and/or a process innovation over the period 1998–2000 (which assumes values 1 for firms answering yes and 0 otherwise).

The second set (ii) of independent variables accounts for the firms' location. Four dummies have been constructed, according to whether the firm is located in the north-west of Italy (Piemonte, Val d'Aosta, Lombardia,

Liguria); in the north-east (Veneto, Friuli, Trentino, Emilia); in the center of the country (Marche, Umbria, Toscana, Lazio); or, finally, in the southern regions of Italy (Abruzzo, Molise, Campania, Basilicata, Calabria, Puglia, Sicilia, Sardegna).

Finally, the third set (iii) of independent variables includes the sector of activity of the firm. All sectors of the economy are covered, from extraction activities to business services. We took great care in defining the sectoral dummies, especially for the service sector, trying to preserve homogeneity both in terms of numerosity and, by and large, of technological characteristics. Among services, for instance, we have constructed a dummy for firms belonging to Computer and related, R&D and KIBS, that is to say the (three-digit level) sectors of architectural and engineering services and technical consultancy. Other business services include legal and accounting services, marketing, cleaning, and security.

5.4.2 Econometric results

Table 5.3 reports the results of the MPM estimation carried out on the full sample of 15 512 firms.[14] It shows the results of the nine separate equations for each of the obstacles evaluated by the sampled firms, as a function of the regressors listed in Appendix A.5.1 and distinguished by location and firm-specific characteristics (the coefficients of the sectoral dummies are not included in the table). The reference categories against which the coefficients have to be read are reported in the table.

The specification of the model chosen proved quite effective in characterizing the evaluation of obstacles by firms. The coefficients of the independent variables related to location are significant for certain type of obstacles, mostly of internal-organizational nature. The dummy for innovativeness is systematically significant across different obstacles. Finally, the variable related to the type of firm also seems to be significantly associated to the evaluation of obstacles. All estimations include sectoral fixed effects.

Overall, a visible pattern of 'regional innovative divide' in terms of obstacle perception emerges, which places the north and the center of Italy as locations where firms tend to evaluate the obstacles to innovate significantly less frequently than those in the South. Firms in the north and center tend to evaluate the lack of financial resources (as an important impediment to their innovative activity) significantly less than firms located in the South. The same applies when we account for obstacles related to the information on technology and markets. While most of the obstacles are perceived as less important by firms located in the north–center of Italy (as compared to the reference category of southern firms), the lack of skilled personnel seems on the contrary to represent a more serious difficulty for firms in the north-east (significance at 1 per cent). Interestingly, the perception of regulation rigidities is also significantly lower for firms located in the north-west. Although probably not fully representative of the variety of local innovation models, these results still support the traditional north–south distinction of the Italian innovation system.

Table 5.3 Multivariate Probit (full sample) – dependent variable: dummy variable for firms perceiving obstacles as important or very important

	Excessive financial risk	Innov. costs too high	Lack of financial sources	Lack of org.al flexibility	Lack of skilled personnel	Lack of info tech.	Lack of info markets	Regulat. rigidities	Lack of clients' responsiveness
Independent variables: location of firm									
North-West	−0.025	−0.012	−0.148***	−0.022	0.015	−0.116***	−0.122***	−0.081**	−0.02
North-East	0.008	0.05*	−0.122***	0.067*	0.13***	−0.023	−0.023	−0.043	0.035
Center	−0.075**	−0.023	−0.113***	0.022	0.026	−0.089**	−0.098**	−0.046	−0.005
South	ref	ref	ref	ref	ref	ref	ref	ref	ref
Independent variables: firm specific									
Innovative	0.358***	0.347***	0.357***	0.194***	0.332***	0.343***	0.308***	0.258***	−0.103***
Size	−0.01	−0.01	−0.042***	0.065***	0.01	0.018	0.003	0.029***	0.036***
Italian group	−0.083**	−0.109***	−0.105***	−0.098***	−0.134***	−0.069**	−0.072**	−0.057*	−0.108***
Foreign group	−0.152***	−0.12**	−0.261***	0.066	−0.122**	−0.102*	−0.018	−0.150***	−0.121**
Single Italian firms	ref	ref	ref	ref	ref	ref	ref	ref	ref
Constant	−1.016***	−0.845***	−0.678***	−1.502***	−1.388***	−1.538***	−1.425***	−1.26***	−1.168***
Observations	15,512								
Log likelihood	−47470.083								

Notes: * significant at 10 per cent; ** significant at 5 per cent; *** significant at 1 per cent.

The coefficients of the dummies of the type of firm by ownership also indicate a robust and clear picture on the differences in the perception of obstacles as dependent on whether the firm belongs to a group, and, particularly, to a foreign group. Firms belonging to a foreign group tend to evaluate the obstacles to innovation as important or very important significantly less than the reference category (single domestic firms). This holds across every type of obstacle, except the lack of organizational flexibility. Interestingly, the coefficients of the dummy 'Foreign group' are significantly lower when compared to those of the 'Italian group'. This result holds also in the case of regulation rigidities, which on the contrary one could have expected to be a more relevant obstacle for foreign-owned firms. Yet, in general, it seems that the actual strong difference in the perception of obstacles arises between firms belonging to a group (i.e. foreign and Italian MNEs) and single domestic firms, rather than between firms according to the nationality of the ownership.

The structural association between the innovativeness of firms and their perception of the obstacles to innovation turns out to be very much in line with the previous empirical literature summarized in Section 5.2.2. Specifically, our results confirm that the more a firm is likely to introduce a product or process innovation, the higher the probability that it indicates as relevant or very relevant the problems faced to innovate. This relationship turns out to be stronger for economic/financial-related obstacles (coefficients between 0.35 and 0.36) and similarly significant for internal-organizational factors and regulatory rigidities (coefficients between 0.19 and 0.34). We find, though, that this link does not hold when firms evaluate the importance of the lack of clients' responsiveness to innovative products as an impediment to carry out innovation activity (the significant coefficients being negative). Framed otherwise, the market response to the introduction of new products/services turns out to be an actual barrier for non-innovative firms, potentially affecting the decision whether to innovate or not. The interpretation of this result, also in the light of the existing literature on the relationship between innovation and impediments perception, leads us to infer that the risk of not meeting the clients' interest and, therefore, of failing to increase market shares, actually prevents firms from carrying out innovation activities. At the micro-level of analysis, this result might be stylized in a 'Schmooklerian' framework, according to which the decision to invest in innovation is somehow 'demand-led'.

As to the role of size, we find that while financial obstacles are considered as important factors hampering innovative efforts more by small than by large firms, the reverse is the case for internal-organizational impediments.

5.5 Summary and conclusions

The major finding of this study is that important differences on how obstacles to innovation are perceived by firms occur both across regions and across types of firms. Overall, as compared to the reference category, which is a non-innovative single domestic firm, located in the south of Italy and operating in the real estate sector, firms located in the north and in the center of the country and belonging to either foreign- or Italian-owned groups tend significantly less frequently to perceive obstacles to innovation as relevant.

In line with the bulk of empirical contributions specifically devoted to this issue, the most interesting result is the structural association between firms' perception of obstacles to innovation and their propensity to innovate. As expected, this relationship turns out to be positive, leading us to conclude that the evaluation of obstacles as relevant is a symptom of a higher awareness of innovative firms of the problems encountered when carrying out innovation activities. Hence, innovative firms, and particularly MNEs, seem to be more aware of the problems encountered as compared to non-innovative firms. Our interpretation is that we should relate the perception of obstacles to the experience and learning processes of firms when they actually carry out innovation. Such processes are relatively faster in MNEs, as they experience various business cultures and institutional environments and face different types of barriers to innovation, leading to higher awareness of problems (and therefore higher 'rating').

As stated above, the fact that the evidence seems to suggest that innovative firms – relative to non-innovative ones – show a higher awareness of the factors that in principle should represent a deterrent for innovation *does not imply*, in our view, that such higher awareness can be used also as a measure of the ability to overcome the obstacles. The policy implications of the two views are symmetrically opposite. Should one attach to the obstacles relevance in firms' assessment the meaning of an indicator of ability to overcome the problems, policy-makers would have no reason whatsoever to remove or release the constraint of the obstacle. On the contrary, should the positive association between innovativeness and perception of obstacles' relevance be interpreted as a measure of (innovative) firms' higher awareness of the problems encountered (not reversing the original meaning of the obstacle as a barrier), this would consistently call for policy action aiming to remove or release such constraints to innovation.

More generally, the integrated modes of knowledge creation within contemporary multinational firms suggest that the quite widespread host location obsession with MNE footloose and crowding-out potential has become obsolete. To disregard MNE transition and its dynamic scope may lead to myopic policy perspectives, which fail to address the possibility for mutual knowledge enrichment for both MNEs and territorial systems, therefore missing fundamental opportunities for local growth. However, how to attract asset-seeking and knowledge-producing foreign investment, or how to promote innovation-conducive environments (i.e. fostering the local capacity to generate, absorb and diffuse knowledge), is something still rather unclear, and further research is needed to provide a sounder base for public intervention.

Notes

1 This chapter is the outcome of a research collaboration between the Italian National Institute of Statistics (ISTAT) and the Department of Informatica and Sistemistica of the University of Rome "La Sapienza", within the Project MIUR-COFIN on 'Foreign Direct Investments in Productive Activities and R&D with Localised Technological Spillovers: Effects on the Host Country'. The authors wish to thank Valeria Mastrostefano, Giulio Perani and Giovanni Seri from ISTAT for having provided access to the data used in this work. We are also grateful to Grazia Ietto-Gillies, Lionel Nesta and Mario Pianta for comments on previous drafts of the chapter. The usual disclaimers apply.

2 See, for example, the very recent and detailed study of the UK CIS by Frenz and Ietto-Gillies (2005).

3 It should be noted that only the sub-sample of innovative firms – those who declared to have introduced at least a product or process innovation over the three years 1998–2000 – is required to answer the majority of the CIS questionnaire. The question on obstacles to innovation is instead addressed to the whole sample of respondent firms, whether they are innovative or not.

4 In particular, the CIS questionnaire includes: excessive financial risk, innovation costs too high, lack of financial sources (economic/financial obstacles); lack of organizational flexibility, lack of qualified personnel, lack of information on technology, lack of information on markets (organizational/internal obstacles); rigidities of regulation and normative standards; lack of customer responsiveness to new products and services (other obstacles).

5 The sample is not stratified by region. ISTAT has conformed to the (standardized) sampling criteria imposed by EUROSTAT, according to which the sample stratification by region is not compulsory and is left to the choice of the single National Institute of Statistics. The descriptive frequencies by macro-region reported respectively in Table 5.1 and Table 5.2 have therefore to be taken cautiously, as the numbers might not be entirely representative.

6 Although not all Italian firms belonging to groups are multinationals and not all single Italian firms are uni-national, it is reasonable to assume that the proportion of firms that are multinational is considerably higher in the case of firms belonging to groups than in the case of single firms. We thus consider Italian firms belonging to groups as a proxy for Italian MNEs. Unfortunately, our data set does not allow the distinction between the Italian groups entirely located in Italy and those whose affiliates/subsidiaries are also located abroad. See Frenz and Ietto-Gillies (2005) for the most detailed categories of firm types in the case of the UK CIS.

7 The location is referred to the enterprise's main headquarters in the national territory, and not to other locations (in the case of multi-plant firms).

8 It should be noted, though, that also in previous rounds of CIS, relating to the 1992–94 and 1994–96 period, only about one-third of Italian (single) firms had declared to have introduced at least one product or process innovation over the period covered by the survey, which might represent therefore a sort of threshold of the Italian industrial tissue. In this case, the percentage of 35 per cent averages the response by both manufacturing and service sectors.

9 In line with these results, Cantwell and Iammarino (2003) found that the location of technological activities of foreign-owned MNEs tends to be even more agglomerated at the sub-national level than that of their domestic counterparts (large nationally-owned MNEs), and that a geographical hierarchy of regional centers in Italy could be established on the basis of different types of agglomeration

forces across the national space. These findings again supported the fact that the majority of Italian regions lags behind, not only in terms of domestic innovative activity, but even more in terms of the absolute level of foreign-owned innovation that they are able to attract.

10 The dependent variable used in the estimation procedures is a dichotomous variable for firms which have evaluated each category of obstacle as important or very important (evaluation 2 and 3 of the Likert scale proposed in the CIS questionnaire). A dummy variable has been thus created, which takes value 1 for firms that have responded 2 or 3 when asked to evaluate the importance of the obstacle, and 0 otherwise. The use of the dummy as dependent variable does not affect the robustness of the results as compared to the use of the (discrete) values of the obstacles evaluation (i.e. the multinomial ordered multivariate probit model).

11 Firms were asked to 'grade the importance of any hampering factor to technological innovation activity which the enterprise has experienced'.

12 An alternative method would have envisaged a regrouping of the obstacles according to their nature (i.e. economic–financial; organizational; others) as in Galia and Legros (2004), Mohnen and Rosa (2000) and Mohnen and Röller (2001), all pointing out the complementarities among obstacles. Yet, we believe that exploring complementarities among sets of obstacles that are already grouped in sets within the questionnaire could be tricky and lead to biased results as firms might tend to evaluate similarly obstacles grouped under the same heading already in the questionnaire.

13 Cappellari and Jenkins built up the STATA algorithms to calculate multivariate Normal probability distribution functions using simulation Maximum Likelihood.

14 The estimation was carried out first on the full sample of responding firms. A second set of estimations has been conducted respectively on the sub-sample of foreign MNEs, Italian MNEs and single domestic firms, to allow for a more in-depth exploration of regional differences within each type of firms. Results are available from the authors.

References

Archibugi, D., Cesaratto, S. and Sirilli G., 'Sources of innovative activities and industrial organisation in Italy', *Research Policy*, 20 (1991) 299–313.

Arundel, A., 'Enterprise strategies and barriers to innovation', in Arundel, A. and Garrelfs, R. (eds), *Innovation Measurement and Policies* (EIMS Publications, European Commission, 1997) pp. 101–8.

Asheim, B.T. and Isaksen, A., 'Regional innovation systems: the integration of local "sticky" and global "ubiquitous" knowledge', *Journal of Technology Transfer*, 27 (2002) 77–86.

Baldwin, J. and Lin, Z., 'Impediments to advanced technology adoption for Canadian manufacturers', *Research Policy*, 31 (2002) 1–18.

Birkinshaw, J., 'How multinational subsidiaries mandates are gained and lost', *Journal of International Business Studies*, 27(3) (1996) 467–96.

Cantwell, J.A., *Technological Innovation and Multinational Corporations* (Oxford: Basil Blackwell, 1989).

Cantwell, J.A., 'The internationalisation of technological activity and its implications for competitiveness', in Granstrand, O., Håkanson, L. and Sjölander, S. (eds),

Technology Management and International Business. Internationalisation of R&D and Technology (Chichester: Wiley, 1992).

Cantwell, J.A., 'The globalisation of technology: what remains of the product cycle model?', *Cambridge Journal of Economics*, 19 (1995) 155–74.

Cantwell, J.A. and Iammarino, S., 'EU regions and multinational corporations: change, stability and strengthening of technological comparative advantages', *Industrial and Corporate Change*, 10(4) (2001) 1007–37.

Cantwell, J.A. and Iammarino, S., *Multinational Corporations and European Regional Systems of Innovation* (London and New York: Routledge, 2003).

Cantwell J.A., and Piscitello L., 'Accumulating technological competence: its changing impact on corporate diversification and internationalisation', *Industrial and Corporate Change*, 9(1) (2000) 21–51.

Cantwell, J.A. and Piscitello, L., 'The location of technological activities of MNCs in European regions: the role of spillovers and local competencies', *Journal of International Management*, 8 (2002) 69–96.

Cappellari, L. and Jenkins, S.P., 'Multivariate Probit Regression using simulated maximum likelihood', *The STATA Journal*, 3(3) (2003).

Castellani, D. and Zanfei, A., 'Multinational experience and the creation of linkages with local firms: evidence from the electronic industry', *Cambridge Journal of Economics*, 26(1) (2002) 1–25.

Chesnais, F., 'Technical co-operation agreements between firms', *Science Technology Industry Review*, 4 (1988) 57–119.

Cooke, P., 'Regional innovation systems, clusters, and the knowledge economy, *Industrial and Corporate Change*, 10(4) (2001) 945–74.

Dicken, P., 'The Roepke Lecture in economic geography: global–local tensions: firms and states in the global space economy', *Economic Geography*, 70(2) (1994) 101–28.

Dunning, J.H. and Narula, R., 'The R&D activities of foreign firms in the United States', *International Studies of Management and Organisation*, 25 (1995) 39–73.

Dunning, J.H. and Wymbs, C., 'The geographical sourcing of technology-based assets by multinational enterprises', in Archibugi, D., Howells, J., and Michie, J. (eds), *Innovation Policy in a Global Economy* (Cambridge: Cambridge University Press, 1999).

Evangelista, R., Iammarino, S., Mastrostefano, V. and Silvani, A., 'Measuring the regional dimension of innovation: lessons from the Italian innovation survey', *Technovation*, 21 (2001) 733–45.

Evangelista, R., Iammarino, S., Mastrostefano, V. and Silvani, A., 'Looking for regional systems of innovation. Evidence from the Italian innovation survey', *Regional Studies*, 36 (2), (2002) 173–86.

Evangelista, R., Perani, G., Rapiti, F. and Archibugi, D., 'Nature and impact of innovation in manufacturing industry: some evidence from the Italian innovation survey', *Research Policy*, 26 (1997) 521–36.

Frenz, M. and Ietto-Gillies G., 'Does multinationality affect the propensity to innovate? An analysis of the third UK Community Innovation Survey', paper presented at the Conference on 'Life after Longbridge: Mature Industrial Clusters and Regional Development', London South Bank University, June 2005.

Frost T.S., 'The geographic sources of foreign subsidiaries' innovations', *Strategic Management Journal*, 22 (2001) 101–23.

Galia, F. and Legros, D., 'Complementarities between obstacles to innovation: evidence from France', *Research Policy*, 33 (2004) 1185–99.

Greene, W.H., *Econometric Analysis*, 4th edn (Upper Saddle River, NJ: Prentice Hall, 2000).

Iammarino, S., 'An evolutionary integrated view of regional systems of innovation. Concepts, measures and historical perspectives', *European Planning Studies*, 13(4) (2005) 495–517.

Iammarino, S., Prisco, M.R. and Silvani, A., 'The geography of production and innovation: how regional styles play in the global scenario', *Regional Science Review*, 18 (1998) 31–45.

Ietto-Gillies, G., *Transnational Corporations. Fragmentation Amidst Integration* (London: Routledge, 2001).

Kline G.J. and Rosenberg, N., 'An overview of innovation', in Landau, R. and Rosenberg, N. (eds), *The Positive Sum Strategy: Harnessing Technology For Economic Growth* (Washington, D.C.: National Academy Press, 1986) 275–306.

Lundvall B.Å., 'Innovation as an interactive process: from user–producer interaction to the National System of Innovation', in Dosi, G., Freeman, C., Nelson, R., Silverberg, G. and Soete, L. (eds), *Technical Change and Economic Theory* (London: Pinter Publishers, 1988).

Malmberg, A., Sölvell, O. and Zander, I., 'Spatial clustering, local accumulation of knowledge and firm competitiveness', *Geografiska Annaler*, 78B(2) (1996) 85–97.

Mohnen, P. and Röller, L., 'Complementarities in innovation policy', *CEPR Discussion Paper Series*, No. 2712 (London: CEPR, 2001).

Mohnen, P. and Rosa, J., 'Les obstacles à l'innovation dans les industries de services au Canada', *CIRANO Scientific Series, 2000s-14*, (2000).

Niosi, J., 'The internationalization of industrial R&D: from technology transfer to the learning organization', *Research Policy*, 28(2–3) (1999) 107–18.

Pearce, R.D., *The Internationalisation of Research and Development by Multinational Enterprises* (New York: St Martin's Press, 1989).

Pearce, R.D. and Papanastassiou M., 'Overseas R&D and the strategic evolution of MNEs: evidence from laboratories in the UK', *Research Policy*, 28 (1999) 23–41.

Petit, M.L. and Sanna-Randaccio, F., 'Endogenous R&D and foreign direct investment in international oligopolies', *International Journal of Industrial Organization*, 18 (2000) 339–67.

Simmie, J., 'Innovation and urban regions as national and international nodes for the transfer and sharing of knowledge', *Regional Studies*, 37(6–7) (2003) 607–20.

Tourigny, D. and Le, C.D., 'Impediments to innovation faced by Canadian manufacturing firms', *Economics of Innovation and New Technology*, 13(3) (2004) 217–50.

Veugelers, R. and Cassiman, B., 'Foreign subsidiaries as a channel of international technology diffusion: some direct firm level evidence from Belgium', *European Economic Review*, 48 (2004) 455–76.

Von Hippel, E., *The Sources of Innovation* (Oxford: Oxford University Press, 1989).

Zanfei, A., 'Transnational firms and changing organisation of innovative activities', *Cambridge Journal of Economics*, 24(5) (2000) 515–42.

Appendix

Table A.5.1 List of variables included in the empirical analysis

Variables	Notes
Dependent variable	
Excessive financial risk	Dummy for firm evaluating the obstacle as important or very important*
Too high innovation costs	
Lack of appropriate sources of finance	
Lack of organisational flexibility within the enterprise	
Lack of qualified personnel	
Lack of information on technology}	
Lack of information on markets	
Insufficient flexibility of regulation and normative standards	
Lack of customer responsiveness to new goods and services	
Independent variables: firm specific	
Size	Number of employees in 1998 (log value)
Foreign groups	Dummy for firm belonging to a foreign group
Italian groups	Dummy for firm belonging to an Italian group
Single Italian firm	Dummy for firm not belonging to a group (Italian)
INN	Dummy for firm introducing a product or a process innovation during 1998–2000 (yes = 1; no = 0)
Total R&D expenditure per employee	Total R&D expenditure per employee (log value)
Innovative firms with no R&D expenditure	Dummy for firm introducing an innovation during 1998–2000 and no R&D expenditure (yes = 1; no = 0)
Independent variables: location of firm	
North-west	Dummy for firm located in the North West (Piemonte, Val d'Aosta,Lombardia, Liguria)
North-east	Dummy for firm located in the North East (Veneto, Friuli, Trentino, Emilia)
Center	Dummy for firm located in the Center (Marche, Umbria, Toscana, Lazio)
South	Dummy for firm located in the South (Abruzzo, Molise, Campania, Basilicata, Calabria, Puglia, Sicilia, Sardegna)

Continued

Table A.5.1 Continued

Independent variables: sectoral affiliation	
Extraction	Dummy for firm belonging to each sector
Food, beverage and tobacco	
Textile, clothing and leather	
Wood, paper, printing and publishing	
Coke, oil, nuclear, chemicals	
Plastic and non metal products	
Metals	
Machinery and equipment	
Electrical machinery, electronics and optical	
Transport goods	
Other manufacturing	
Energy, gas and water	
Trade	
Hotels and restaurants	
Transport services and communication	
Financial services	
Real estate	
Computer, R&D, KIBS**	
Other business services	

Notes: * Evaluation on a Likert scale: 0 (not relevant); 1 (low importance); 2 (medium importance); 3 (high importance).
Dummy variables have been created that take value 1 for evaluation 2 and 3 and 0 otherwise.
** KIBS include engineering and technical consultancy.

Part II

Clusters and Industrial Development

6

Evolution of Industry Clusters through Spin-offs and the Role of Flagship Firms

Manuel P. Ferreira, Ana Teresa Tavares and William Hesterly

6.1 Introduction

The manner in which clusters emerge and evolve is important for public policy and corporate strategy. Understanding how and why clusters emerge and develop provides insights into agglomeration phenomena, innovation capacity, location advantages, and may influence local governments' investments. Extant research has traced the origins of clusters to historical trajectories and to reasonably random and undefined exogenous events (Hendry et al., 2000). In other instances, scholars identify life-cycle patterns to show that clusters are living entities that thrive and fail in response to environmental shifts (Swann, 1998). Others focus on unusual entrepreneurial dynamism and innovation in some regions (Saxenian, 1994; Pinch and Henry, 1999; Feldman, 2001; Zander, 2003). Extant research has two main perspectives: first, the discussion on how innovation may or may not occur as the density of firms in the cluster increases; second, the idea that firms' co-location induces isomorphic behavior.

In this chapter we follow our recent work (Ferreira et al., 2006) in advancing a complementary approach to entrepreneurship in industry clusters, and especially those clusters that have a regional base. Specifically, we posit that some clusters may emerge and develop more endogenously than usually considered in the literature, through the gestation of new firms. The latter are entrepreneurial spin-offs by employees that exit their parent firm, or previous employer, to start their own businesses. By focusing on entrepreneurial issues at the genesis of at least some clusters, we contribute primarily to research on the evolution of clusters as a function of entrepreneurial activity by entrepreneurs that are insiders to the cluster, and also to the literature on the value of networks to support entrepreneurial dynamism and success (Coviello and Munro, 1995; Sorenson and Audia, 2000). Moreover, we also formulate some insights on the possible role of multinational enterprises

(MNEs) in originating clusters, further complementing existing clusters' research. MNEs may be leader firms whose business focus, strategies, and business ties inside and outside of the cluster may prove a fertile ground for individuals' entrepreneurial actions.

The intersection of research on clusters with that on entrepreneurship permits a novel approach to social networks in regionally bounded clusters. Based on studies on parent-progeny relations (genealogical trees), advantages and hazards (Brittain and Freeman, 1980; Saxenian, 1994; Klepper, 2001; Phillips, 2002), we suggest a format of new firms' foundation relying mainly on the network benefits emerging for both spin-offs and parents. Our approach departs from extant studies observing spin-offs as parasites that feed from knowledge acquired at the parent, or from the 'brain-drain' hypothesis (Stinchcombe, 1965; Brittain and Freeman, 1980; Phillips, 2002). We suggest that beyond the argument that spin-offs imitate the parent's routines, structures, markets, products, and technologies, there is a more appealing argument based on the social embeddedness emerging in some spin-offs. The latter are spawned from leader flagship firms that will drive the cluster's evolution. These may be domestic or foreign-owned. In this chapter we clarify our concept of 'flagship'. We further articulate some considerations on the effects of the flagships being an indigenous versus a foreign entity leading the cluster's genesis and evolution.

We advance that the relationships of the new spin-offs with parent firms, other siblings, and the broader business network of the parent carry substantial network benefits. These parent–progeny relationships can be crucial to overcome liabilities of newness and smallness (Stinchcombe, 1965; Hannan and Freeman, 1977), and the absence of legitimacy (Carroll, 1984) that often characterize small new spin-offs. Not only do these relationships increase spin-offs' likelihood of success because they help to identify opportunities, secure start-up resources (Singh et al., 1986; Oviatt and McDougall, 1995; Uzzi, 1997), and reduce resource dependence uncertainties, they also provide an advantage to insider spin-offs vis-à-vis outsider start-ups. We illustrate this model of cluster emergence and evolution (termed 'motherhood model'), and the respective network benefits arising for both spin-offs and parents in an exploratory case study of a plastic molds cluster in Portugal. In sum, we build on the idea that only by observing the genesis of the ties between firms are we better able to understand the formation and evolution of the inter-firm organizational forms that preside in clusters, thus enabling us to understand how clusters develop.

The remainder of the chapter is organized as follows. Section 6.2 reviews relevant work on clusters' dynamism, focusing on parenting processes, and explains the endogenous evolution of some clusters based on entrepreneurial activity and network benefits. Section 6.3 focuses on flagship firms, specifies how we use the 'flagship' concept, and highlights a debate on the relevance of the flagship being a domestic company, or a foreign-owned subsidiary.

Following that, we provide a case-study illustration. We conclude and discuss policy issues related to the themes addressed.

6.2 The motherhood model

6.2.1 Parenting spin-offs

Research on industry clusters identified a specific model of clusters' development in which the clusters' composition can be traced back to one, or a few, original firm(s). This work has been rooted in entrepreneurship studies and mainly in the observation that clusters might start by accident through the presence of focal firms that generate spin-offs (Arthur, 1990; Hendry et al., 2000). Some case studies found evidence that, indeed, the origin of the clusters is often due to one or a small group of companies (Saxenian, 1994; Dalum, 1995). The composition of those clusters is thus strongly influenced by a parenting process (Klepper, 2001), and the ties of the firms in the clusters, is found in the original firm's genealogy. For example, Dalum (1995) noted how in the radio communications cluster in Denmark most firms' genealogy was tied to one manufacturer of off-shore radio equipment. Saxenian's (1994) work on Silicon Valley uncovered that a large part of the dynamism of that area was induced by employees that exited Fairchild in what she called the 'family tree' of Fairchild Semiconductors (Saxenian, 1994). The offspring that originated the Springfield metalworking sector (Forrant and Flynn, 1998) is another example of an original firm and a family tree. Not all these spin-offs maintain ties to the parent; some may even compete with it, or the entrepreneur may have left the firm disgruntled. However, generally, spin-offs tend to locate near the parent, possibly to avoid the burden of relocating families and moving away from social contacts (Zander, 2003). Thus, these spin-offs will be able to benefit from previous social and business contacts.

Research on parent–progeny or parenting processes in entrepreneurship and clusters research has mostly focused on the transfer of parental practices, routines, blueprints and skills to the new firm (Brittain and Freeman, 1980; Carroll, 1984). These studies suggest that the prior history and work experience of the new entrepreneur have a strong influence on spin-offs' survival (Helfat and Lieberman, 2002). In essence, this work proposes that new firms' capabilities are largely determined by founders' prior experience in the parent that they may leverage in identifying new business opportunities and in managing operations (Shane, 2000; Burton et al., 2002). Other studies emphasized the inheritance effects accruing to spin-offs and whether they mimic the organizational practices and forms of their parents (Phillips, 2002).

Much of the research on parenting processes at the origin of the cluster, and in entrepreneurship more broadly, stresses the negative impact of entrepreneurial spin-offs on parents. These studies question whether spin-offs are parasites that deplete the parent's knowledge, and exploit business opportunities that are stolen from the mother (Phillips, 2002). Mimicking behaviors, the transfer of knowledge, routines, organizational practices, blueprints, skills, procedures, and market orientations from the parent to spin-offs are the central issue analysed (Stinchcombe, 1965; Brittain and Freeman, 1980; Carroll, 1984; Phillips, 2002). In this view, spin-offs are a dreadful outcome that should be avoided. It is well known that many US firms bind the workers in employment contracts that prevent them from using knowledge acquired at the parent in setting up their own companies or in working for competitors. In sum, parenting models and processes as they have been examined are based on the imprinting conditions upon spin-offs' that affect the role and development of the spin-off (Stinchcombe, 1965; Baum et al., 2000).

A major drawback of parenting processes, mimicking behaviors and inheriting business models and practices is the slower pace of innovation that may lead to the stagnation/decline of the entire industry or cluster. The ties that bind also blind, by isolating spin-offs from external agents. Mimicking behaviors may lead spin-offs to rely only on current relationships, restricting them to local search, and to ignore opportunities outside their visible landscape (Gulati and Gargiulo, 1999). In other words, parenting processes may induce rigidities (Leonard-Barton, 1992). For instance, when firms in a cluster display similar actions/behaviors it is less likely that clustered firms will make significant efforts in innovating (Ferreira et al., 2003). Thus, arguments against parenting processes are mainly based on the view that isomorphism may lead to substantial inertia and a low level of innovation, and that spin-offs act as mere replicators of parents' structures and practices. This view thus posits that parenting events generate substantial losses to the parent, moreover creating added competition in the industry and jeopardizing the industry's future evolution because spin-offs and parents will tend to innovate less, to waste resources, snatching markets and opportunities from each other.

6.2.2 Mothering spin-offs: a more positive perspective

Following our recent work (Ferreira et al., 2005, 2006), we suggest a complementary approach to the negative view advanced in former parenting studies, proposing a 'motherhood' model to differentiate from prior research on parent–progeny relations. Our conceptualization focuses on network benefits to both mother and spin-off. We highlight the importance of social capital for spin-offs' success. This approach is consistent with Burt's (1992) argument that entrepreneurs' success is partly determined by the structure of their relational networks, and with Abrahamson and Rosenkopf's (1997) view that to understand firms' founding, behavior, performance, and innovation we need to comprehend the social network in which firms originate. It is also attuned to Sedaitis's (1998) suggestion that the entrepreneur's pool of social ties to incumbents prior to the founding of spin-offs is crucial to new ventures' success.

The mothering process we present is similar to the parenting processes above in that mothers generate spin-offs as employees spot an opportunity and exit. However, it highlights a different set of benefits as we depart from the view of the spin-off as a parasite to examine the benefits for both spin-offs and parent firms. It is worth noting at the outset that the benefits to spin-offs are, at least partly, a function of who the mother firms are: here we refer to the more fertile mothers as 'flagship firms' (more considerations on flagships will be provided in the next section).

In a brief explanation (for detail check Ferreira et al., 2005, 2006), our mothering process highlights a large number of new (typically small) firms that are founded by employees exiting to found their own firm (Freeman, 1986; Saxenian, 1994; Pinch and Henry, 1999; Hendry et al., 2000; Phillips, 2002). These spin-offs have stereotypical liabilities (lack of resources, clients, financing, undefined reputation) that they need to overcome in order to exchange with other firms. They access the resources needed by transacting with other firms in the market. However, other firms may be reluctant to exchange with new spin-offs (Hite and Hesterly, 2001; Podolny, 2001). The mothering rationale we advance comes into play now. By being gestated inside an established firm, spin-offs gain access not only to the parent and siblings but also, and more broadly, to the mother's network of social/business contacts. Progeny ties to the mother and its network reduce uncertainties felt by other firms in exchanging with the spin-off, as ties provide identification and 'borrowed reputation'. Similarly, offspring from the same mother share a sense of identification and cohesiveness among them (Hite and Hesterly, 2001). These are essentially social network benefits that accrue to spin-offs and increase their ability to survive.

In sum, the dominant features of the mothering process rely on spin-offs by employees with tangible or intangible ties to the mother and to its network. However, some mothers will gestate a larger number of spin-offs. These are the parents that have certain characteristics that increase the spin-offs' likelihood of success. Hence, some parents can transmit more valuable social resources to spin-offs, guaranteeing that spin-offs will be able to obtain the resources and markets required (financing, employees, customers, interacting with public and private institutions, etc.).

As regards consequences we suggest that, instead of being stuck in isomorphic behaviors and imitations that could lead to decline, a cluster characterized by a mothering type of evolution is able to realign and to adapt to market/technological shifts. Although spin-offs may mimic some structural features of the parent, those features are not usually what determines the employee's exit. When the social benefits involved can be identified, it is the opportunity or market/technological gap detected, and the advantage over other spin-offs with a smaller social capital inheritance that drives the exit. Some spin-offs may compete with the mother or siblings (as some parenting models advance) but that is not a requisite. What is required is an opportunity, probably in a competitive or in a complementary niche. In this way, there is substantial space for innovation or at least for the commercial application of an innovation not currently exploited by the mother. In any case, the industry evolves. Hence, not surprisingly the cluster develops towards the areas of most intense entrepreneurial activity, because these are the ones where more new spin-offs are occurring, possibly as a result of technological and market changes.

Cluster evolution and realignment may be driven by the expansion of leader firms and their activities across technological and market boundaries. As leaders explore across boundaries their product-market scope enlarges and new resource needs emerge (Hite and Hesterly, 2001). Employees inside leaders may detect opportunities before others do. This is a fertile ground for spin-offs. The question to be addressed is: who are the leaders able to drive cluster realignment? We call them 'flagships' and explain subsequently what we mean by this concept, debating the advantages/disadvantages of these flagships having foreign versus domestic ownership.

6.3 Flagship firms: multinationals versus domestic firms as cluster leaders

6.3.1 Definition of flagship firms

Here, flagships are mother firms that generate a larger number of entrepreneurial spin-offs because they offer greater network benefits to their offspring. The quality of the social network benefits voluntarily or unintendedly transferred to the spin-offs is an indicator of which firms are flagships. Because these are the firms that gestate a disproportionate amount of new and more successful spin-offs, they are likely to be the primary drivers of the cluster's evolution. We suggest that flagships are easy to identify and may be domestic or foreign-owned (multinational) firms.

The designation 'flagship firm' was used in the international business (IB) literature (D'Cruz and Rugman, 1993; Rugman and D'Cruz, 2000) to reflect those companies providing strategic leadership for its partners in the business system. They developed this concept (always assuming that the flagship is a MNE) within their 'Five Partners Business Network' model. Rugman and D'Cruz (2000) stated that the flagship leads a network of virtual relationships. In short, they see the flagship as the strategic leader of a multi-actor network existing around MNEs, comprising other key partners (suppliers, customers, competitors, and key partners in the business infrastructure).

Despite some commonalities, highlighted below, we do not reproduce the Rugman and D'Cruz meaning of flagship here. We use a more minimalist conceptualization – *the flagship is the company that leads the cluster*. This is more in line with the objective dictionary definition of flagship: '(1) the ship that carries the commander of a fleet and flies his flag; (2) the chief one of a related group; "it is their flagship newspaper" '.[1]

When referring to the cluster, the flagship is easily identifiable as the main firm, the one at the cluster's genesis. Hence, a key difference vis-à-vis the IB literature noted is that we do not focus on fully developed clusters, but rather on the starting or growth phases of the cluster (a complementary, not antagonistic, standpoint).

Moreover, the flagship needs not to be foreign-owned, not even a domestic MNE. It may be a pure domestic firm (as occurs in the case analysed later in this chapter). We also do not posit the need for the flagship to formally co-ordinate the network/cluster, or even to establish frequent commercial relationships with spin-offs. Being spawned from the leader is enough for network benefits to materialize, as previously explained (see also Ferreira et al., 2003). However, though we do not exclude a scenario of more or less formal coordination, we do not see it as a requirement. Moreover, we do not, as Rugman and D'Cruz (2000) do, posit a need for multi-directional information flows coordinated by the flagship, even if that could enhance network spillovers.

We use the term flagship applied to a firm, and not, as the previously cited authors do, as a strategy (that they envisage based on long-term collaboration and learning, especially helpful to overcome internal and/or environmental constraints, e.g. in the internationalization process). Our model focuses on a decentralized configuration, rather than on a centralized one. Furthermore, Rugman and D'Cruz (2000) postulate that the flagship network is internationally oriented. We do not see it as a requisite and suggest that the network can be entirely domestic. Last but not least, in a flagship strategy *à la* Rugman and D'Cruz, no partner is rival of the others; conversely, we accept possible interfirm rivalry. Thus, some of the unique configuration traits proposed by Rugman and D'Cruz for the flagship are not essential from our perspective. Hence, the view that has been adopted in IB is somewhat different, and more maximalist, than the straightforward flagship notion used here.

Our conceptualization of flagship has, however, some commonalities with former uses of the concept in the IB literature. A common perspective is that the flagship is at the core of the cluster, acting (even if indirectly or involuntarily) as a 'strategic centre' (Lorenzoni and Baden-Fuller, 1995). Another point of convergence is the strategic asymmetry between the flagship and its network partners (i.e. the spin-offs), at least at the beginning of the spin-off process that generates the cluster.

As already noted, we do not see a need for the flagship to be foreign-owned. Yet, in the next section we examine that case, highlighting some nuances arising when the flagship is a foreign MNE in contrast to a pure domestic company.

6.3.2 Multinationals and clusters

The role of MNEs in clusters has been explored (Birkinshaw and Hood, 2000; Holm et al., 2002). Nonetheless, previous research departs usually from the perspective of a MNE subsidiary setting up in an extant industrial cluster; we focus on the moment of the inception of the cluster.

MNE-focused scholars such as Enright (2000) and Rugman and Verbeke (2003) have recognized that in most clusters MNEs are leading participants. One of the most extreme articulations of this view is Rugman and D'Cruz (2000) mentioned above, suggesting that MNEs act as flagships to lead, co-ordinate, and manage strategically the value-added activities of partner firms in a network, including key suppliers, customers, competitors, and the non-business infrastructure.

A commonality between an important body of the IB literature and our research is the fact that repeated interactions breed trust, thus reducing transaction/information costs (Gulati, 1995; Casson, 1998), an idea that is also at the heart of Marshallian (Marshall, 1920) ideas of industrial agglomeration, and present in several chapters of this book (especially those focused on linkages). These interactions and lower transaction costs facilitate business in general and exploitation of opportunities, as they ease exchange and contracting out work to other firms that have complementary resources and skills; in turn, inter-firm linkages and co-operation tend to generate spillovers, leading to a self-reinforcing process and to a virtuous dynamic.

The presence of MNEs in clusters is often investigated in terms of the impact (spillovers) of MNEs' activities; we do not delve into this rich strand of research (for a comprehensive review see Tavares and Young, 2005). Rather, even if spillovers are not the theme focused here, when arguing in favor of network benefits, we are admitting that spillovers exist even if originated in domestic firms (if they generate business opportunities thereby leading to spin-offs).

The scenario of having a MNE as the flagship may be interesting from a host economy perspective: especially for those without a very consolidated/sophisticated industrial base, a foreign subsidiary may be a source of novel technologies and business models. Thus, if the foreign MNE operates in an industry not present in the host economy, or if such subsidiary is a step above local counterparts in the same or related industries, clearly the host economy may have scope to take advantage of, for instance, relevant demonstration and learning effects. However, the local subsidiary does not always undertake high value-added activities, nor more sophisticated, than those of indigenous firms. In these cases, spillovers may be limited, as they probably are when the subsidiary has a too large technological gap vis-à-vis local firms.

A MNE subsidiary, especially when it is large and high-tech, also tends to provide employment opportunities to talented staff, who may learn effectively the 'tricks of the trade', and how to perceive opportunities. Employees may use the experience acquired (Burton et al., 2002; Helfat and Lieberman, 2002) in the MNE to set up their own spin-off (as occurred in the software industry in Ireland, and in India).

Nonetheless, having a MNE as a flagship bears an additional risk, as MNEs may be more footloose (less rooted) than domestic firms. Foreign-owned subsidiaries tend to have fewer local strings and carry out divestments and/or relocations more easily, with potential dramatic effects on the local economic and social environment. In turn, one could also counter-argue that, because MNEs have by definition their international network and are used to operating flexibly they have usually greater survival probability than less flexible non-internationalized firms. This argument is harder to defend when the domestic-owned firm would itself be a MNE, in which case it could arguably have a similar potential advantage of a foreign-owned firm.

Another important element to this discussion has to do with the level of autonomy enjoyed by the subsidiary (vis-à-vis its MNE group). It has been argued that autonomous subsidiaries tend to be more locally embedded, and that local embeddedness may create uniqueness that underlies getting more status, thus autonomy, from the group. The importance of the local context, and in particular the dynamic of the local business environment, has also been discussed in IB research (Birkinshaw and Hood, 2000; Holm et al., 2002), particularly looking at the strength/relevance of the local context versus corporate control as determinants of subsidiary evolution. The chapters by Pedersen and by Andersson and Persson in this volume address the theme of the determinants of subsidiary evolution.

Regarding autonomy, the usual hypothesis is that subsidiaries less integrated in their groups tend to generate greater network benefits than highly integrated ones. However, as Chapter 15 of this book notes, there seems to be a U-shaped relationship whereby too much autonomy may mean that the subsidiary is not sharing much with the group, limiting technological exchange, and other forms of information/knowledge-based interactions.

A relevant facet of local embeddedness may be that MNEs, as Hymer (1976) defended, are burdened by the 'costs of foreignness', i.e. they are at a disadvantage vis-à-vis their domestic counterparts in such aspects as establishing networking ties with local firms. However, these costs may be offset by reputation and other 'ownership advantages' (Dunning, 1977) that MNEs may hold that make local network stakeholders want to relate to the foreign subsidiary.

Finally, it is worth noting Birkinshaw and Sölvell's (2000) conclusion that subsidiaries in leading edge clusters are more embedded than in other sectors. It is thus likely that the quality of local clusters is a determinant of the type of subsidiary an economy can attract. However, we come again to the conclusion that IB research is too centered in already existing clusters, making our conceptualization in this chapter a valuable call for a need to research further cluster's inception. Moreover, the roles of foreign subsidiaries vary considerably across clusters and are associated to the dynamism of the cluster and the overall level of foreign ownership (Birkinshaw and Hood, 1998).

To conclude, despite the factors highlighted, there is not a conclusive answer to what may be the best situation: a domestic firm or a foreign-owned subsidiary acting as a flagship. In a contingent view we consider that both can be important, with the MNE having more scope for radical change although with greater risks of instability. In this section we discussed some of the nuances involved in both situations, which we will be back to when debating possible policy options (in the last section of this chapter). Next, we present an application of our model to a specific empirical case.

6.4 Illustrative case: plastic molds cluster in Marinha Grande, Portugal

The plastic molds cluster, located near Marinha Grande, Portugal, provides an interesting exploratory example of how industrial clusters emerge through essentially an entrepreneurial process fed by insider spin-offs. The dynamism of this cluster illustrates the geographically bounded concentration of firms in related and supporting industries (Porter, 1998). It also shows that spin-offs do not need to compete with each other. In fact, although some competition may exist, co-operative ties seem to prevail, especially at the startup stage of the spin-offs' 'lifecycle' (Hite and Hesterly, 2001). Co-operation not only seems to be socially determined, as individuals/firms co-locate in a small area, but it may also explain how the industry has been capable of responding to changing market needs. A look at the industry's exports renders evident a continuous shift from low margin, low tech, to high tech manufacturing required to serve sophisticated domestic and foreign clients.

6.4.1 Historical background

The history of the molds industry in Portugal dates back to the seventeenth century and to a glass manufacturing plant established in Marinha Grande by the Stephen family. In the twentieth century the first molds for glass were produced, and a history of highly specialized human resources, with very specific capabilities, initiated. In the 1940s the small firm Aníbal H. Abrantes started the shift from manufacturing molds for glass (mainly for bottles) to the production of molds for the plastic industry. Aníbal H. Abrantes is one of the most reputed plastic molds firms, and in the milieu it is even referred to as 'the university of the molds'. In the mid 1950s Aníbal H. Abrantes initiated exports, and from then on the internationalization of the industry cluster kept growing. The orientation to foreign markets is still a major characteristic of the cluster, with about 90 per cent of the production being exported. As new entrepreneurs established in the region, essentially entrepreneurs that foresaw an opportunity to spin-off from the most reputed and visible firms, a rich geographically bounded cluster emerged. The growth of the cluster was remarkable, and currently comprises about 200 firms employing 5000 workers, and exporting to 50 countries. The cluster's exports have tripled in the last 14 years, and the cluster is recognized world-wide to be at the forefront of technological advances in the industry.

6.4.2 Historical incidents and entrepreneurship

With the Portuguese war in the former African colonies in the 1960s a shortage of human resources led firms to explore new organizational models. To overcome personnel shortage, molds firms started hiring their own employees to perform highly specialized tasks in a peculiar outsourcing arrangement. These employees either used the firm's facilities and equipment for their overtime/independently contracted work, or the firm would set up in the employees' garage the equipment necessary for the activities outsourced. The employees' overtime work was independently contracted. As the co-operation with the employees was maintained they started creating their own spin-offs, frequently in their own garages. Most of these firms persist today and this founding process where employees leave the parent firm to constitute their own spin-offs seems to be a major driver of the cluster expansion, as our own initial studies (Ferreira et al., 2005, 2006) predicted.

6.4.3 Innovation in the plastic molds cluster

As authors reviewed above advanced, the relative dangers that insider entrepreneurs could adopt mimicking behaviors, possibly leading the cluster to maturity, stagnation and decline, is not the case in this cluster. The isomorphic tendencies that could emerge when entrepreneurial employees simply mimic prior experiences are not prevailing and substantial innovations and complementarities dominate the cluster. Perhaps this is due to the cluster's industry structure, exclusively composed by small and medium firms (SMEs), price-takers, and seeking to minimize transaction costs that could emerge if opportunism were triggered. The innovative capacity and the ability to rethink the industry were obvious in the implementation of the latest communication technologies that permit joint product development with clients. These new technologies allow firms to keep permanent contact not only with clients, but also with other network actors, e.g. suppliers. Moreover, the shift of production from the toys industry to the far more demanding customers in the electronics and auto industries, are good examples of the ability to search beyond local boundaries.

The continuous generation of spin-offs induces important changes in the cluster's configuration. As incumbents expand, their product-market scope enlarges and new resource needs emerge. This is a fertile ground for entrepreneurial employees to constitute spin-offs. Simultaneously, over time, the firm becomes less resource dependent on the mother (and the latter's ties) in favor of exchange ties with the self-gestated spin-offs. We note, however, that the umbilical ties do not completely fade away, and the hubs identifiable in the cluster are largely co-operative. However, it would not be astonishing if, in periods of high uncertainty and change, co-operation were replaced by some degree of competition among firms (and hubs).

Innovative ability is also possible due to low competition among firms, an interesting finding in small firms' networks. Competition is reduced because firms lean towards specialization, hence, effective differentiation, in terms of mold size (weight), design, material used, and client industry. This co-operation is fuelled by an institutional framework with little parallel (in this industry) in other locations: two associations (CEFAMOL – Associação Nacional da Indústria de Moldes and CENTIMFE – Centro Tecnológico da Indústria de Moldes e Ferramentas), multiple relations to Portuguese and foreign universities, and to an established international 'brand' image of the region (perceived as a European and world leader in some types of molds). Furthermore, an industry structure populated by SMEs leaves many opportunities for outsourcing. SMEs need to rely on outside firms for the majority of the activities required: from transport to accounting, packaging to equipment, to distribution and after-sales service. This generates a special dynamic and an active engagement of all stakeholders involved. Besides, the relative specialization of SMEs generates chances for new spin-offs up and down the product value chain in untapped segments (see Figure 6.1).

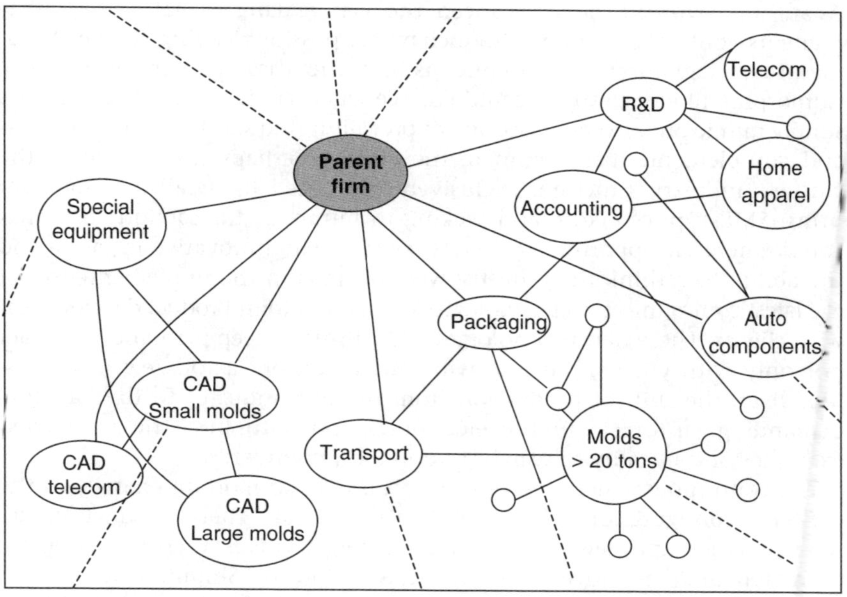

Figure 6.1 Specialization in the network
Source: Authors.

6.4.4 Strengths

The cluster's strength appears to lie in the peculiar founding circumstances, in which incumbent firms continuously gestate new spin-offs. This is, in our opinion, an interesting model of industry development. Moreover, it seems to be applicable to a multiplicity of clusters in various industrial sectors and countries.

The 'motherhood model', as we designate it, is based on employees exiting the parent to constitute their own spin-offs, with the support of the mother. This model also illustrates an effective vehicle for the flow of information, and innovation diffusion, with clear bandwagon effects (Granovetter, 1973). As the spin-off grows, a hub of self-gestated spin-offs gradually emerges around it, as its employees also start exiting to found their own spin-offs. In consequence, the cluster expands. The innovator (initiator) firm gradually gestates new spin-offs (followers) Figure 6.2. The continuous repetition of this process originates a highly inter-linked network of largely co-operative firms with fluid hubs around central, high status flagship firms.

In the Portuguese plastic molds cluster the high level of co-operation among firms assumes diverse forms: employee mobility and exchange, technical assistance, co-operation in special and/or large projects, participation in research projects and conferences, share of clients' orders, pooling of complementary assets for a client's order, and contribution to trade fairs, that we have identified so far. Such interactions between incumbents increase information flows, foster the development of more cohesive ties, facilitate innovation diffusion, and widen employees' tacit and explicit knowledge. Then co-operation between firms provides opportunities for the discovery of business opportunities or untapped markets by entrepreneurial employees to constitute their spin-offs. The more frequent, the broader the forms of co-operation, and the more such collaborations involve the transfer of tacit knowledge, the greater are the opportunities for generating new spin-offs. Hence, firms that co-operate more are also more likely to be the grounds for the gestation of more spin-offs.

This cluster is further supported by personal cohesive ties of the new entrepreneurs to former employers. These are the ties that *prima facie* provide legitimacy to the new spin-off, opening an array of opportunities to engage in exchange with other firms in the area. Sedaitis (1998) suggested that the network of relationships of the new firm depends essentially on the social density of the founding network. The prevailing tie in the cluster appears to be acquaintanceship, i.e. an 'advice tie' originated in the prior employment relation. This cohesive tie provides a source of familiarity and trust (Kale et al., 2000), essential for the successful gestation of spin-offs, and crucial to create a more efficient environment through the reduction of transaction costs.

In short, this case provided an illustration of how our motherhood model works in practice. This model may also be instrumental in explaining other

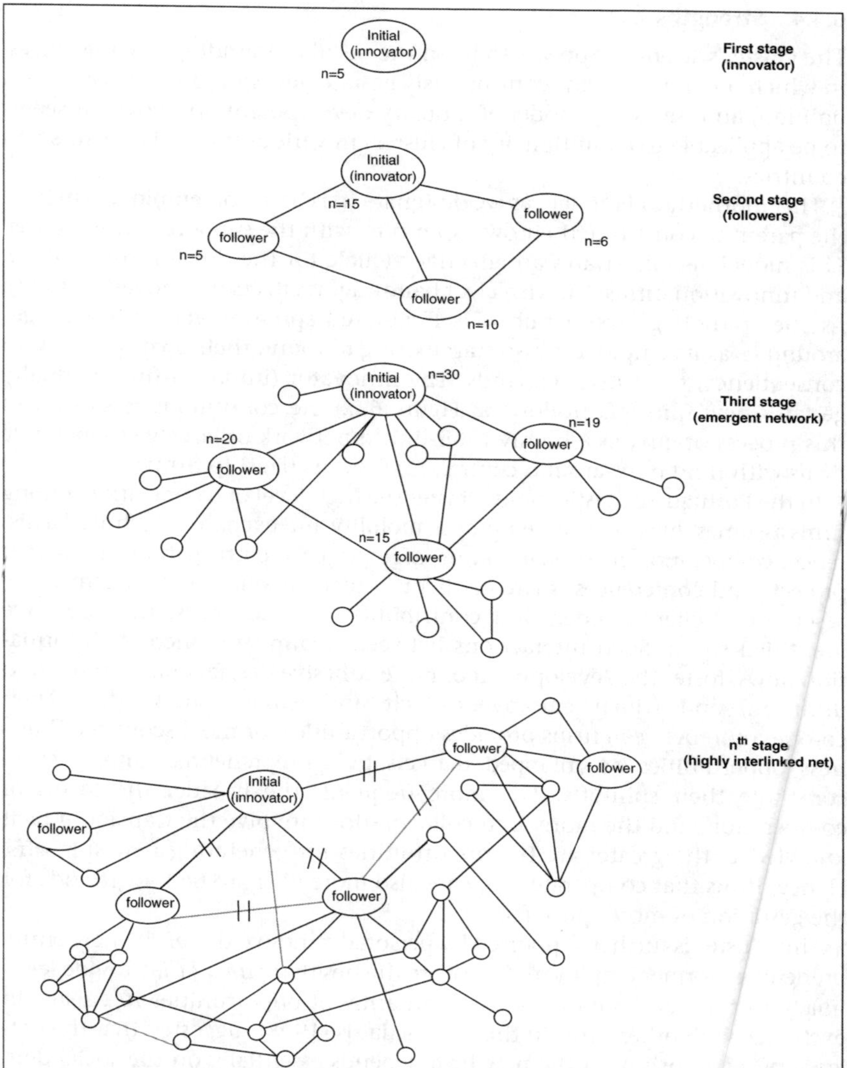

Figure 6.2 Snapshots of the evolution of a mothering-based network

Source: Authors.

sectors and clusters in other regions and countries. In future research, we expect to venture more applications to other industrial clusters, as well as deepening the one presented in this chapter through a clinical, in-depth case study depicting and explaining the networks of ties briefly highlighted above.

6.5 Concluding remarks and policy issues

6.5.1 Conclusions

This chapter sheds light on the phenomenon of cluster emergence and evolution, focusing on an endogenous process by which a flagship firm gestates entrepreneurial spin-offs. In contrast to other parenting processes envisaging spin-offs through a negative lens (dissatisfied employees leaving), the 'motherhood model' presented here offers a rather positive perspective. It focuses essentially on employees exiting because there is a promising entrepreneurial opportunity, often complementary to the parent's activity. The odds of success of the spin-offs are nurtured by the intended or unintended inheritance from the mother firm, near whom they co-locate.

Our model highlights network benefits and reputation spillovers arising from being gestated by a prestigious firm, that help spin-offs to overcome the stylized liabilities of newness and smallness plaguing their founding phase, and also to secure resources and reduce uncertainty. We made the case that these benefits put entrepreneurial spin-offs by insider employees in a position of advantage vis-à-vis outsider startups.

We also specify the meaning of flagship adopted in this chapter. Our model is applicable no matter if the flagship is a domestic or foreign firm. Still, we elaborate on some advantages and disadvantages of having a foreign subsidiary versus an indigenous company as a flagship. Even if the model applies to both cases there is some specificity concerning a potential impact on the host economy. This issue is further analysed in the subsequent policy section.

There is ample empirical evidence in different countries and industries that this motherhood process is a realistic model explaining the emergence and development of several clusters. To illustrate this perspective we examined the case of the plastic molds cluster of Marinha Grande, Portugal – an international leader in various niches of the industry. We observed that this cluster's inception was driven by a flagship firm that led to the set-up of several entrepreneurial spin-offs and that these spin-offs have benefited considerably from the expertise gathered by founders/entrepreneurs at the flagship company, and from the network benefits inherited, that help relating not only to the mother and to other offspring, but also to external entities, as we posited.

6.5.2 Policy reflections: how to stimulate dynamic, innovative, and entrepreneurial clusters

This chapter argues in favor of the frequent applicability of a positive perspective of cluster emergence and evolution. We believe that there is scope for policy-making to improve the situation created by the firms' dynamic. In this section we address policy issues to promote entrepreneurial spin-offs and to stimulate cluster dynamism and innovation. This policy discussion, that does not make recommendations *tout court*, but offers as a set of suggestions and reflections on relevant themes, is rather topical nowadays. Its relevance is justified by the failure of several cluster initiatives that have been attempted in many countries (where governments, for example, threw subsidies without a systems view of the cluster or of the networks at stake). It is also justified because regions increasingly seek to figure out ways of promoting endogenous growth and development though inward investment, namely, stimulating innovation. Agglomerations are currently perceived as a critical factor in helping to promote this endogenous dynamism. In a necessarily short manner, the critical issues at stake, from a policy viewpoint, are:

- Can policy-makers stimulate the emergence of clusters through spin-offs from flagship firms?
- How to spot and nurture effective flagships?
- If the flagship is a MNE, how can the 'right type' of firm be attracted? If the MNE subsidiary is already operating in the country, what can be done to stimulate its role in the formation and evolution of the cluster? How can inter-firm ties be enhanced?
- How to promote innovative activities, and innovation diffusion, in the cluster?

There are no easy answers, or 'right recipes', to above questions. Notwithstanding, the following suggestions offer a starting point for debate.

First and foremost, very often, cluster-related policies have failed. A vast body of literature, e.g. Enright (2000), questioned how appropriate and sufficient are cluster-stimulating measures. Research has noted the lack of a systems view, and the crucial understanding that clusters cannot be forced from night to day; they often take long to emerge, through gradual trust-building among actors. These trust-related bonds distinguish a cluster from a mere agglomeration of firms. Our model highlights exactly the importance of these bonds.

It is also important to note that for a flagship-led cluster to emerge, the first requirement is the very existence of the flagship. Policy-makers should thus look at the industrial landscape seeking to identify which are the flagships with more 'fertility potential'; in Ferreira et al. (2005, 2006) we argued that flagships are likely to have some identifiable traits: large, older, and central, and more connected. Government should thus first take stock of the companies offering more potential for spin-offs, and take proactive/selective policies to stimulate what appear to be the 'right' ones. This does not mean the former obsession of some countries with 'picking winners' and giving them subsidies, but rather more sophisticated and targeted measures. For instance, promoting innovation through funding selected projects; promoting ties with universities, either via match-making (decreasing transaction costs), supporting joint projects with universities and other research institutions, and so forth.

When the flagship is foreign-owned there are two situations: first, whether a potential flagship is not yet an investor in the economy; and second, in the case when the MNE is there, what then should be done. In the first case, and taking into account the lack of multilateral FDI regulation (Young and Tavares, 2004) and the race for FDI (Oxelheim and Ghauri, 2003) in which nearly all countries are involved, support arises for proactive FDI attraction policies – but these are likely to be effective only if very selective – to the point of identifying a specific company and trying to persuade it to invest (as Ireland and others did). The 'right' type of firm to chase would vary according to the initial conditions of the country, the expected upgrade, extant resources and capabilities of workers; in short, would presuppose a serious evaluation of the match between the economy's locational determinants (Dunning, 1977) and the firm's characteristics. Sectoral targeting is very common, as is targeting attending to technology levels and the potential for creating jobs.

Another question arises when the firm is already in the country, in which case a set of measures can be enacted: promote linkages, material and intangible/knowledge-based (as highlighted by several chapters in this volume) – hopefully with spin-offs; trying to upgrade extant innovative activities of the flagship – hence stimulating a 'learning environment'; and making sure labor laws stimulate the design of contracts that do not prevent the use of previously accumulated knowledge. Employment contracts are important regardless of whether the flagship is foreign or domestic. Restrictive employment contracts can seriously hinder entrepreneurship.

Further recommendable measures relate to stimulating human capital formation and on-the-job training to help employees learn as much as possible before spinning off more equipped to understand markets and technologies, thus enhancing spin-offs' survival prospects. Other survival-enhancing policies could be to provide selected business services that spin-offs have no resources for establishing on their own.

Any selective measure to stimulate networking (suppliers' networks, purchasing centrals, information exchange, increased use of information technology, joint collaborative projects) would potentially bear fruits. So would in principle innovation-targeted policies, with an underlying long-term vision of what is needed for the economy's industrial strategy. For instance, intellectual property measures such as assisting SMEs in applying for patents may be an idea. Also, helping companies to engage in university/R&D collaboration and acquire consultants' services, e.g. for strategy definition, and for technology development and commercialization. Strategies favoring niches that would not cause entropy or isomorphism should receive priority.

Special attention should always be given to policy implementation and evaluation, also having the courage to correct mistakes. This was not done often when cluster-related policies became fashionable, and when recipes proved ineffective. One thing is certain: any policy is as good as its implementation; policy experimentation should occur, and the old tactic of subsidy-throwing avoided, but always very selectively and carefully planned/evaluated and with a long-term, systemic, perspective.

Note

1 worldreference.com site.

References

Abrahamson, E. and Rosenkopf, L., 'Social network effects on the extent of innovation diffusion: a computer simulation', *Organization Science*, 8(3) (1997) 289–309.

Arthur, W., 'Silicon Valley locational clusters: when do increasing returns imply monopoly?', *Mathematical Social Sciences*, 19 (1990) 235–51.

Baum, J., Calabrese, T. and Silverman, B., 'Don't go it alone: alliance network composition and startups' performance in Canadian biotechnology', *Strategic Management Journal*, 21 (2000) 267–94.

Birkinshaw, J. and Hood, N., 'Multinational subsidiary evolution: capability and charter change in foreign-owned subsidiary companies', *Strategic Management Journal*, 23(4) (1998) 773–95.

Birkinshaw, J. and Hood, N., 'Characteristics of foreign subsidiaries in industry clusters', *Journal of International Business Studies*, 31(1) (2000) 141–54.

Birkinshaw, J. and Sölvell, Ö., 'Leading-edge multinationals and leading-edge clusters', *International Studies of Management and Organization*, 33(2) (2000) 3–9.

Brittain, J. and Freeman, J. 'Organizational proliferation and density dependent selection', in Kimberly, J. and Miles, R. (eds), *The Organizational Life Cycle* (San Francisco: Jossey Bass, 1980) pp. 291–338.

Burt, R., *Structural Holes: The Social Structure of Competition* (Cambridge, MA: Harvard Business School Press, 1992).

Burton, M., Sorensen, J. and Beckman, C., 'Coming from good stock: Career histories and new venture formation', in Lounsbury, M. and Ventresca, M. (eds), *Research in the Sociology of Organizations* (New York: Elsevier, 2002) pp. 229–62.

Carroll, G., 'Organizational ecology', *Annual Review of Sociology*, 10 (1984) 71–93.

Casson, M., *Information and Organization* (Oxford: Oxford University Press, 1998).

Coviello, N. and Munro, H., 'Growing the entrepreneurial firms: networking for international market development', *European Journal of Marketing*, 29(7) (1995) 49–61.

D'Cruz, J. and Rugman, A., 'Developing international competitiveness: the five partners model', *Business Quarterly*, Winter (1993).

Dalum, B., 'Local and global linkages. The radiocommunications cluster in Northern Denmark', *Journal of Industry Studies*, 2 (1995) 89–109.

Dunning, J., 'Trade, location of economic activity and the MNE: a search for an eclectic approach', in Ohlin, B., Hesselborn, P. and Wijkman, P. (eds), *The International Allocation of Economic Activity* (London: Macmillan, 1977) pp. 395–418.

Enright, M.J., 'The globalization of competition and the localization of competitive advantage: policies towards regional clustering', in Hood, N. and Young, S. (eds), *The Globalization of Multinational Enterprise Activity and Economic Development* (London: Macmillan, 2000) pp. 303–31.

Feldman, M., 'The entrepreneurial event revisited: firm formation in a regional context', *Industrial and Corporate Change*, 10(4) (2001) 861–91.

Ferreira, M., Li, D. and Tallman, S., 'Innovation, knowledge sharing, and clustered firms' ability to capture rents from innovation', Paper presented at the Academy of Management meeting (New Orleans, 2003).

Ferreira, M.P., Hesterly, W. and Tavares, A.T., 'Industry clusters evolving: a new perspective on parenting spin-offs', in Fai, F. and Morgan, E. (eds), *Managerial Issues in International Business* (London: Palgrave Macmillan, forthcoming 2006).

Ferreira, M.P., Hesterly, W., Tavares, A.T. and Armagan, S., 'Network and firm antecedents of spin-offs: mothering spin-offs', Paper presented at the Economics of Networks Conference (Budapest, 2005).

Forrant, R. and Flynn, E., 'Seizing agglomeration's potential: the greater Springfield Massachusetts Metalworking sector in transition, 1986–1996', *Regional Studies*, 32(3) (1998) 209–22.

Freeman, J., 'Entrepreneurs as organizational products: semiconductor firms and venture capital firms', *Advances in the Study of Entrepreneurship, Innovation, and Economic Growth* (Greenwich, JAI Press Inc., 1986) pp. 33–52.

Garvin, D., 'Spin-offs and the new firm formation process', *California Management Review*, January (1983) 3–20.

Granovetter, M. 'The strength of weak ties', *American Journal of Sociology*, 78 (1973) 1360–80.

Gulati, R. and Gargiulo, M., 'Where do networks come from?', *American Journal of Sociology*, 104(5) (1999) 1439–93.

Gulati, R., 'Does familiarity breed trust? The implications of repeated ties for contractual choice in alliances', *Academy of Management Journal*, 38(1) (1995) 85–112.

Hannan, M. and Freeman, J., 'The population ecology of organizations', *American Journal of Sociology*, 83 (1977) 929–64.

Helfat, C. and Lieberman, M., 'The birth of capabilities: market entry and the importance of prehistory', *Industrial and Corporate Change*, 11(4) (2002) 725–60.

Hendry, C., Brown, J. and DeFillipi, R., 'Regional clustering of high technology-based firms: opto-electronics in three countries', *Regional Studies*, 34(2) (2000) 129–44.

Hite, J. and Hesterly, W., 'The evolution of firm networks: From emergence to early growth of the firm', *Strategic Management Journal*, 22 (2001) 275–86.

Holm, U., Malmberg, A. and Sölvell, Ö., 'MNEs impact on local clusters: the case of foreign owned subsidiaries in Sweden', *IIB Working Paper* 02/11 (Stockholm: Stockholm School of Economics, Institute of International Business, 2002).

Hymer, S., *The International Operations of National Firms: A Study of Direct Foreign Investment Abroad* (Cambridge, MA: MIT Press, 1976).

Kale, P., Singh, H. and Perlmutter, H., 'Learning and protection of proprietary assets in strategic alliances: Building relational capital', *Strategic Management Journal*, 21 (2000) 217–37.

Klepper, S., 'Employee startups in high-tech industries', *Industrial and Corporate Change*, 10(3) (2001) 639–74.
Leonard-Barton, D., 'Core capabilities and core rigidities: A paradox in managing new product development', *Strategic Management Journal*, 13 (1992) 111–25.
Lorenzoni, G. and Baden-Fuller, C., 'Creating a strategic centre to manage a web of partners', *California Management Review*, 37(3) (1995) 146–63.
Marshall, A., *Principles of Economics*, 8th edn (London: Macmillan, 1920).
Oviatt, B. and McDougall, P., 'Global start-ups: entrepreneurs on a worldwide stage', *Academy of Management Review*, 9(2) (1995) 30–43.
Oxelheim, L. and Ghauri, P., *European Union and the Race for Foreign Direct Investment in Europe* (Oxford: Elsevier, Pergamon, 2003).
Phillips, D., 'A genealogical approach to organizational life changes: the parent–progenitor transfer among Silicon Valley law firms, 1946–1996', *Administrative Science Quarterly*, 47 (2002) 474–506.
Pinch, S. and Henry, N., 'Paul Krugman's geographical economics, industrial clustering and the British motor sport industry', *Regional Studies*, 33(9) (1999) 815–27.
Podolny, J., 'Networks as pipes and prisms of the market', *American Journal of Sociology*, 107(1) (2001) 33–60.
Porter, M., 'Clusters and the new economics of competition', *Harvard Business Review*, (1998) 77–90.
Rugman, A.M. and D'Cruz, J.R., *Multinationals as Flagship Firms: Regional Business Networks* (Oxford: Oxford University Press, 2000).
Rugman, A.M. and Verbeke A., 'Multinational enterprises and clusters: an organizing framework', *Management International Review* 43(3) (2003) 151–9.
Saxenian, A., *Regional Advantage: Culture and Competition in Silicon Valley and Route 128* (Cambridge: Harvard University Press, 1994).
Sedaitis, J., 'The alliances of spin-offs versus startups: Social ties in the genesis of post-Soviet alliances', *Organization Science*, 9(3) (1998) 368–81.
Shane, S., 'Prior knowledge and the discovery of entrepreneurial opportunities', *Organization Science*, 11(4) (2000) 448–69.
Singh, J., Tucker, D. and House, R., 'Organizational legitimacy and the liability of newness', *Administrative Science Quarterly*, 31(2) (1986) 171–94.
Sorenson, O. and Audia, P., 'The social structure of entrepreneurial activity: geographic concentration of footwear production in the US, 1940–1989', *American Journal of Sociology*, 106 (2000) 324–62.
Stinchcombe, A., 'Social structure and organizations', in March, J. (ed.), *Handbook of Organizations* (Chicago: Rand McNally, 1965) pp. 142–93.
Swann, P., 'Clusters in the US computing industry', in Swann, P., Prevezer, M. and Stout, D. (eds), *The Dynamics of Industrial Clustering: International Comparisons in Computing and Biotechnology* (Oxford: Oxford University Press, 1998).
Tavares, A.T. and Young, S., 'FDI and multinationals: patterns, impact and policies', *International Journal of the Economics of Business*, 12(1) (2005) 13–16.
Uzzi, B., 'Social structure and competition in interfirm networks: the paradox of embeddedness', *Administrative Science Quarterly*, 42 (1997) 35–67.
Young, S. and Tavares, A.T., 'Multilateral rules on FDI: do we need them? Will we get them? A developing country perspective', *Transnational Corporations* (2004) 1–29.
Zander, I., 'The micro-foundations of cluster stickiness – entering the mind of the entrepreneur', Paper presented at the 4th IB research forum at Temple University, Philadelphia (2003).

7
Promoting Industrial Clusters: Evidence from Ireland[1]

Anne Marie Gleeson, Frances Ruane and Julie Sutherland

7.1 Introduction

New economic geography focuses on the importance of the location of industries within economies and emphasizes the role of centripetal and centrifugal forces that affect the spatial concentration and sectoral specialization of industry. The tendency for enterprises conducting innovative economic activity to cluster has been found to be higher in industries where external economies of scale and new economic knowledge play an important role. When this knowledge is tacit, geographical boundaries will exist (Audretsch and Feldman, 1996). Enterprises will locate within close geographic proximity to each other in order to benefit from potential knowledge externalities and to gain access to skilled workers endowed with a high level of industry-specific human capital. The spatial concentration and sectoral specialization of enterprises thus have significance for regional and industrial policy as the establishment of high-technology clusters has the potential to increase economic activity and enhance regional development.

Ireland's success in attracting foreign direct investment (FDI), evidenced by the strong presence of multinational enterprises (MNEs) in its manufacturing sector, has been well documented (Barry et al., 1999; Ruane and Uğur, 2005). Consistent public policy instruments promoting Ireland as an export platform for manufacturing FDI over the past four decades have integrated regional and industrial policy objectives through proactive and selective support for MNE projects. Specifically, the focus of industrial policy has been to attract MNEs in the electronics (NACE 30–33) and chemical and pharmaceutical sectors (NACE 24). In the former case, the policy has sought to encourage MNEs across the range of upstream–downstream industrial activities so that an effective industrial clustering of MNEs would result (Görg and Ruane, 2001). For example, the Industrial Development Authority (IDA), which is charged with promoting Ireland as an industrial location for FDI, has enticed leading enterprises in these high-tech sectors (Intel, Dell, IBM, Hewlett Packard) in the hope of generating a 'contagion' effect that encourages other

MNEs in electronics to locate in Ireland with upstream and downstream links to these sector leaders (Krugman, 1997). In the case of chemicals and pharmaceuticals, policy has focused on creating a geographic clustering of upstream activities, centered primarily on the creation of a serviced site in Cork, as well as encouraging the location of pharmaceutical plants in downstream activities more widely distributed throughout Ireland.

The success, or otherwise, of policies encouraging spatial and sectoral industrial clustering of MNEs in Irish manufacturing is relatively little analysed or understood. Moreover, the correlations between spatial and sectoral MNE clustering and the concentration of local (Irish-owned) enterprises (LEs) have not been measured. This chapter compares the spatial and sectoral concentration of manufacturing enterprises in the Irish economy in 1985 and 2001 in order to establish the impact of industrial and regional policy objectives on the clustering of enterprises during this time. Entropy indices are estimated as indicators of clustering patterns for MNEs and LEs separately. To examine the clustering relationships, if any, between enterprises in Irish manufacturing, correlation coefficients are estimated to measure the co-location patterns of LEs and MNEs in 1985 and 2001. The combination of entropy indices and correlation coefficients permits analysis of the impact of stated industrial and regional policy goals on the overall sectoral and spatial clustering of manufacturing within Ireland.

The remainder of this chapter is organized as follows. Section 7.2 describes previous literature on the use of entropy measures as a means of measuring industrial clustering. Section 7.3 reviews Irish industrial and regional strategy over the past 40 years with an emphasis on the increasingly high-tech focus of policy since the late 1980s. In Section 7.4 the data set used is described and entropy indices and correlation coefficients of manufacturing at Irish county and sectoral level for 1985 and 2001 are reported. Section 7.5 focuses on the implications of the results obtained for clustering policy in Ireland.

7.2 Context

The analysis of enterprise clustering presented in this chapter is an application and extension of the *new economic geography* literature that explains the spatial concentration and dispersion of economic activity within countries or regions (Krugman 1979, 1980, 1991) in the context of the new theories of international trade and endogenous growth (see Fujita and Thisse, 1996 and Ottaviano and Puga, 1997 for comprehensive literature surveys). The basic model of new economic geography has its origins in the new international trade theory literature developed by Krugman (1979, 1980), which, in turn, has its foundations in the Dixit–Stiglitz (1977) model of monopolistic competition. Increasing returns to scale (IRS) and product differentiation play a crucial role in explaining the spatial distribution of production by acting as centripetal forces that depend on transportation costs, with the best locations being those that are close to the consumer market and to suppliers. Hirschman's (1958) forward and backward linkages come into play, leading to a spatial concentration of people and production reinforced by higher wages in large market locations that attract industrial labor, enlarging the market further through linkage effects. The model thus explains how a core–periphery industrial pattern can emerge as a function of transport costs, economies of scale and industrial labor relocation.

Krugman (1993) suggests further grounds for concentration that can result in a spatial 'lock-in' effect, which creates a locational path-dependence favoring the growth of clusters built on one or two large enterprises that act as leaders or anchor enterprises and feed the growth of numerous smaller enterprises (Davies and Lyons, 1996). This can occur by accident or as a result of 'first nature' location conditions, such as the location of specific raw materials or other geographical advantages. Industry locates in a particular area in order to utilize such specific local resources and in so doing attracts further inflows of capital and labor. A pattern of spatial concentration and agglomeration emerges which persists even after the original factors have been exhausted or the initial industries have declined or exited.

Krugman and Venables (1995) and Helpman (1998) extend the Krugman (1991) model by examining the impact of costs associated with increasing agglomeration, such as increased housing and other congestion costs, which limit the degree of centripetal pull in the face of falling transportation costs. This ultimately leads to a dispersal of production to the periphery where wages and expenses are lower, suggesting the existence of a U-shaped relationship between production costs and spatial concentration. Costs of concentration initially fall due to the benefits of internal and external economies of scale resulting from expansion of the market. Over time, the increase in congestion and other related costs outweigh the original benefits of spatial concentration and as a consequence, with the assistance of decreased transport costs, economic activity begins to disperse.

Focusing on specialization, Krugman and Venables (1996) show how regional specialization may be more intense where enterprises have strong forward and backward linkages with certain types of suppliers and buyers where pecuniary externalities arise through market transactions. In this two-sector model both sectors are imperfectly competitive and enterprises are horizontally linked so that there is a greater proportion of exchange between enterprises in the same sector than with enterprises in other sectors. The result is that if one more firm locates in the region it will benefit same sector enterprises through linkage effects, but the resulting increased demand for labor and other inputs, and consequent rising factor prices, may harm all enterprises. Increasing factor demand and prices in the region may result in the non-linked enterprise dispersing to areas where costs are lower.

Thus decisions regarding enterprise production and location depend on many factors, including the location of specific resources, economies of scale and linkage effects. Once located, there is an interconnectedness between space and sector, which manifests itself in spatial and sectoral clusters that reflect different degrees of spatial concentration and sectoral specialization. Different indicators are used to measure the degree of specialization and concentration and many of the indices are a variant of, or related to, the Hirschman–Herfindahl index. In this chapter we use entropy indices based on Aiginger and Davies (2004) to measure spatial concentration (at county level) and sectoral specialization (at NACE 4-digit level).

Following Aiginger and Davies (2004), and focusing on the number of enterprises (E), the entropy index of sectoral specialization ($SPEC$) is a measure of the extent to which a given geographical area (Irish county, $r = 1,\ldots, K$) specializes in a number of industrial (NACE) sectors ($i = 1,\ldots, N$).

$$SPEC_r = -\sum_{i=1}^{N}\left(\frac{E_{ir}}{E_r}\right)\ln\left(\frac{E_{ir}}{E_r}\right) \tag{7.1}$$

$SPEC_r$ is an inverse measure of sectoral specialization. If manufacturing sectors are of equal size in a county, then $(E_{ir}/E_r)=1/n$ for all sectors, and $SPEC = \ln(n)$. If a county is completely specialized in one sector only, then $SPEC = \ln 1 = 0$. The value of $SPEC$ increases the more evenly enterprises in a county are spread across sectors. Similarly, spatial concentration ($CONC$) in Irish manufacturing is measured as:

$$CONC_i = -\sum_{r=1}^{K}\left(\frac{E_{ir}}{E_i}\right)\ln\left(\frac{E_{ir}}{E_i}\right) \tag{7.2}$$

which defines the extent to which manufacturing activity in a given sector is concentrated in a particular Irish county. If there is equal spatial dispersion of industry then $CONC = \ln(r)$, and if there is total concentration, $CONC = \ln 1 = 0$. Weighted averages or typical levels of sectoral specialization and spatial concentration are calculated with county and sectoral shares of total Irish manufacturing, respectively, as weights. Thus average sectoral specialization ($AVSPEC$), using county shares of Irish manufacturing as weights, is defined as:

$$AVSPEC = \sum w_r \cdot SPEC_r, \quad \text{where } w_r = \frac{E_r}{E} \tag{7.3}$$

Similarly, average spatial concentration ($AVCONC$), using industry shares of Irish manufacturing as weights, is defined as:

$$AVCONC = \sum v_i \cdot CONC_i, \quad \text{where } v_i = \frac{E_i}{E} \tag{7.4}$$

The two indices can be re-expressed as

$$AVSPEC = -\left\{\sum\sum\left(\frac{E_{ir}}{E}\right)\cdot \ln\left(\frac{E_{ir}}{E}\right)\right\} - \left\{-\sum\left(\frac{E_r}{E}\right)\cdot \ln\left(\frac{E_r}{E}\right)\right\} \quad (7.5)$$

$$AVCONC = -\left\{\sum\sum\left(\frac{E_{ir}}{E}\right)\cdot \ln\left(\frac{E_{ir}}{E}\right)\right\} - \left\{-\sum\left(\frac{E_i}{E}\right)\cdot \ln\left(\frac{E_i}{E}\right)\right\} \quad (7.6)$$

where the second terms in both equations are also entropy indices. The second term in equation (7.5) is the spatial entropy index at national level, Irish Concentration (*IRCONC*). Similarly, the second entropy term in equation (7.6) is the sectoral entropy index at national level, Irish Specialization (*IRSPEC*).

In Section 7.4 we present estimates of equations (7.5) and (7.6) (*AVSPEC* and *AVCONC*) for all manufacturing in Ireland and for MNEs and LEs separately between 1985 and 2001, a period of significant change in Ireland's industrial policy and performance landscape.

7.3 Ireland's industrial and regional policies

Ireland's policy of promoting FDI dates back to the late 1950s, when it started to promote itself as an export-platform manufacturing base for the European market. Beginning in the early 1970s, and in anticipation of joining the EU, policy towards FDI became increasingly selective, seeking to attract MNE investment specifically into modern high technology (high-tech) sectors. It was widely accepted that Irish entrepreneurs were not well placed to move into the production of high-tech products and the strategy was designed to fill this entrepreneurial gap through MNE investments, specifically in electronics and pharmaceuticals. Such high-tech products were identified as being readily suited to an export-platform island economy because of their low per-unit-value transportation costs.

The strategy of establishing high-tech MNE based sectors had two dimensions, sectoral and spatial. The aim of the sectoral dimension was to entice key international manufacturing enterprises by emphasizing Ireland as an ideal platform for exporting into the EU. This required a project centered approach to attracting MNE investment that worked through an identification process comprising four steps. The first step identified niche markets with global growth potential, especially European growth potential. The second step generated information on enterprises in these sectors, typically in the US, which were already exporting large volumes into Europe and which looked, in terms of a Vernon-type product cycle, likely to be considering a European production base. The third step was to persuade these enterprises to consider Ireland as an investment base; such persuasion was only credible for high value-to-volume products, as low value/volume products would not find an island such as Ireland a competitive long-term location choice. The final step involved agreeing an incentives package that would secure the investment and at the same time ensure significant employment benefits to Ireland as a host country. Irish policy-makers recognized the heterogeneity of potential MNEs and their different host-country potentialities at a very early stage.

The project-based approach established the framework for developing the spatial dimension of the strategy, namely the high-tech cluster concept that developed in the 1980s, which had two intended outcomes. First, high-tech MNE clusters would generate Marshallian external economies and optimize the positive cumulative effects of attracting FDI enterprises. These effects would be reflected in (i) the development of potential Hirschman-type production linkages between foreign enterprises; (ii) the growth of a skilled labor pool to which all entrants and incumbents would have access; and (iii) the possibility of higher levels of linkages and technology transfer from MNEs to LEs as LEs clustered next to MNEs. Secondly, the creation of clusters raised the possibility of regional markets emerging with critical mass in terms of labor and service supply; it was argued such regional markets would allow greater numbers of enterprises to locate successfully outside Dublin and Cork, the largest Irish cities. This approach replaced the earlier policy, operating in the 1960s and 1970s, of distributing individual manufacturing plants in a highly dispersed way across small-town clusters in Ireland.

The successful implementation of this sectoral and spatial approach to cluster creation required the identification of key sectors and associated locations, and their promotion to potential inward FDI enterprises. In the case of electronics it also built on an active program (in the 1980s and 1990s) that fostered production links between enterprises, thereby building a vertically integrated electronics sector to service the European market. With the location of Intel, Microsoft, Dell and Hewlett Packard as the hub, the spokes were quickly populated by dozens of smaller electronics and software enterprises, all of which wanted to interconnect with these key industrial leaders. As Krugman (1997) explained, the Irish economy was a significant beneficiary from the process of clustering and also of some good luck. Part of this luck was 'made' in the consistent enterprise-centered approach developed over the previous 25 years, and in the management by policy-makers of the process of rapid cluster building. For example, policy has been highly active in addressing skill needs (including specialized skills) and in managing a good industrial relations environment for incoming investors.

It is now widely recognized that consistency in public policy, through proactive and selective support for MNE projects in a manner that has integrated regional and industrial policy objectives, has been crucial to the success of building an MNE export-platform base in Ireland. However, the consequences of this policy success for the concentration and specialization of MNEs and LEs are not clear.

Most FDI during this period has been in high-tech industries and in the form of greenfield export-platform investment, reflecting the fact that the small Irish domestic market is not the primary attraction for most FDI investment; US MNEs locating in Ireland export 94 per cent of their output, 76 per cent of which is bound for EU destinations (IDA, 2002). Consequently, the spatial distribution of Irish consumers is not relevant to the location choices of MNEs, which are driven by production rather than market considerations. The absence of a significant local market means that in principle MNEs are more likely to be able to respond positively to incentives created by regional policy in terms of their location choices. As such, there is a natural contrast between the situation of export-focused MNEs and the primarily domestic focused LEs. Also, the question of whether the location pattern of LEs mirrors that of MNEs is of particular interest where evidence of LE/MNE dualism is apparent.

This chapter explores how the spatial and sectoral clustering of Irish manufacturing has changed since the mid-1980s in response to the proactive industrial and regional clustering policies described. In the improved economic climate of the late 1990s, and with the effects of globalization becoming ever more evident, the policy emphasis on spatial dispersion outside large centers, sectoral specialization, and cluster creation has grown significantly, raising two important questions in the context of Irish manufacturing industry: Are MNEs becoming more or less sectorally specialized and spatially concentrated? To what extent is there evidence that LEs are mirroring the clustering patterns of MNEs?

7.4 Evidence of clustering for Irish manufacturing

Our empirical study is based on sectoral data derived from a census of manufacturing plants in Ireland with three or more employees. The data on enterprise numbers are collected as part of the Census of Industrial Production conducted annually by the Central Statistics Office of Ireland. Data between 1985 and 2001 are used to measure changes in spatial and sectoral clustering at the Irish county and 4-digit NACE sector level respectively for MNEs and LEs using the entropy indices described in Section 7.2. In order to take account of the restructuring of the Irish economy towards high-tech industry, we disaggregate manufacturing into two sectoral groups, namely high tech (chemicals and pharmaceuticals, NACE 2411–2470, and electronics, NACE 3001–3350) and low tech (enterprises in the remaining NACE categories).

Table 7.1 reports entropy indices for all Irish manufacturing enterprises. Columns 1–3 show estimates for *AVCONC* and Columns 4–6 show estimates for *AVSPEC*. The entropy index for total enterprises (column 1) is on a downward trend over the period 1985 to 1992/93, whereafter it rises. Since the entropy index is an inverse measure of concentration, this implies an inverted

U-shaped pattern of spatial concentration over the period, with average spatial concentration across counties showing an increasing trend until 1992–93 before declining; by 2001 enterprises were less spatially concentrated than they were in 1985. This pattern is also evident in the average spatial concentration indices for both MNEs and LEs (columns 2 and 3), but notably the turning point for MNEs is rather later than for LEs (1995 compared with 1992). While the patterns over time are quite similar, there is a striking level difference between MNEs and LEs – the former are significantly more spatially concentrated than the latter, with entropy indices for MNEs around two-thirds the level of those for LEs. Thus while the spatial patterns over the period are similar, there has been no reduction in the gap between MNEs and LEs in terms of the degree of spatial concentration.

Focusing next on changes in sectoral specialization between 1985 and 2001, the index for all enterprises in Irish manufacturing on average across counties (*AVSPEC*) is trended upwards (column 4) over the period. This implies that sectoral clustering has declined continuously. However, this average result for all enterprises masks differences in sectoral specialization between MNEs and LEs. The decline in *AVSPEC* for MNEs (column 5) indicates that they have become more sectorally clustered over the period. This MNE trend toward sectoral specialization contrasts with that for LEs – the upward

Table 7.1 Entropy indices for manufacturing enterprises, 1985–2001

	AVCONC			*AVSPEC*		
Year	(1) TOTAL	(2) MNE	(3) LE	(4) TOTAL	(5) MNE	(6) LE
1985	2.236	1.488	2.209	4.036	3.334	3.869
1986	2.217	1.474	2.188	4.052	3.342	3.881
1987	2.217	1.483	2.190	4.072	3.360	3.890
1988	2.195	1.454	2.167	4.084	3.345	3.899
1989	2.194	1.455	2.165	4.091	3.330	3.912
1990	2.188	1.450	2.155	4.092	3.301	3.914
1991	2.181	1.435	2.148	4.106	3.269	3.931
1992	2.179	1.432	2.145	4.081	3.209	3.910
1993	2.190	1.428	2.155	4.084	3.144	3.928
1994	2.201	1.423	2.176	4.096	3.195	3.935
1995	2.193	1.411	2.172	4.107	3.195	3.943
1996	2.200	1.421	2.173	4.123	3.181	3.963
1997	2.216	1.452	2.187	4.140	3.196	3.984
1998	2.203	1.470	2.171	4.131	3.172	3.972
1999	2.224	1.466	2.196	4.128	3.125	3.979
2000	2.244	1.462	2.219	4.126	3.108	3.991
2001	2.249	1.488	2.220	4.116	3.163	3.959

Source: Authors' own estimates derived from CSO data.

trend in column 6 highlights the constant decline in sectoral specialization for LEs. While the gap between the indices for MNEs and LEs widens significantly over the period, it is noteworthy that the difference in sectoral specialization between the two groups (measured by *AVSPEC*) is much less than the difference in spatial concentration (measured by *AVCONC*).

As a result of the emphasis in Irish policy shifting the economy from low tech to high tech, we present entropy indices for both sectoral groups in Table 7.2.

Table 7.2 indicates that the spatial distribution of the high-tech sectors is significantly more concentrated than that of the low-tech sectors throughout the period, reflecting perhaps their general need to access more location-specific factors, and hence their lower ability to be spatially footloose. However, this difference became much less marked as enterprises in the high-tech sector become increasingly spatially dispersed and the low-tech sectors became more spatially concentrated over the early part of the period. In terms of sectoral specialization, the high-tech sectors became less specialized over the period, reflecting the growth in activities in entirely new high-tech sectors in the early period; the downward trend in recent years is perhaps some evidence of sectoral consolidation and clustering among the high-tech enterprises. The upward trend in the *AVSPEC* index for the low-tech sectors,

Table 7.2 Entropy indices for total manufacturing by sectoral groups, 1985–2001

	AVCONC		*AVSPEC*	
Year	(1) **High tech**	(2) **Low tech**	(3) **High tech**	(4) **Low tech**
1985	1.835	2.292	2.655	3.816
1986	1.840	2.273	2.670	3.827
1987	1.870	2.269	2.668	3.848
1988	1.858	2.248	2.689	3.856
1989	1.856	2.248	2.671	3.865
1990	1.895	2.236	2.663	3.865
1991	1.881	2.233	2.705	3.870
1992	1.905	2.226	2.679	3.848
1993	1.925	2.234	2.656	3.856
1994	1.932	2.245	2.670	3.869
1995	1.910	2.242	2.686	3.877
1996	1.909	2.251	2.702	3.893
1997	1.945	2.263	2.716	3.909
1998	1.949	2.248	2.681	3.904
1999	1.969	2.269	2.635	3.907
2000	1.972	2.290	2.642	3.908
2001	2.025	2.287	2.632	3.895

Source: Authors' own estimates derived from CSO data.

reflecting greater sectoral dispersal, is consistent with structural adjustment in low-tech sectors in response to increased competition on the domestic market and new opportunities on export markets as the EU single market opened up.

The combination of results in Tables 7.1 and 7.2 suggests that we look beyond the degree of specialization/concentration measured by the entropy indices to focus on how the actual patterns of location by sector (NACE 4-digit) and space (county) have changed between 1985 and 1991. We use correlation coefficients to examine the co-clustering of MNEs and LEs both spatially and sectorally, and to establish whether the clustering policy in operation has led to LEs mirroring the clustering patterns of MNEs. Spatial correlation coefficients are calculated using $MNE\ Share = MNE_{k,r}/MNE_{k,\ Total}$ and $LE\ Share = LE_{k,r}/LE_{k,\ Total}$, where r = county and k = high- or low-tech sectors. Sectoral correlation coefficients for high-tech and low-tech sectors are calculated in a similar manner as $MNE\ Share = MNE_{k,i}/MNE_{k,\ Total}$ and $MNE\ Share = LE_{k,i}/LE_{k,\ Total}$, where i = sector and calculated for high-tech and low-tech sectors, as reported for 1985 and 2001 in Table 7.3.

We find very high correlation coefficients for the spatial concentration of all enterprises over the period, with virtually no change in the coefficient values between 1985 and 2001. This aggregate result also reflects what has occurred spatially in the high- and low-tech sectors, implying a strong positive relationship between the spatial co-location MNEs and LEs in all sectors at county level. In other words, the spatial centers of high-tech clustering of MNEs are very similar to those of LEs.

By contrast, sectoral correlation across all enterprises is much lower and has fallen significantly over the period. This indicates that at the 4-digit level, there is low and declining sectoral clustering of MNEs and LEs – at this level of disaggregation, MNEs and LEs are not operating in the same sectors. To take account of sectoral restructuring and the focus of shifting activities increasingly into the high-tech sectors we decompose the total of manufacturing into high-tech and low-tech groups and look at sectoral correlations within those groups. Not unexpectedly, the correlation coefficients for both

Table 7.3 MNE and LE correlation coefficients

	Spatial		**Sectoral**	
	1985	**2001**	**1985**	**2001**
Total	0.954	0.959	0.301	0.122
High tech	0.955	0.964	0.578	0.594
Low tech	0.941	0.947	0.487	0.229

Source: Authors' own estimates derived from CSO data.

groups are higher than for manufacturing as a whole. However, what is striking is the marked difference in the patterns for high-tech and low-tech sector groups. The correlation for high tech is stable at close to 0.6, showing that the MNEs and LEs are to a considerable degree operating in the same high-tech sectors. By contrast, the coefficient values are very small for the low-tech sector, and reduced by over 50 per cent between 1985 and 2001, indicating a diminishing relationship between MNEs and LEs. Thus there is little evidence of MNE–LE sectoral clustering when looking at total manufacturing and the low-tech sectors but there does appear to be some tendency toward clustering between MNEs and LEs in the high-tech sectors.

The dominance of the low-tech sectors, in terms of enterprise numbers, should be noted in order to aid the interpretation of the above results. The LE share of the low-tech sectors rose from 88 per cent in 1985 to over 90 per cent in 2001, at a time when the low-tech share of total manufacturing enterprises in Ireland fell from 88 to 86 per cent. Over the same period, the LE share of manufacturing enterprises in the high-tech sectors increased from 50 to 55 per cent, while the high-tech share of total manufacturing has increased from 12 to 14 per cent. These results are in line with the industrial policy for restructuring of the Irish economy where the focus has been on attracting high-tech MNEs to foster growth in the indigenous high-tech sector.

7.5 Conclusions for policy

This chapter has analysed the spatial concentration and sectoral specialization of both host-country and foreign enterprises in an economy that has experienced extraordinary growth in export-platform FDI over the past two decades. The use of entropy indices as indicators of spatial and sectoral clustering in Irish manufacturing for both MNEs and LEs separately, as well as for high-tech and low-tech sectors, combined with correlation coefficients to estimate the co-location patterns of LEs and MNEs, allow investigation of the overall impact of industrial and regional policies on Irish manufacturing.

LEs and MNEs exhibit a similar trend in spatial concentration, growing more clustered from 1985 until the early 1990s before trending toward greater spatial dispersion up to 2001. Throughout the period, the level of spatial concentration among MNEs remained far greater than that of LEs across counties. This trend from more to less spatial concentration since the late 1990s in part reflects the spatial policy direction implemented at the time, which focused on developing the regional areas of Ireland while at the same time reducing congestion in the main cities of Dublin and Cork. However, the effect can only partly be due to policy as the real emphasis on spatial dispersal came late in the 1990s when the economy reached full employment, while it is evident that the process of spatial dispersal had begun in the early 1990s. This suggests that there were market forces already driving enterprises out of the more concentrated locations so that while policy reinforced those pressures, it cannot be said to have led them.

The pattern of sectoral specialization for MNEs and LEs provides more of a contrast, as MNEs have become more sectorally concentrated over the period while LEs have become less concentrated. Since policy in attracting MNEs is deliberately selective, with the IDA focused on key sectors in manufacturing, such a pattern for MNEs is to be expected. The less concentrated sectoral pattern among LEs is consistent with general restructuring, and given the high sectoral correlations coefficients for high tech MNEs and LEs, may well be evidence that LEs are following MNEs into the same sectors. Further sectoral level analysis is required to address this issue.

The analysis in this chapter highlights the overall impact of industrial and regional policy on the sectoral and spatial development of the Irish economy. Our evidence on the timing of spatial dispersion suggests that recent policies have intensified rather than initiated the spatial changes that were already under way in Irish manufacturing from the early 1990s, and these effects are likely to be further intensified as EU membership increasingly constrains the use of grant aid for manufacturing in the greater Dublin area. The much higher sectoral correlation for MNEs and LEs in high-tech sectors suggests that the promotion of sectoral clustering in these sectors is successful. The downward trend in sectoral specialization for LEs and for the low tech sector generally suggests that sectoral restructuring is still occurring on a significant scale in Ireland. In Section 7.3, we asked two questions: Are MNEs becoming more or less sectorally specialized and spatially concentrated? To what extent is there evidence that LEs are mirroring the clustering patterns of MNEs? The answers are clear: MNEs are becoming more sectorally specialized, less spatially concentrated, and LEs are mirroring the clustering patterns of MNEs in spatial terms but not the sectoral patterns. Further analysis at regional level is required to investigate the spatial consequences of such restructuring and the patterns of clustering that are emerging between foreign and local enterprises in Ireland.

Note

1 The authors acknowledge the assistance of the Central Statistics Office (CSO), Ireland in allowing access to the micro-data used in this chapter. To facilitate the research conducted the CSO gave the authors controlled access that was at all times within the CSO's premises under stringent conditions. We are grateful for financial support from the Irish Research Council for the Humanities and Social Sciences and DEPFA Bank in undertaking this research. Helpful comments were received from Austin Gleeson, Tom Pius Moloney, Ana Teresa Tavares, Aurora Teixeira and participants at the workshop in Porto in April 2005, as well as conference attendees at the Irish Economics Association Annual Conference in May 2005.

References

Aiginger, K. and Davies, S., 'Industrial specialisation and geographic concentration: two sides of the same coin? Not for the European Union', *Journal of Applied Economics*, VII(2) (2004) 231–48.

Audretsch, D. and Feldman, M., 'R&D spillovers and the geography of innovation and production', *American Economic Review*, June, 86(3) (1996) 630–40.

Barry, F., Bradley, J. and O'Malley, E., 'Indigenous and foreign industry', in Barry, F. (ed.), *Understanding Ireland's Economic Growth* (London: Macmillan, 1999) pp. 45–74.

Davies, S. and Lyons, B. (eds), *Industrial Organisation in the European Union: Structure, Strategy and the Competitive Mechanism* (Oxford: Oxford University Press, 1996).

Dixit, A. and Stiglitz, J., 'Monopolistic competition and optimum product diversity', *American Economic Review*, 67(3) (1977) 297–308.

Fujita, M. and Thisse, J.F., 'Economics of agglomeration', *Journal of the Japanese and International Economies*, 10 (1996) 339–78.

Görg, H. and Ruane, F., 'Multinational companies and linkages: panel-data evidence for the Irish electronics sector', *International Journal of the Economics of Business*, 8 (2001) 1–18.

Helpman, E., 'The size of regions' in Pines, D., Sadka, E. and Zilcha, I. (eds), *Topics in Public Economics* (Cambridge: Cambridge University Press, 1998).

Hirschman, A., *The Strategy of Economic Development* (New Haven, Conn.: Yale University Press, 1958).

IDA Ireland, *Annual Business Survey* (Dublin: Government Publications, 2002).

Krugman, P., 'A model of innovation, technology transfer, and the world distribution of income', *Journal of Political Economy*, April, 87 (1979) 253–66.

Krugman, P., 'Scale economies, product differentiation, and the pattern of trade', *American Economic Review*, 70 (1980) 950–9.

Krugman, P., *Geography and Trade* (Cambridge, MA: MIT Press, 1991).

Krugman, P., 'First nature, second nature and metropolitan location', *Journal of Regional Science*, 33 (1993) 129–44.

Krugman, P., 'Good news from Ireland: a geographical perspective' in Gray, A. (ed.), *International Perspectives on the Irish Economy* (Dublin: Indecon Economic Consultants Public Policy Series, 1997).

Krugman, P. and Venables, A., 'Globalisation and the inequality of nations', *Quarterly Journal of Economics*, 110 (1995) 857–80.

Krugman, P. and Venables, A., 'Integration, specialisation and adjustment', *European Economic Review*, 40 (1996) 959–67.

Ottaviano, G. and Puga, D., 'Agglomeration in a global economy: a survey', *CESPRI Working Paper No. 97* (1997).

Ruane, F. and Uğur, A., 'Trade and foreign direct investment in manufacturing and services', in Newman, C. and O'Hagan, J. (eds), *The Economy of Ireland: National and Sectoral Policy* (Dublin: Gill & Macmillan, 2005), pp. 162–87.

8

Emerging Business Models for Biotechnology Firms and Clusters: Policy Responses in Peripheral Regions of the EU

Ewen Peters and Stephen Young

8.1 Introduction

The aim of this chapter is to identify the emerging policy challenges[1] facing the governments of peripheral regions in Europe where biotechnology has been targeted as a growth sector. Specifically, the research question is whether it is advisable or feasible for peripheral areas to target this sector, where financial demands and risks (but also potential returns) are high. In answering this question, the relevance of clusters and innovative milieux[2] to the biotechnology sector is examined in the light of the new post-genomic business models that are emerging.

Comparisons are made between two peripheral European economies, Ireland and Scotland, which for over two decades have been attempting to nurture life-science activity; and two leading firms, respectively Elan Corporation plc and Cyclacel plc. Issues that emerge include, first, given that biotechnology is *prima facie* a global industry, that there are fundamental localization challenges concerning the anchoring of key elements of business activity, particularly headquarters functions and research and development (R&D) within the local economies. And, second, that of providing the appropriate (but inevitably risky) forms of public support for key biotechnology companies to remedy market failures, which in turn requires in-depth understanding of likely successful business models in a fast evolving sector.

8.2 Business models in the life sciences

Within biotechnology, converting a discovery into a commercially viable product is a high-risk process that is often highly complex, challenging, very

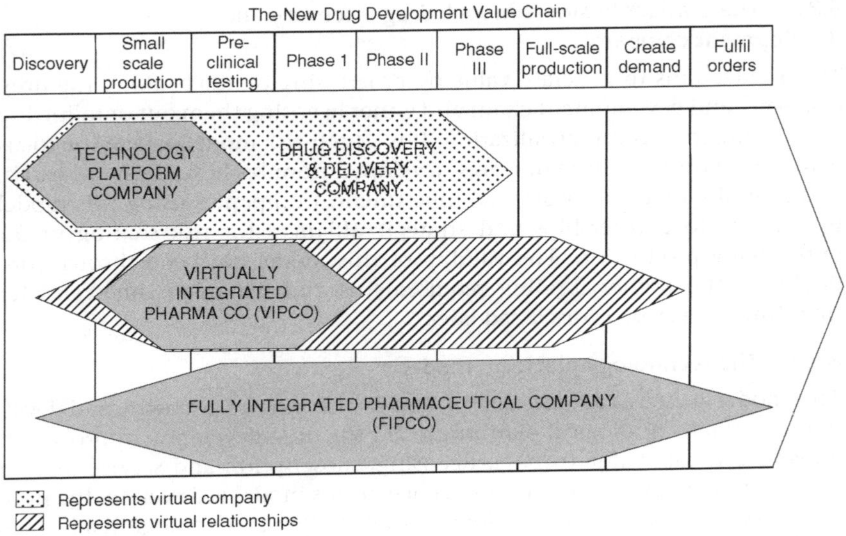

Figure 8.1 Business models and the new drug development value chain
Source: Adapted from Arthur D. Little (2002).

time consuming and expensive – where obtaining development funding is crucial to success. Hence, considerable importance is attached to the validity of the new company's business model – how it is resourced, how it creates value and how returns are realized. A range of business model options have been identified, closely tied to the new drug development value chain, and these are illustrated in Figure 8.1 and discussed below.

8.2.1 The vertical model

This is a fully integrated organizational model, in which the company possesses internal development, manufacturing and marketing capabilities. The main attraction is the potential for value generation throughout the entire value chain, albeit operating within a high-risk environment. The model offers the potential for high returns on research investment, and strong profit margins similar to those of the pharmaceutical sector. This model has proved successful with some of the largest biotech companies, but others have found difficulty in obtaining the levels of finance required to support such a wide range of activities.

When a company outsources most of its operations, while still managing the chain internally, it is referred to as a *virtually* integrated pharmaceutical company; or in the terms of D'Cruz and Rugman (2000) the 'flagship pharma firm'.

8.2.2 The product business model: drug discovery and development company

This model aims to generate value by progressing products along the drug discovery and development chain and either licensing them out or following them through to commercialization. It is a high-risk model, where the focus is placed upon a pipeline of drugs developed from basic scientific research, and typically requires substantial investment. Companies using this model may resort to partnerships and strategic alliances for risk sharing or for outsourcing production. The validity of this model has been thrown into doubt by the higher funding requirements and extended timescales for investment returns, as discussed later.

8.2.3 The technology platform model

This model is used as a research tool, and focuses on the discovery and development of a technological platform to aid the drug development process. It aims to generate value through licensing, subscription and service fees for the technology platform. The emergence of this model is driven by the need to reduce the risks in drug development by applying technological advances to drug discovery. However, under the best circumstances the technology platform company can only uniquely remove one bottleneck in the drug discovery process. Some critics point out that the technology on which these companies rely risks being commoditized (Lim, 2003).

8.2.4 Hybrid business model

Some companies that were initially technology-platform businesses have tried to establish a sustainable value base by gradually turning the technology platform into a proprietary drug discovery effort. This would ultimately transform the company into an emerging pharmaceutical company. The hybrid business model is thus a fusion of the product and platform business models that may be capable of generating a pipeline of products. It gives investors the benefits of reduced risk and near-term revenue generation, without compromising the potential for greater value creation. Therefore it is attractive as a business model in offering long-term drug development with near-term technology commercialization potential.

8.3 A new approach to business models in biotechnology

The business models reviewed above have emerged from the work of Fisherman (2001, 2002) and the Harvard Biotechnology Group (2002). A feature of all of these is their emphasis on the technology and product dimensions of biotechnology. This is overly simplistic from both business competitiveness and public policy perspectives, since no attention is paid to the *location* of value chain functions, where localization/globalization dynamics are highly influential; and to the significance of *networks and relationships* in business strategy domestically and internationally. To achieve a fuller understanding for policy purposes, other key dimensions of a business model need to be incorporated. Figure 8.2 presents a model that incorporates two other important vectors, namely, the internationalization of the firm and business relationship modalities. The former is both a marketing requirement in this global sector, and a source of opportunities for knowledge sourcing, for attracting key skills and for accessing process competences, specialized capital and complementary assets internationally for different elements of the drug discovery and development process. Business relationship modalities focus upon the wide range of options for strategic alliances that exist between market-based transactions and greenfield ventures or mergers and acquisitions (M&As) (Inkpen, 2001), and that may be either domestic or international.

Table 8.1 develops the framework of Figure 8.2 by highlighting the relationships among value chain activities, internationalization and business modalities. Reflecting the resource-based view of the firm (Penrose,

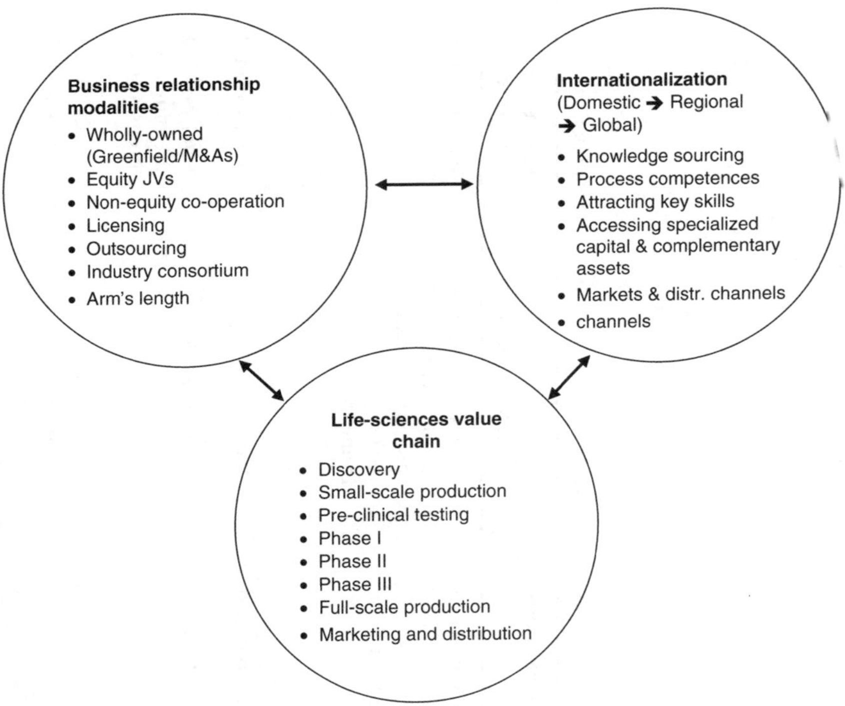

Figure 8.2 Dimensions of enhanced life-sciences business model

Source: Authors.

Table 8.1 Relationships among value chain activities, internationalization and business modalities

Life sciences value chain	Internationalization	Business relationship modalities	Comments
HQ functions: Decision-making; HRD; Technology; Procurement	Domestic, with access to innovative milieu; but potentially mobile (to US)	Wholly-owned	Strong network ties to private and public organizations and companies necessary to anchor HQ to home base.
Discovery	Domestic	Wholly-owned	R&D may be part-funded by alliance partners, in return for rights in other value chain activities e.g. testing, manufacturing, marketing.
Small-scale production	Domestic	Wholly-owned	Early scale-up usually integrated with R&D.
Pre-clinical testing	Domestic and/or international	Wholly-owned Licensing-in or licensing-out	Domestic, wholly-owned activities crucial for core drug discovery and development companies. Licensing-in common to expand portfolio.
Phase I	Domestic and/or international	Wholly-owned Licensing-in or -out	Ditto

Phase II	Domestic and/or international	Wholly-owned Licensing-in or -out	Ditto
Phase III	Domestic and/or international	Wholly-owned Licensing-in or -out	Ditto
Full-scale production	Global – least cost locations	Strategic alliances (contract manufacturing arrangements)	Large firms with manufacturing capabilities may offer sub-contract production facilities
Marketing & distribution	Domestic base, but potentially mobile (closer to main markets)	Small firms commonly collaborate with large drug/biotech firms	Small firms need partnerships for international marketing; but latter may be core competence
Markets	Global, but US market critical	As above	
Company growth	Global	M&As common, together with range of strategic alliances	Patterns depend upon business model pursued. Integration and management of MAAs represent major challenge

Notes: HRD = Human resource development; M&As = Mergers & acquisitions; MAAs = Mergers, acquisitions & alliances.

Source: Authors.

1959; Barney, 1991), the core competences of most biotechnology companies will reside in their R&D operation within the drug discovery and development process, suggesting wholly-owned and domestically-based research and development. Early small-scale development is closely linked to R&D technically and, therefore, locationally; but thereafter much of the value chain can be potentially outsourced. Markets are global (albeit dominated by the US), although international marketing (the role, for example, of an international sales force) may represent a core competence for some enterprises.

In respect of modality, a wide range of options exists as regards both outward and inward internationalization, the out- and in-licensing of drugs being especially significant for widening markets and extending the product range respectively. Also important are alliances with the major pharmaceutical companies, and wider linkages with the international life-science community, and the international financial community. These networks have a variety of objectives, including accessing scarce factor inputs, accessing non-core competences and complementary assets, and accessing final markets.

To address these issues we believe that policy-makers will have to acquire, first, a much greater appreciation of the type of business models suggested by the framework outlined earlier; and, second, a deeper understanding of the key market failures that are likely to characterize the future operating environment. In particular, it will be important to understand the extent to which the existence of local innovative milieux will strengthen and support these models and where the limits to local support may lie.

8.4 New business models, market failure and public policy

8.4.1 Capital market failure, business growth and clusters

The need to achieve 'critical mass' is an oft-stated aim and objective of cluster strategies for aspirant life-science regions. However, access on a long-term basis to large amounts of specialized risk capital is the single most important factor in determining business survival and growth for drug discovery and development companies.

Following the technology bubble bursting in 2001, public markets in Europe and North America remained virtually closed to biotechnology companies until late 2003. Underlying market sentiment has remained volatile subsequently, with companies responding by being forced to take 40–50 per cent discounts or abandoning Initial Public Offering (IPO) plans. A number of factors explain the price weaknesses, but especially important is the fundamental change in investor attitudes toward risk. This provides a basis for arguing that a new funding gap has emerged and that a new market failure now exists.

The biotechnology and life-science sector is now talking about the emergence of a revised funding model. Previously, venture capitalists (VCs) would typically look to invest $< \$100$ million and expect to exit in 5–7 years, with a 30 per cent return; the equivalent figure currently is $250–$300 million of venture and development funding. While expected returns would be commensurately higher, early investors would only be able to exit after 10–12 years (ING, 2004). This could force a significant restructuring of the supply-side of the early-stage capital market with fewer business angels and VCs willing or able to play such a long game of higher risk/higher return.

If such a model comes to dominate, the underpinning market failure gives policy-makers a powerful rationale for new intervention. However, the extent of this failure (a maximum of around $150 million for every drug which gets to Phase III trials) represents a very high cost, which governments may be unlikely to be willing or able to meet directly. From a policy perspective it seems that emerging clusters, especially in peripheral regions, could have only an indirect role to play, with external linkages being more significant.

8.4.2 Information and co-ordination failures, transaction costs and clusters

Strategic linkage with the pharmaceutical industry remains one of the major determinants of success for specialist biotechnology companies: more than half of the biopharmaceutical drugs approved in 2000 in the United States were either co-developed or marketed by pharmaceutical companies. Delivering shareholder value going forward will be a major challenge as the next ten years will see a series of patent expiries on drugs whose current annual sales are worth over $40bn (Ernst and Young, 2001). This will require pharmaceutical companies to grow at an annual rate in excess of 10 per cent to compensate, and, therefore, the development of three or four new drugs per year (while the average pharmaceutical firm currently only produces one new drug per year from in-house development).

Two basic responses have been apparent. First, Big Pharma has downsized and spun-out non-core activity, pursued a series of mergers and acquisitions, and become more active in corporate venturing. Secondly, Big Pharma has outsourced much of its R&D to more productive biotechnology companies, either as licensing deals, alliances or as contract research and manufacture.

The restructuring described above, combined with current technology and organizational trends (Peters and Hood, 2002), is creating an increasingly complex and challenging operating environment for drug discovery companies where business is multimodal: a dense web of relationships now support the operation of collaborative networks, license deals, arms-length contracts etc. The consequence is that biotechnology firms face a future of rapidly rising transaction costs. This would represent a significant market failure and a growing source of economic inefficiency which public policy may need to address.

Enright (2000) has rightly observed that cluster strategies have a potentially important role to play in rectifying market failures such as those described above. Unfortunately, cluster strategies tend to be applied more as recipes by the public sector than attempts to address specific market failures (Peters and Hood, 2000). Moreover, as drug discovery businesses internationalize, there will be a growing need to develop international networks and linkages in order to access markets (both product and factor) and to harness and deploy a wider range of specialized complementary assets than would normally be available in an emerging local industry cluster. Information and co-ordination failures that exist along the international businesses interface will require a different set of policy responses.

8.5 Company cases

The remainder of this chapter compares in detail the different strategies pursued by two peripheral countries in Europe – Ireland and Scotland – that have targeted biotechnology for a number of years. Two company case studies are presented to highlight the relevance of the framework developed above and to consider the implications for policy. The cases illustrate the different strengths of each of the local milieux and their respective policy approaches, distinguishing the 'flagship firm' approach pursued by Ireland, as represented by Elan Corporation plc, which draws on Ireland's established strengths in high volume/high value manufacturing and emerging strengths in development; and the policy challenges presented for the more fragmented structure, but greater number of specialized biotechnology companies which has evolved in Scotland. Here the product-based company Cyclacel plc, drawing on the recognized excellence of Scotland's life-science base and the growing strength of Scotland's drug development expertise, has the potential to emerge as a significant player.

8.5.1 Ireland and Elan Corporation: biotechnology policy in Ireland

Irish BioScience

Total Irish expenditure on biotechnology R&D in the period 1994–98 has been estimated at Euro 46.3mn (Martin, 1999), only a very small proportion of which supported basic research, until the late 1990s. Recent years have seen a significant increase in the level of investment that Ireland is making in biotechnology-related R&D: according to Enterprise Ireland over Euro1bn has been invested through combined funding. This is set to rise further under if the Irish government meets the ambitious target of increasing gross expenditure on R&D from 1.4 per cent to 2.5 per cent of GNP by 2010 (see Inter-Departmental Committee on Science and Technology, 2004); and adopts recommendations to give a high priority to life science.

Commercialization of Irish BioScience

Biotech policies have been in place in Ireland since 1987 when the National Biotechnology Program was set up to develop commercially-oriented research in Irish universities. In 2002 Enterprise Ireland launched a new biotechnology strategy specifically focused on technology transfer and campus-based initiatives. This was complemented by the establishment of Enterprise Ireland's Biotechnology Directorate in 2003 that is seen as having a crucial role in improving Ireland's commercialization chain.

Over many years the Irish Industrial Development Authority (IDA) has successfully pursued a strategy of attracting and retaining healthcare-related direct investment from overseas, especially large-scale pharmaceutical plants. The current IDA strategy now complements the work of Enterprise Ireland by focusing on the attraction and retention of R&D projects in the life sciences. The other national organization playing a major role in the funding of biotechnology research is Teagasc, the development agency responsible for agriculture and food research.

A set of initiatives was put in place in the period 1994–99 to address the shortage of venture capital for biotechnology firms. However, the biotechnology industry in Ireland did not benefit to any great extent, with the exception of the Campus Companies Venture Capital Fund and the Millennium Entrepreneur Fund. In 2002, Enterprise Ireland began to address the special needs of the biotechnology sector in the form of a dedicated seed-fund. It involves a partnership with financial institutions in order to share the risks. Moreover, tax credits for incremental R&D undertaken by companies was introduced in the Finance Act, 2004. Ireland now has over 170 companies employing 35 000 people in the pharmaceutical, chemical, biopharmaceutical, medical devices and diagnostic sectors (Irish Minister for Trade and Commerce, 2005). These companies are largely engaged in manufacturing and production activities. By comparison, just over 10 per cent of these people are employed in 41 companies (23 indigenous firms and 18 multinational enterprises), engaged in activities that are primarily based on biotechnology research and processes (InterTrade Ireland, 2003). As such, the biotechnology sector remains a very young and relatively small component of the economic activity related to the life sciences in Ireland.

As one of only two European companies to achieve annual sales in excess of $1 billion in the biotechnology domain, Elan Corporation has been the major beneficiary of policy initiatives in Ireland. The latter include the favorable planning and health and safety regulations that existed during the 1960s and 1970s; and the relatively low and attractive rates of corporation tax available in Ireland during the 1980s and 1990s favored the volume manufacture of high-margin low-bulk products for export. This policy environment helped Elan to achieve 'flagship status' and to do so in the absence of Ireland having an internationally-recognized science-base. As the above analysis suggests, the Irish government is well aware that the policy approach, which supported Elan's main development through the 1980s and 1990s, is no longer sustainable.

Elan Corporation plc

Corporate history

Elan was incorporated as a private limited company in Ireland on 18 December 1969 and became a public limited company on 3 January 1984. Prior to 1987, Elan funded much of its R&D under agreements with a number of larger pharmaceutical companies. These companies conducted the necessary clinical and regulatory procedures, were responsible for all aspects of marketing and paid Elan royalties reflecting the contribution to the product development program. Elan was essentially a contract research organization at this stage. After 1987, Elan decided to market its products directly, and its new strategy integrated the formulation, clinical testing, registration, manufacturing and the marketing phases in the drug life cycle. Its aim was to become an integrated bio-pharmaceutical company, a strategy reinforced during the 1990s. Elan pursued an increasingly active strategy of growth-through-acquisitions and joint ventures during the 1990s, in order to build a stronger and more mature product pipeline (strategic-asset-seeking) and to gain quick access to new international markets, especially the United States (market-seeking). Most of the acquisitions were made through the stock markets due to the cash rich position of Elan and the tax exemptions received from the Irish government. In retrospect, Elan seriously over-extended itself in this period. In 2002 the company suffered a number of serious setbacks, namely, the Phase II failure of an Alzheimer's Disease vaccine; the announcement of an investigation by the SEC into accounting practices at the firm; and the company's largest product went generic. Market capitalization fell from $19 billion to $1–2 billion, forcing large-scale restructuring. About 50 wholly-owned and joint ventures were divested and entire business lines sold off.

Current profile

Elan is listed on the New York, London and Dublin Stock Exchanges and now positions itself as a neuroscience-based biotechnology drug company engaged in discovering, developing, manufacturing and marketing advanced therapies in neurology, autoimmune diseases, and severe pain. According to its latest Annual Report, Elan generated revenues of $464m in 2004 ($1.16bn in 2002) and employed 1899 people (4400 in 2002). Of the total employed in 2004, 575 were in R&D, 571 in manufacturing, 314 in sales, and 439 in administration. Pre-tax losses in 2004 amounted to $403m. Following recent criticism from Ireland's financial authorities, and as part of its major restructuring, Elan has recently been reorganized into two discrete business units.

- *Biopharmaceuticals*, which discovers, develops, manufactures and sells its own drugs;
- *Global Services & Operations (GS&O)*, which provides a broad range of product development and sub-contract manufacturing services, and provides

technology platforms to assist drug companies bring their products to the market.

This may portend a possible future break-up of the business were Elan's performance to fail to recover.

Drug discovery and development

Elan's product pipeline. The company has a range of products at different stages in the drug discovery and development pipeline; but the only marketed products in the US are drugs for pain relief and for the treatment of infections. In autoimmune diseases, a significant setback occurred in February 2005 with the voluntary suspension of the marketing of Tysabri following the untimely death of sufferers using the drug (which is used as a therapy for multiple sclerosis). With the loss of products from divestments and the problems with Tysabri, Elan now has a rather narrowly-based product portfolio and drug pipeline.

Drug development and delivery

Elan has a long track record of innovation and expertise in drug optimization and delivery from solving problems of poor solubility to customizing release profiles for oral dosage forms. This has helped to enhance the performance of numerous drugs that have subsequently been marketed world-wide. Its primary focus, however, is on using its scientific expertise to provide superior technology platforms that offer innovative, high-value technologies to address difficult drug delivery and life-cycle challenges which drug companies face. This expertise is embodied in Elan's patented drug delivery and delayed release technologies. These can be used alone or in combination to create intelligent drug delivery systems. For example, Elan has been able to develop a superior range and depth of solid oral dosage technologies which are patented and unrivalled in the drug delivery industry. And a significant number of marketed products have been developed using Elan's patented technologies.

Manufacturing and other value chain facilities

Elan's proprietary technology platforms represent important assets as is the co-location of drug development and manufacturing activities on certain sites: the latter allows for more efficient scale-up and transfer to commercial manufacturing activities. The company's main manufacturing facilities are in Ireland, with its other manufacturing operation in the US (Georgia). R&D is also undertaken in Ireland, as well as in California and in Pennsylvania. Finally, product development takes place in the UK and at another Californian site. Elan provides a turn-key manufacturing service to drug companies across a wide range of products. Its core manufacturing competence lies in meeting customer requirements involving the development of drugs whose delivery is characterized by solid, oral dosage. Customers can select from a broad range of manufacturing options to fit their precise product requirements. Manufacturing is thus complementary to the company's drug development and delivery activities.

Hierarchies, markets and alliances

Between 1996 and 2001, Elan pursued strategic alliances with biotechnology, drug delivery and pharmaceutical companies. Most of these were with other drug companies, where Elan collaborated in their development, scale-up, manufacturing and commercial launch activities. Elan's collaborative approach helped to spread the high risk of drug development, and its home-base in Ireland has been an important launch pad for this strategy, with 17 of the top 20 global pharmaceutical companies being represented in Ireland. Moreover, Elan's open, service-oriented approach helped facilitate the inward transfer of scarce specialized knowledge and intellectual capital. However, the bulk of these strategic alliances were eliminated as part of the post-2002 restructuring. Notwithstanding the central importance of Elan's external linkages, the significant growth in the company's scale and scope that took place during the 1990s was largely achieved through an aggressive acquisition program. From 1997–2000 alone, the company made around 13 major acquisitions at a cost of over $3bn. All except one of these were in the US, and the majority were R&D-oriented companies, reflecting Elan's objective of extending its drug discovery capabilities. Once again, most were divested following the corporate crash of 2002.

Internationalization

Headquartered in Dublin, Ireland, Elan generated its largest world-wide revenues of $1.8 bn in the year 2001. Revenue growth was especially dynamic between 1997 and 2001, helping to make Elan one of Europe's largest bio-pharmaceutical companies. However, during this period, two markets, Ireland and the United States, accounted for over 90 per cent of total revenues generated. And by 2004, of total revenues of $464m, the United States accounted for $321m (70 per cent) and Ireland for $130.5m (28 per cent). Of the latter, export sales from Ireland totaled $111.1m, with the US as the major export market ($67.9m). As such, Elan appears to resemble a narrow, bi-national enterprise rather than a broad-based, multinational enterprise with global ambition.

Conclusion

In European terms, Elan continues to represent a significant international business of scale. Its serious under-performance, following various major set-backs and exacerbated by the bursting of the technology bubble in 2001, has resulted in large-scale restructuring and retrenchment. The creation of two new business units may well indicate that, under investor pressure, Elan as a single company may have a limited future. If this future materializes, the likelihood would be that two separate businesses would emerge: one focused on own-drug discovery, development, manufacturing and marketing eventually, the other focused on sub-contract manufacturing and product development support through to commercialization of other drug companies' products.

The nature of Elan has thus changed radically since its ambitions of becoming an integrated bio-pharmaceutical company were thwarted by a variety of internal and external problems (not all of which have yet been resolved). In European terms, Elan's products and processes continue to exhibit significant scale and scope. In international terms, however, Elan lacks true international breadth and depth, and its products and processes (and the technologies that underpin them) lack dominance. Elan's partnership approach that was a major feature of its strategy has in effect been abandoned; and in reality this strategy, together with that of acquisitions, created problems of management (especially risk management) and co-ordination, which led *inter alia* to the unraveling of the enterprise after 2002. Big challenges now exist for Irish government policy given Elan's 'flagship' status as a national champion in biotechnology.

8.5.2 Scotland and Cyclacel

The traditional importance of the life sciences to Scotland is perhaps best reflected in its science base: according to Scottish Enterprise, for example, while Scotland has less than 10 per cent of the UK's population, life-science researchers in Scottish universities and institutions currently attract over 14 per cent of funding from all sources available in the UK. Life-science researchers in Scotland have a world-class reputation in many important health areas. Dolly, the world's first cloned sheep, was born in Scotland (later recognized world-wide as the scientific 'Breakthrough of the Year' in 1996).

Biotechnology policy in Scotland

The UK context. Life science (and its commercialization) plays a key role in UK public policy and has been positioned in the vanguard of the UK's emerging knowledge economy (DTI, 2003). The general direction of life-science policy in a devolved Scotland is still largely influenced by this wider UK policy context, and Scotland stands to benefit from the Government's (DES, DTI, HM Treasury 2004) commitment to increase UK R&D investment as a proportion of national income to 2.5 per cent by 2014. Life science companies in Scotland will also continue to be affected by changes to UK fiscal policy, such as the proposed further extension of R&D tax credits to SMEs and larger firms.

Scottish life science. The life sciences in Scotland have been a priority for economic development since 1975 when the Scottish Development Agency – Scotland's first economic development body – was established. They have remained a major focus for the sectoral work of Scottish Enterprise (SE), established in 1991 as successor to the SDA.

Since the millennium, SE's life-science initiatives have focused on:

(i) improving access to, and the availability of, risk funding to support start-up activity and early-stage business growth and development in the life sciences;
(ii) enhancing the research/clinical trial/industry interface and thereby raising the level of early-stage commercialization activity based on innovative life-science applications with international market appeal;
(iii) improving skills availability and the science and technology infrastructure, often through property-led initiatives; and
(iv) attracting direct investment in high-value R&D projects from overseas.

Excluding the basic pharmaceutical sector, there are now over 550 organizations in Scotland's life-science community – employing over 26 500 people – and Scotland is home to 15 per cent of the UK's life-sciences companies. Of these, over 50 academic institutions – and 80 companies – are engaged in drug discovery and early stage development including Phase I and Phase II clinical trials (Scottish Enterprise, 2005). In this context, Cyclacel makes an especially important contribution to Scotland's drug discovery and development effort: in-house research undertaken by Scottish Enterprise (2005) shows that in Scotland 26 clinical trials are currently being conducted by five companies (Ardana, Amarin, Crusade Lab, Cyclacel and ProStrakan). When ranked by the value of therapeutic area being targeted, Cyclacel commands first and fourth place. The importance of Cyclacel's position is further emphasized by the fact that the science and discovery on which the pipelines of ProStrakan and Amarin are based are primarily sourced outside of Scotland: ProStrakan's pipeline is largely acquired and its main R&D facility is in Paris. Moreover, London-headquartered Amarin also has a pipeline that is largely acquired; this contrasts markedly with the Cyclacel position: of the nine research programs underway, seven are based on Scottish science and Dundee-based 'in-house' research – only two programs are licensed-in.

The success of Cyclacel going forward is an important bell-wether for policy-makers in their on-going monitoring and assessment of the effectiveness of public policies to stimulate and support product-based indigenous life-science firms operating in the high-risk domain of drug discovery and development (see, for example, Life Sciences Industry Advisory Group for Scotland, 2005).

Cyclacel

Corporate profile

Established in 1997 as a spin-out from the University of Dundee in Scotland, Cyclacel is a public limited drug company which designs and develops small molecule drugs to stop uncontrolled cell division in cancer and other similar diseases. Reflecting its scientific roots, Cyclacel is Dundee-based and headquartered. Cyclacel's proprietary discovery engines integrate cell cycle biology expertise with a large library of gene-based targets, RNAi functional genomics, chemo-genomics and clinical biomarker technologies to deliver new drugs. Fundamentally, this approach to drug discovery and design aims to improve the efficiency of drug targeting in the early stages of discovery and development in order to reduce compound attrition rates during the later, more expensive stages of drug development. As at December 2004, Cyclacel had total assets of around £13.5 million, the bulk of this being cash in hand. Cyclacel currently employs around 60 staff the majority of whom work in drug discovery and pre-clinical development. Reflecting the nature of its business, Cyclacel's workforce is highly graduate/post-graduate intensive: over half of the staff have Ph.Ds. The bulk of Cyclacel's senior managers and scientists were recruited from outside Scotland. Since its inception in 1997, Cyclacel has spent around $80 million on IP protection. The breakdown of patents issued/pending is as follows:

- 26 issued patents
- 66 issued under licence
- Around 70 pending in the US and EU
- Around 70 pending in other countries
- Another 73 pending where Cyclacel has a license or option to licensee.

Cyclacel's Senior Management Team, Board of Scientific Advisers and staff scientists are highly experienced and internationally respected. This is reflected in Cyclacel's ability to raise significant VC funding during a major downturn in the capital market.

Funding history

With the securing of venture capital finance in December 2003, Cyclacel in total has raised over $103 million of early-stage funding from investors since inception in 1997. This is the largest amount of venture capital ever raised by a university-spin out in Europe. Following this, Cyclacel's fundraising focused on the possibility of attracting first-time support from the public markets. Accordingly, the company registered its intention in January 2004 to raise capital through an IPO launch on both the New York Stock Exchange (Nasdaq) and London Stock Exchange. This was abandoned at the last minute because of stock-market volatility.

Products, processes and technologies

Cyclacel's drug pipeline. Cyclacel currently has nine drugs in its pipeline. CYC 202, its lead candidate drug (targeted at lung and blood cancer and kidney disease), is in Phase II clinical trials. Because of the poor efficacy of existing therapies, these markets are regarded as open to new entrants offering improved therapies. Cyclacel currently estimate the combined annual worth of these markets at around $1 billion globally, were these markets to be fully captured. CYC682, an oral nucleoside analogue in Phase Ib trials, has been shown in pre-clinical tests to have superior efficacy to alternatives.

Process know-how. Cyclacel's core process competence is based on:

- Its specialized understanding of cell life cycles based on the science developed by its founder.
- Its use of structure-based drug design and use of biomarkers that involves a proprietary genes-to-drugs approach. This approach to drug discovery and design aims to improve the speed, efficiency and yield of drug target selection in the early stages of the process in order to decrease compound attrition rates during the later, more expensive stages of drug development.

Hierarchies, markets and alliances

Hierarchies. As noted above, Cyclacel's core product IP, related process competences, and proprietary platform technology have mostly been developed in-house, to date. If the international capital markets remain highly risk averse, Cyclacel may need to consider a trade sale of its lead candidate drug in about 18 months time. The company's strategic intent, however, is to retain ownership of promising follow-up candidate drugs and eventually to try to market these directly on an international basis. The current expectation, therefore, is for Cyclacel to become much larger and more vertically integrated, forward to the market; although such a strategy has been thrown into doubt by the new funding model and the need for faster revenue generation.

Markets. Having completed four rounds of venture funding, it is clear, current difficulties notwithstanding, that the international market for public capital will remain a potentially important source of future development funding for Cyclacel – and the company will need to be investor-ready. Likewise, the international labor market is also (and will remain) an important source of key management expertise and specialist skills. This is reflected in Cyclacel's wage bill that is about a third higher than the average for companies in the Scottish life-science sector. Cyclacel also uses specialist outside vendors to support key processes such as structure-based drug design and the undertaking of enzyme assays.

Alliances and partnerships. However, in pursuit of the partnership model which is central to the future development of its business (Cyclacel F-1 SEC Registration Statement), Cyclacel has entered into corporate alliances with AstraZeneca, CV Therapeutics, GlaxoSmithKline, Lorus and Sankyo, all in the oncology field. Cyclacel also has important scientific and academic links with a number of universities in the UK, as well as in France and the Czech Republic. A number of these external linkages (especially but not exclusively at Scottish universities) are key to strengthening and supporting the development of the Scottish-based discovery and development competences that lie at the heart of Cyclacel's business.

Internationalization

For a young drug discovery company such as Cyclacel, the tangible and intangible dimensions of internationalization need to be considered, with the latter being predominantly important at this stage of business development. The main operational aspects of its internationalization are the clinical trials that Cyclacel runs for drugs that have entered Phase 1 and beyond. For least-cost reasons, all of Cyclacel's clinical trial work is currently conducted elsewhere in Europe. It is a process that is managed virtually by a handful of specialist managers. At this stage, what is perhaps more strategically vital, are Cyclacel's linkages with the international life-science community and the international financial community. In playing such a long game, it is in those communities of interest where Cyclacel must seek to forge and maintain increasingly effective ties and relationships. Similarly, it is in these communities where reservoirs of social capital, market capital and goodwill for the future must be built up in order to maintain and enhance Cyclacel's positive international profile and reputation. The failed IPO notwithstanding, Cyclacel still has world-class assets to deploy in the members of the Board of Cyclacel and its Senior Management Team, and its Board of Scientific Advisers.

Conclusion

Cyclacel's business approach and strategy is largely modeled on that of the American company Vertex (the US company widely recognized as a star in the biotechnology firmament (The Economist Group, 2001)). The model is based on superior science and world-class business know-how applied to drug discovery and development and embodied in both proprietary products and processes. It is a 'pipeline model' where dynamic innovation is required to produce sufficient numbers of promising candidate drugs which lend real strength and depth to the pipeline, thereby de-risking the company for potential investors. The cash-burn associated with such a model is extreme. And in the current climate where there are fewer lead-specialist investors and more risk-averse attitudes, raising sufficient funds to reach the required regulatory milestones – even for the fittest of young drug developers – is now becoming a much steeper uphill struggle. In particular, greater risk aversion is now reflected in the preference of lead specialist investors for biotechnology companies seeking an IPO to be well partnered and to have lead candidate drugs in Phase II b or Phase III of clinical trial – a step beyond where Cyclacel now sits. Moreover, unlike its US mentor, Cyclacel neither has significant presence in, nor immediate access to, the full range of resources, specialized complementary assets, and infrastructure of the world's largest and most dynamic innovative life-science milieux (i.e. Boston and San Diego). The challenge Cyclacel's management team face in driving the business forward on the basis of this model requires that new sources of long-term risk capital are found. In this, capital markets alone may be insufficient and other forms of financial, including government, support may be needed.

8.6 Discussion and conclusions

The research question addressed in this chapter is whether it is advisable or feasible for peripheral areas to target the biotechnology sector, where financial demands and risks are high but so are potential returns given the industry's role as a gateway to the knowledge economy.

Both Ireland and Scotland have invested heavily in biotechnology over an extended period. Their respective government strategies have been quite different historically, reflecting their country capabilities and attractiveness, but are now converging as Irish policy shifts towards the development of its science-base. Scotland still possesses much stronger science foundations and drug discovery pipelines than Ireland or, indeed, other peripheral areas in Europe. Yet neither country can be said to possess anything like a fully developed cluster or innovative milieu as Table 8.2 reveals (but see Peters and Hood, 2002, on Scotland). Moreover, the intensification of globalization in the biotech domain is changing the nature and extent of related localization process. Thus the market failures, in respect of higher funding requirements and transaction costs in internationalization and business networks, may significantly limit both the scale and the scope for development of vertically integrated downstream activity in 'local' innovative milieux that are life-science based. To provide simply one example concerning skilled labor, it has proved difficult in Scotland to attract talent from the Cambridge/London cluster, let alone internationally; while leading figures have exited Scotland for improved prospects abroad. Ireland is better placed in this regard because of its better integrated global diaspora.

On the corporate side, given these policy challenges going forward, product-based firms (the drug delivery and development company in Figure 8.1) could struggle in both Scotland and Ireland; while service-based companies (the technology platform and hybrid business models) may be better adapted for long-term survival. Question marks will therefore remain over Cyclacel's ability to survive in its current form – merger or acquisition seems likely – and it is uncertain whether any newcomer would be prepared to maintain Cyclacel's deep pipeline drug discovery approach and the HQ and R&D hub at Dundee. Similarly, question marks remain over Elan's ability to survive in its present form. The restructuring of Elan suggests a service-based company (providing solutions for delivering drugs to the market) with an Irish HQ may emerge; and a product-based company with a US HQ and questionable prospects may also emerge. Under this scenario it would be debatable whether such outcomes represent good value for the taxpayer in either Ireland or Scotland.

The example of Ireland highlights the challenges for public policy in taking a position on the structure of the biotech sector, and seeking to nurture a flagship firm. There are attractions to this approach that can be integrated closely with a cluster model of development, but the risks of picking

Table 8.2 Local innovative milieux in Ireland and Scotland

Features of local innovative milieux	Ireland	Scotland
Strong science base in universities as a core prerequisite for building drug discovery and development activity	−−	+++
Basic research on convergent technologies, with efficient mechanisms for spillovers between research institutions	−−	++
Attraction of R&D labs of Big Pharma may provide alternative way into research (but core science base still critical)	++	−−
Environment encouraging start-up activity from university science	−	++
Institutions and complementary assets to support pre-market, early-stage commercialization	−	+
Presence of Big Pharma to support commercial activity from spin-offs and spin-outs	++	−−
Integrated healthcare system with large patient base for clinical trial activity	−−	+
Access to large amounts of risk capital	−−	−−−
Access to international competences and complementary assets, and business networks: global and virtual milieux	+	−−

Notes: + weakly positive, +++ strongly positive; − weakly negative; −−− strongly negative.
Source: Authors.

winners are high. In Scotland, while the number of players suggests a buoyant, innovative industry, most are small start-ups. The Scottish approach may be termed a 'nursery-model', where an environment of early-stage development revolving around university science is encouraged. The high costs and risks of progressing along the value chain suggest high failure rates for such firms and a high propensity for M&As. Successful companies will need to pay great attention to boosting cash flows by providing services (e.g. clinical trials) for international partners. Policy-makers are faced with high funding demands for only a handful of embryonic dominant players (such as Cycaclel) that may emerge from the pack and have the potential to go to the market.

The evidence from this chapter is that it is indeed advisable for peripheral areas to target the biotech sector, especially where the basic preconditions exist, namely, a strong science-base; but the jury is still out on the feasibility of this strategy. In particular, policy-makers have not resolved the fundamental challenge of how to promote the successful emergence and development of significant international businesses of scale. In part the answer lies in filling the gaps in the local milieux. Even more important for Scotland in particular is access to the global and virtual milieux. If there is one fundamental conclusion for Scotland, it concerns the requirement for better internationally experienced management teams to build on the science strengths and get closer to the market. The conclusion for Ireland is the reverse; namely, to build the science-base through funding university research and attracting R&D labs associated with Big Pharma.

Notes

1 The focus of this chapter is public policy in Ireland and Scotland. There is, however, a significant EU dimension to biotech policy, as represented, for example, by the Sixth EU Research Framework Programme (FP6 2003–2006) which aims to create a European Research Area by focusing on selected priority areas of research (including biotechnology and lifescience). See also European Commission (2002).

2 Space does not permit a discussion of the literature on clusters and innovative milieux. The latter notion, which gives primacy to knowledge creation and diffusion rather than traditional economies of agglomeration, is a more relevant focus for policy intervention in biotech because of the centrality of the drug discovery and development process. On regional innovative milieux, see Aydalot (1986); Maillat (1995); Malmberg et al. (1996). The tendency for biotechnology and lifescience activity to cluster is a well-understood and well-documented European phenomenon. The top 10 regions in Europe (6.8 per cent of the total) account for 60 per cent of the patent total; and a relatively small number of local clusters account for the large majority of biotechnology firms and research centers in Europe (Pammolli and Riccaboni 2001).

References

Arthur D. Little, *Biotechnologies: Parlons Vrai, Overview of Biotechnology Sector and Business Models* (London, 2002).

Aydalot, P., *Milieux Innovateurs en Europe* (Paris: Gremi, 1986).

Barney, J.B., 'Firm resources and sustained competitive advantage', *Journal of Management*, 17 (1991) 99–120.

D'Cruz, J.R. and Rugman, A.M., *Multinationals as Flagship Firms. Regional Business Networks* (Oxford University Press, 2000).

DES, DTI, HM Treasury (Department of Education and Skills, Department of Trade and Industry, Her Majesty's Treasury), *Science and Innovation: Working Towards a Ten-Year Investment Framework* (London, March 2004).

DTI – Department of Trade and Industry (UK), *Improving National Health, Increasing National Wealth, Bioscience Innovation and Growth Team* (London: DTI, 2003).

The Economist Group, 'Drugs Ex Machina', *The Economist Technology Quarterly*, 22 September (2001) 36–8.

Enright, M.J., 'Regional clusters and multinational enterprises', *International Studies of Management and Organization*, 30(2) (2000) 114–38.

Ernst and Young, *Integration, European Lifesciences Report 2001* (London, 2001).

European Commission, *Lifesciences and Biotechnology: A Strategy for Europe* (Brussels, 2002).

Fisherman, J.M., *Biostrategy Seminar: Biotechnology Business Models* (MIT 14 November, 2001).

Harvard Biotechnology Group, *Five Broad Business Models for Biotechnology* (McKinsey and Company, 2002).

ING, *Biotech Valuation* (December 2004).

Inkpen, A.C., 'Strategic alliances', in Rugman, A.M. and Brewer, T.L. (eds), *The Oxford Handbook of International Business* (Oxford University Press, 2001), pp. 402–27.

Inter-Departmental Committee on Science and Technology, *Building Ireland's Knowledge Economy: The Irish Action Plan for Promoting Investment in R&D* (Dublin, July 2004).

InterTrade Ireland, *Mapping the Bio-Island* (Dublin, 2003).

Irish Minister for Trade and Commerce as reported by Finfacts Ireland Business News (June, 2005) <www.finfacts.com>.

The Life Sciences Industry Advisory Group for Scotland, *Scottish Lifescience Strategy: Achieving Critical Mass for Sustainable Growth – 2020 Vision* (Glasgow, 2005).

Lim, L., 'Bio-business models', *Cambridge Manufacturing Review*, Winter (2003).

Maillat, D., 'Territorial dynamics, innovative milieu and regional policy', *Entrepreneurship and Regional Development*, 7 (1995) 157–65.

Malmberg, A., Sölvell, Ö. and Zander, I., 'Spatial clustering, local accumulation of knowledge and firm competitiveness', *Geografiska Annaler*, 78(2) (1996) 85–97.

Martin, P., *The National Report of Ireland in Inventory of Public Biotechnology R&D Progress in Europe, Vol. 2* (Brussels: Office for Official Publications of the European Commission, 1999).

Pammolli, F. and Riccaboni, M., 'Geographical clusters in the biotechnology industry', *Working Paper* EPRIS Project (2001).

Penrose, E.T., *The Theory of the Growth of the Firm* (Oxford: Basil Blackwell, 1959).

Peters, E. and Hood, N. (2000), 'Implementing the cluster approach: Some lessons from the Scottish experience', *International Studies of Management and Organization*, 30(2) (2000) 68–92.

Peters, E. and Hood, N., 'Scotland's biotechnology cluster: Strategic issues and responses', in Hood, N., Peat, J., Peters, E. and Young, S. (eds), *Scotland in a Global Economy: The 2020 Vision* (London: Palgrave Macmillan, 2002), pp. 91–111.

Scottish Enterprise Web Site <www.scottish-enterprise.com> (2005).

Part III

Linkages between Multinationals and Local Firms

9
Determining Factors of Subsidiary Development

Torben Pedersen

9.1 Introduction

The literature on subsidiary development has greatly expanded over the last 20 years, evaluating the dynamics behind the evolution of subsidiary roles. The literature emerged out of a critique of the traditional 'center–periphery' view where the firm-specific advantages are developed and controlled by the parent company, while the foreign subsidiaries are the long-arm of the parent company in exploiting the firm-specific advantages in the local market. The observation that some subsidiaries have a strategic role in the multinational corporation (MNC) that reaches beyond their local undertakings (e.g. Étemad and Dulude, 1986; Bartlett and Ghoshal, 1989; Gupta and Govindarajan, 1994) was the starting point of research on different subsidiary roles and the evolution of subsidiary roles.

A common understanding in recent literature is that MNCs consist of dynamic networks of more or less specialized subsidiaries linked to each other in multiple, complex patterns. It is generally argued that the new picture reflects the response of the MNCs to increasing global competition and rapid technological development, and a consequent shift in focus from exploitation of firm-specific advantages (Hymer, 1976; Buckley and Casson, 1976; Hennart, 1982) to global accumulation and dissemination of knowledge within MNCs (Kogut, 1993; Malnight, 1996; Doz et al., 2001). An important element in the new picture is that some subsidiaries are specialized in certain skills, capabilities or fields of operations of importance for the whole corporation and are therefore gaining a more strategic role than other subsidiaries. This process of subsidiaries enhancing their resources and competences is denoted 'subsidiary development'.

Subsidiary development is an important phenomenon for several reasons. The first is that multinational corporate evolution is driven in large part by the changes that occur in foreign subsidiaries. The process of subsidiary development is itself indicative of the much discussed trend towards greater international dispersal of value-adding activities in MNCs. The second reason is that the subsidiary draws from, and contributes to, the development of its host economy. Much has been written about the importance of technological spillovers from foreign direct investment (FDI) and how subsidiaries are building relationships with local counterparts and tapping into local clusters in various ways. Clearly, in countries where foreign-owned subsidiaries make up a large part of the industrial sector, the phenomenon of subsidiary development is not an academic exercise: rather, it lies at the heart of fundamental economic questions such as national competitiveness, investment attraction and job creation.

This chapter will explore the factors determining subsidiary development and provide a better understanding of which factors matter for different dimensions of subsidiary development. A central question is to what extent is subsidiary development determined by MNC headquarters (HQ) vs. own subsidiary initiative? And how important is the dynamism of the local business environment (e.g. leading-edge clusters, or other favorable conditions) for subsidiary development? The policy implications follow from the insights into how the different factors are influencing subsidiary development, which provide guidance on how policy-makers can target their efforts in order to facilitate upgrading and development of the foreign-owned subsidiaries.

9.2 Subsidiary development

Historically, MNCs used their subsidiaries abroad mainly for the purpose of the adaptation of products developed in their home countries to local tastes or customer needs, and the adaptation of processes to local resource availabilities and production conditions. In this case subsidiaries were dependent on the competence of their parent companies, and accordingly their role was essentially 'home-base exploiting' (Kuemmerle, 1999). In recent years instead, linked to the closer integration of subsidiaries into international networks within the MNC, some subsidiary operations have gained a more creative role, e.g., to generate new technology in accordance with the comparative advantage in innovation of the country in which the subsidiary is located. This transformation has led to subsidiaries gaining new roles in the MNC.

The new subsidiary roles have been expressed in several different terms used to characterize such subsidiaries; world product mandate (Étemad and Dulude, 1986), strategic leader (Bartlett and Ghoshal, 1989), global innovator or integrated player (Gupta and Govindarajan, 1991), and center of excellence (Holm and Pedersen, 2000). All of these terms point to the fact that, over the past two decades or so, subsidiaries have been evolving out of their traditional role of being the subservient executors of headquarters commands. This process 'through which MNC subsidiaries enhance their resources and capabilities, and in so doing add increasing levels of value to the MNC as a whole' has been called 'subsidiary development' (Birkinshaw and Hood, 1998b).

In a review of the MNC subsidiary management research, three streams of research is identified as shown in Table 9.1 (Birkinshaw and Hood, 1998b). These are the streams of: HQ-subsidiary relationship, subsidiary role and subsidiary development. The basic assumption in the literature on HQ-subsidiary relationship is the hierarchical nature of the MNC where HQ delegates decision rights to subsidiaries, in a way that decision rights are 'loaned, not owned' (Foss and Foss, 2002). Subsidiaries are controlled by HQ which retains the power of veto, i.e., the ability to overrule any subsidiary decision. This implies that the process of subsidiary development is designed by HQ and new subsidiary mandates are decided by the parent firm and subsequently given to the subsidiary.

The starting point of the literature stream on subsidiary roles is the recognition that subsidiaries may obtain very different roles in the MNC. The focus is mainly on identification of typologies capturing the various roles of MNC subsidiaries. A strong emphasis is on the level of subsidiary autonomy and product mandates, while issues pertaining to the integration and collaboration with other MNC units are of minor importance. Product mandates are earned rather than given by headquarters. In this view, subsidiary autonomy is based on unique resources and bargaining power, and is, therefore, fundamentally different from discretion in the sense that it is much more difficult for HQ to revoke. In other words, subsidiaries with some bargaining power and autonomy have a degree of 'ownership' over their decision rights rather than holding them to the whim of headquarters.

Table 9.1 Three streams of research on subsidiary management

	HQ–subsidiary relationship	Subsidiary role	Subsidiary development
Focus on research	Aspects of dyadic relationship between subsidiary and HQ	Internal, corporate and environmental factors explaining different subsidiary roles	Changes in role and activities of subsidiary over time
Assumptions about nature of MNC	Hierarchy: subsidiaries are controlled by HQ	Heterarchy/network: subsidiaries have different roles and have relationships with multiple units inside/outside the firm	
Research approach	Cross-sectional / static		Longitudinal / dynamic
Theoretical foundations	Transaction cost theory, contingency theory	Social network theory	Evolutionary theory, resource-based theory

Source: Based on Birkinshaw and Hood (1998b, p. 5).

The subsidiary development stream builds on the network perspective of the MNC and focuses more on the evolution of subsidiary activities over time. The MNC is conceptualized as an 'interorganizational network' of loosely coupled entities, rather than a hierarchical monolith. This loose coupling gives the subsidiary freedom to develop its own unique resource profile by linking the internal network of MNC units and the external network of local counterparts (e.g. key suppliers and customers). The basic idea is that over time the subsidiary accumulates valuable competences through its network relationships, so other MNC units become dependent on their competences, which leads to an enhanced status and an extension of their activities. It reflects the reality that many subsidiaries have specialized competences on which the rest of the MNC is dependent and this allows the subsidiary to take more control of its own destiny (Pfeffer and Salancik, 1978). Although it is recognized that MNC subsidiaries pursue their own interests and are not a mechanical instrument of headquarters' will (Nohria and Ghoshal, 1994), the emphasis is on MNC interdependence rather than on subsidiary autonomy. The subsidiaries are pursuing their own interest by engaging in the internal MNC network (as well as in the external network) and using their bargaining power to attract more internal resources from the MNC rather than seeking autonomy and isolating themselves from other MNC units.

One critique that has been forwarded against the literature on subsidiary development is that most studies on subsidiary evolution have only approached the issue from an internal perspective. The subsidiary development is viewed as the result of corporate headquarters' assignment or an autonomous process within the subsidiary. However, less attention has been given to the external environment. Although the literature on aggregate economic activity suggests that different roles do evolve, the strategic management literature, with a few exceptions (Benito et al., 2003), lacks research relating subsidiary development to exogenous factors that are not firm specific.

The external environment can have considerable impact on the scope and competence level of subsidiaries. These are often acknowledged in the literature within the rubric of location advantages. Considerable attention has been paid in the literature to the role of location advantages in determining the initial entry decisions of MNCs in any given market. However, once the decision to enter a given market through FDI is taken, the kinds of activity and the level of competence of the subsidiary are also co-determined by the nature of the location advantages of the host location. The host country's location advantages play an important role in determining the level of competence of the subsidiary. This is on at least two levels. First, the level of competence is a function of the quality of the location advantages that the host location can provide. High competence levels require complementary assets that are non-generic in nature, and are often associated with agglomeration effects, clusters, and the presence of highly specialized skills. In other words, firms are constrained in their choice of high competence subsidiaries by resource availability. For instance, R&D activities tend to be concentrated in a few locations because the appropriate specialized resources are associated with a few specific locations. Second, MNCs have been shown to prefer to engage in sequential investment in locations that may provide sub-optimal returns but with which they have prior experience.

9.3 Determining factors

In their seminal paper Birkinshaw and Hood (1998a) argue, based on an extensive literature review, that subsidiary development (or, as they put it, 'subsidiary evolution') is driven by three factors, namely head-office assignment, subsidiary choice, and local environment determinism. These three drivers of subsidiary development also refer to the three actors that have both vested interests in subsidiary development and the necessary power to influence the process of such development. These are subsidiary management, parent company management and host-country policy-makers. All three actors have an impact on the activities and competences of the subsidiary, which in turn may lead to subsidiary development as shown in Figure 9.1.

Only a few studies have really investigated how these factors affect subsidiary development. One notable exception is the paper by Birkinshaw and Hood (1998b) that explores the model based on a clinical study of 13 US-owned subsidiaries in Scotland and Canada. But, to our knowledge, none has tested the model based on large sample data, although a larger quantitative study will

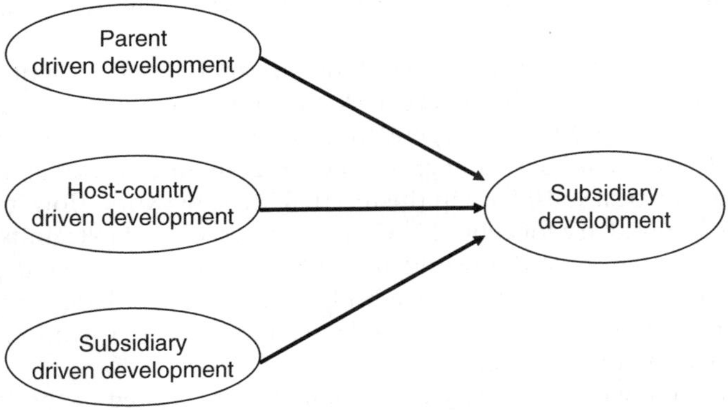

Figure 9.1 Three factors driving subsidiary development

permit a test of the general value of the model. In addition, it will make it possible to explore the more interesting questions of the strengths and interaction between the three drivers. For example, parent company managers and host country policy-makers can either impact the subsidiary activities and resources directly or they can work more indirectly by influencing the actions of subsidiary management. It is an empirical question whether they mainly have a direct or an indirect impact on subsidiary development. However, this is a question with major implications for both policy-makers and parent company managers as it makes an immense difference whether they target subsidiary development more directly (e.g. through regulation and mandates) or indirectly by influencing the behavior of subsidiary managers.

9.4 Dimensions of subsidiary development

Subsidiary development is obviously a multi-dimensional construct, as subsidiaries can develop in many dimensions (e.g. in terms of activities, competences, mandates, and size). Some authors have proposed a typology of subsidiaries, where subsidiary development is equal to gaining more scope or mandates (e.g. White and Poynter, 1984; Jarillo and Martinez, 1990), that is, the breath of activities conducted by the subsidiary. However, as pointed out by Benito et al. (2003), the subsidiary scope is only one of the relevant dimensions describing subsidiary roles and subsidiary development. Scope and mandates may be very broad or more narrowly focused on specific activities where the subsidiary has a high level of competence. Therefore, scope and mandates alone do not necessarily determine subsidiary roles. Benito et al. (2003) argue that the level of competence or the depth of subsidiary activities is an important dimension missing when only looking at subsidiary scope. Consequently, it is useful to look at the combination of scope (breath) and competence levels (depth) when discussing subsidiary development.

A third aspect of subsidiary development that has been put forward recently in the literature is the level of MNC-interdependence (Forsgren and Pedersen, 1998). MNC interdependence is concerned with the extent to which activities that have strategic implications are interrelated and the role played by the focal subsidiary in the internal MNC network. How important is the focal subsidiary for other MNC units and do other MNC units depend on the competences of the subsidiary and *vice-versa?* The more dependent other MNC units are on the focal subsidiary, the stronger bargaining power and the more resources they will be able to appropriate in the internal MNC network (Holm and Pedersen, 2002).

The final model of subsidiary development that is tested in the following is presented in Figure 9.2.

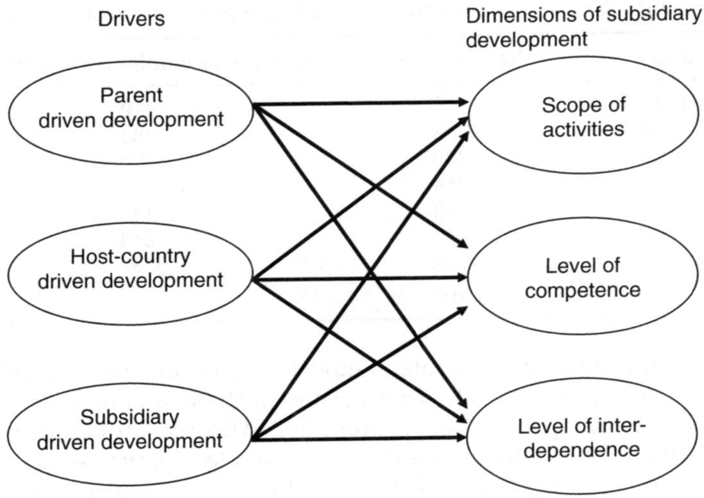

Figure 9.2 Model of subsidiary development

9.5 Data and analysis

The data for this chapter was collected as part of the Centers of Excellence project launched in May 1996 with the purpose of investigating headquarters–subsidiary relationships and the internal flow of knowledge in MNCs. In order to collect comparable quantitative data, a questionnaire that could be applied in all the involved countries was constructed. After several project meetings and extensive reliability tests of the questionnaire on both academics and business managers, this was accomplished (for more information on the CoE project, see Holm and Pedersen, 2000).

For practical reasons, each project member was made responsible for gathering data on foreign-owned subsidiaries within their own country. Thus, all subsidiaries in the database belong to MNCs. In the data gathering, subsidiary managers, rather than headquarters, were respondents.

This chapter is based on empirical data from seven countries: Austria, Denmark, Finland, Germany, Norway, Sweden and the UK. All countries are located in the northern part of Europe. The four Nordic countries and Austria are relatively small, while Germany and the UK are among the largest in Europe. Approximately 80 per cent of the questionnaires were answered by subsidiary executive officers, while financial managers, marketing managers or controllers in the subsidiary answered the remaining 20 per cent. The response rate varies between 20 (UK) and 55 per cent (Sweden), depending on the country of investigation. The quality of the data is quite high with a general level of missing values of not more than 5 per cent.

Table 9.2 Sample size and subsidiary employees in the different countries

Country	Sample size	Subsidiary employees (mean)
Austria	313	318
Denmark	308	284
Finland	238	200
Germany	254	1574
Norway	262	130
Sweden	530	244
UK	202	3787
Total	2107	742

As shown in Table 9.2, the total sample covers information on 2107 subsidiaries, comprising all kinds of subsidiaries in all fields of business. Among countries, the sample size ranges from 202 (UK) to 530 (Sweden). With the exception of Sweden, the size of the sample is rather similar in the six countries. The average number of employees in subsidiaries is 742 and the median is 102. Within the five smaller countries the average size of the subsidiaries is very similar, while Germany and UK comprise substantially larger subsidiaries because of larger market sizes.

9.5.1 Construct analysis

The hypotheses are tested in a LISREL model that allows for simultaneous formation of underlying constructs (the measurement model) and test of structural relationships among these constructs (the structural model). The validity of LISREL models is estimated by the validity of the entire model, i.e., by the nomological validity. But before estimating the nomological validity of the model, with the causal relations specified, it is important to judge the convergent validity, i.e., the homogeneity of the constructs included in the model, and the discriminant validity, i.e., to what extent the constructs are independent. First, we will describe the operationalization of the constructs included and then we will evaluate the different forms of validity. The three drivers of subsidiary development are latent constructs (underlying and not directly measurable) and are therefore measured as reflective measures, while the three dimensions of subsidiary development are manifest dimensions and are therefore measured as formative measures.

Parent company drivers

The parent company drivers mirror the discretion granted to the focal subsidiary by the parent firm. It may be the decision rights granted to the subsidiary on issues of strategic importance for the subsidiary. Based on the scale developed by Roth and Morrison (1992), respondents were asked to identify the level at which certain decisions were made, where 1 = corporate level (HQ), 2 = divisional level and 3 = subsidiary level. Decisions were as follows: entering new markets in host country; entering foreign markets; introduction of new products/services. Taken together these three items indicate the degree of decision rights granted to the subsidiary. The higher the value the more decision rights is granted.

Host-country drivers

Host-country drivers reflect the dynamism of the local business environment. Building on the main elements of Porter's (1990) diamond model, respondents were asked to assess the business environment in which they compete along three dimensions: availability of supply material; quality of suppliers; and demanding customers. The three items are measured on a 7-point Likert-type scale from 1 (very low) to 7 (very high). Together these three host-country factors indicate to what extent the subsidiaries get dynamic impulses from the local environment (i.e. the external network of suppliers and customers).

Subsidiary drivers

This construct measures the focal subsidiary's own initiative and effort in upgrading the activities and competencies. For each of the six activities: development, production of goods and services, marketing/sales, logistics/ distribution, purchasing, and HRM, respondents were asked to 'indicate the level of own investments for the past three years'. We chose a 3-year time frame to eliminate single-year fluctuations. All six items were measured on a 7-point Likert-type scale (1 = very limited and 7 = very large). Taken together the items are forming a construct for the level of own subsidiary initiative.

Scope and competence

In line with Benito et al. (2003), scope and level of competence are manifest variables and, therefore, measured as formative measures. The two variables were operationalized as follows:

$$\text{Scope of tasks} = \Sigma a_i \tag{9.1}$$

where a_i = any given activity i (development, production of goods or services, marketing/sales, logistic/distribution, purchasing, HRM) undertaken by a given subsidiary. Since it takes a value of 1 if an activity is performed, and 0 otherwise, the variable simply sums up the number of activities. Hence, the values range from 1 (i.e., a single-activity subsidiary) to 6 (i.e., the whole range of tasks is carried out).

$$\text{Level of competence} = \Sigma c_i \,/\, \Sigma a_i \tag{9.2}$$

where c_i is a measure of the level of competence of the subsidiary in performing a given activity i, as perceived by the subsidiary managers on a 7-point Likert-type scale (1 = weak competence and 7 = very strong competence). Since the level of competence indicator c_i is counted only for activities a_i actually undertaken by the focal subsidiary, it provides a measure of the average overall level of competence of that subsidiary.

Level of interdependence

This reflects the extent to which the focal subsidiary is integrated in the MNC network, i.e., how dependent other MNC units are on the focal subsidiary and vice versa. The respondents were asked to assess to what extent other MNC units where dependent on them: 'What would be the consequences for other units in the Foreign Company if they no longer had access to the competencies of the subsidiary?' (1 = no consequences, 7 = very significant consequences). In a similar vein, the subsidiary dependence on other MNC units was captured by the following question: 'What would be the consequences for the subsidiary if it no longer had access to the competencies of other MNC units?' (1 = no consequences, 7 = very significant consequences). Taken together, these two items measure the level of interdependence between the focal subsidiary and other MNC units.

Assessing convergent validity

A measurement model was created in order to assess convergent and discriminant validity of the three reflective measures. In Table 9.3, convergent validity is judged by the R^2-values measuring the strength of the linear relationships, the t-values (a significance test of each relationship in the model), and the factor loading for each indicator (Jöreskog and Sörbom, 1993). The constructs in this LISREL model all have good convergent validity, i.e., they are homogeneous constructs. As can be seen in Table 9.3, the strength of the linearity in relations between constructs and items are in all cases very strong (all R^2-values are above the usual threshold of 0.20), which is also a reflection of the very large data set of more than 2100 observations. From Table 9.3 we can conclude that the t-values for all items are highly significant (all above 16) and that their (standardized) factors loadings are strong (all above 0.39).

The second step in the analytical process was to form the structural model by specifying the causal relations in accordance with the proposed model. We tested single causal relations with t-values and factor loadings between the constructs in the model. The Chi-square test statistic is frequently used to evaluate the performance of the entire model, i.e. whether the model is significantly different from the data (nomological validity). However, as a measure of fit the Chi-square has drawbacks limiting its use in this study. Thus, Bentler and Bonett (1980) have made the point that the Chi-square statistic – being a function of the sample size – is very sensitive to large sample sizes. Consequently, a different alternative measure has been proposed. The two goodness-of-fit indices, GFI (goodness-of-fit index) and AGFI

Table 9.3 Constructs and items of reflective measures

Constructs and items	Factor loading*	t-value	R^2-value
Granted decision rights			
Decisions on entering new markets in host country	0.54	20.65	0.29
Decisions on entering markets abroad	0.62	23.59	0.38
Decisions on introduction of new products/services	0.42	16.32	0.22
Local business environment			
Availability of supply material	0.65	24.80	0.42
Quality of suppliers	0.86	32.99	0.75
Demanding customers	0.39	16.56	0.23
Subsidiary initiative			
Investments in development	0.41	19.19	0.25
Investments in production of goods and services	0.43	20.18	0.26
Investments in marketing/sales	0.55	27.88	0.30
Investments in logistics/distribution	0.67	38.16	0.45
Investments in purchasing	0.75	45.97	0.56
Investments in HRM	0.69	39.80	0.47

Note: * All factor loadings are highly significant at $p < 0.01$ with *t*-value above 16.

(adjusted goodness-of-fit index) measure how closely the model fits the data by comparing the predicted and observed covariance matrices. In general, one considers a 0.9 level of GFI and AGFI as satisfactory, and a 0.8 level as the minimum acceptable for a model (Bentler and Bonett, 1980). A related measure is the Norm Fit Index (NFI), which is interpreted as the percentage of variance in the data that the model explains (similar to R-square in OLS-models). As this study applies a very large sample of more than 2100 observations the Chi-square statistics is of limited use and instead we apply the alternative measures of goodness-of-fit of the model.

9.6 Results

Through repeated iterations a LISREL analysis proceeds with the fine-tuning of the model to obtain a more coherent representation of the empirical data. The purpose of the LISREL analysis is to arrive at and confirm a model consisting of specified causal relations. The first step was to test all possible causal relationships simultaneously. However, some of the relations turned out to be insignificant and after omitting the insignificant relations we arrived at the structural model shown in Figure 9.3. The figures given are standardized factor loadings of causal relations with t-values in parentheses. The values of GFI and AGFI are 0.92 and 0.88, respectively, and therefore well above the minimum level (of 0.8), suggesting an acceptable fit with the

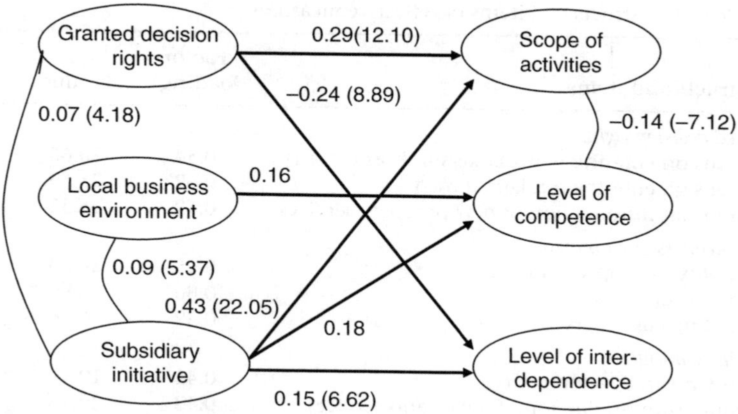

Figure 9.3 LISREL model of subsidiary development

data (the Chi-square value is $1755_{df=75}$ with $p < 0.01$). The NFI is 0.73 indicating that the model is able to explain almost three-quarters of the variance in the data, which is very satisfactory.

It turns out that subsidiary initiative is positively related to all three dimensions of subsidiary development and, in particular, for the scope of activities with a standardized coefficient of 0.43. The local business environment only has a significant relation to the level of competences indicating that the local dynamism in the business environment is an important source of upgrading of subsidiary competences. HQ granting of decision rights to the subsidiary has a positive effect on the scope of subsidiary activities, but a significant negative effect on the extent to which the subsidiary is integrated in the internal MNC network.

When allowing for correlations among the three drivers in the model, it appears that both HQ-granting decision rights and local business environment are (highly) significantly correlated with subsidiary initiative indicating that they have a strong indirect effect on subsidiary development going through the influence on own subsidiary initiatives.

Furthermore, scope of activities is negatively correlated with the level of competences giving support to the argument that breadth of subsidiary activities often comes at the cost of depth in these activities. This implies that the subsidiaries with a high level of competences typically will be relatively narrow in their scope, while subsidiaries with broader scope tend to have lower average level of competence for the activities they are carrying out.

Surprisingly, we do not find a significant correlation between level of competence and level of interdependence as was expected. The reason could be that subsidiaries in some cases develop high levels of competence in the interaction with external counterparts, which is of very little use in other MNC units as it is developed in a different context. The implication is that high levels of competence will only be associated with high levels of interdependence when the competence is useful and important for other MNC units. At the core of this argument is the distinction between autonomous subsidiaries that may have high competences but of little use in other MNC units and Centers of Excellence with high competences that are used in other MNC units (Holm and Pedersen, 2000).

9.7 Discussion and policy implications

The literature on subsidiary development identifies three drivers affecting the subsidiary development process (here operationalized by the dimensions of scope, competences and interdependence): parent-company drivers, host-country drivers and subsidiary drivers.

Parent companies can play an active role, in the sense that a decision can be taken to provide a subsidiary a formal responsibility for a certain task. Such a decision can be based on certain identified potential of the subsidiary. In contrast to this the HQ role can also be more indirect. Over time the activities of the subsidiary within a certain area are recognized at HQ, and eventually the subsidiary is given the responsibility for that area. In the latter case, the main driving force in subsidiary development is the subsidiary's entrepreneurship, which is eventually confirmed by HQ.

The results indicate that HQ mainly has discretion over the scope of subsidiary activities and the extent to which subsidiaries are integrated in the MNC network, while they have less influence on the level of subsidiary competence. Interestingly, it seems like the granting of more decision rights to the subsidiary expands the scope of activities, but at the same time has a negative effect on the degree of interdependence, meaning that the subsidiary will tend to isolate themselves more from other MNC units. In that sense the parent company drivers may seem to have a double character, enhancing one dimension of subsidiary development (scope of activities), but at the cost of another dimension (level of interdependence).

The most important driver of subsidiary development is the subsidiary's own set of entrepreneurial initiatives for upgrading their resources and competences. In contrast to the effect of parent-company drivers, subsidiary drivers have a uniform positive effect on all three dimensions of subsidiary development.

The main direct effect of policy-makers goes through the creation of a dynamic local business environment, which has a strong positive influence on the level of competence. With a coefficient of 0.16 the local business environment is almost as important as own subsidiary initiatives in upgrading the level of competences. However, the direct effect of policy-makers on scope of activities and level of interdependence is very limited.

This chapter brings out some important findings for policy-makers, the most significant being that subsidiaries indeed have some freedom to form their own development path, so it is worthwhile creating a supportive business environment for their upgrading and development. A prudent policy towards 'monitoring' MNC subsidiaries requires an understanding of their development dynamics.

References

Bartlett, C. and Ghoshal, S., *Managing Across Borders: The Transnational Solution* (Boston, MA: Harvard Business School Press, 1989).

Benito, G.R.G., Grøgaard, B. and Narula, R., 'Environmental influences on MNE subsidiary roles: economic integration and the Nordic countries', *Journal of International Business Studies*, 34(5) (2003) 443–56.

Bentler, P.M. and Bonett, D.G., 'Significance tests and goodness of fit in the analysis of covariance structures', *Psychological Bulletin*, 88 (1980) 588–606.

Birkinshaw, J. and Hood, N., 'Multinational subsidiary evolution: capability and charter change in foreign-owned subsidiary companies', *Academy of Management Review*, 23(4) (1998a) 773–95.

Birkinshaw, J. and Hood, N. (eds), *Multinational Corporate Evolution and Subsidiary Development* (London: Macmillan, 1998b).

Buckley, P. and Casson, M., *The Future of the Multinational Enterprise* (London: Macmillan, 1976).

Doz, Y.L., Santos, J. and Williamson, P., *From Global to Metanational: How Companies Win in the Knowledge Economy* (Boston: Harvard Business School Press, 2001).

Étemad, H. and Dulude, L.S. (eds), *Managing the Multinational Subsidiary. Response to Environmental Changes and to Host Nation R&D Policies* (London: Croom Helm, 1986).

Forsgren, M. and Pedersen, T., 'Centres of excellence in multinational networks', in Birkinshaw, J. and Hood, N. (eds) *op. cit.* (1998).

Foss, N. and Foss, K., 'Authority and discretion: tensions, delegation and implications for new organizational forms', *Paper presented at the 2001 LINK conference*, Copenhagen, Denmark (2002).

Gupta, A.K. and Govindarajan, V., 'Knowledge flows and the structure of control within multinational corporations', *Academy of Management Review*, 16(4) (1991) 768–92.

Gupta, A.K. and Govindarajan, V., 'Organizing for knowledge flows within MNCs', *International Business Review*, 3(4) (1994) 443–57.

Hennart, J.F., *A Theory of the Multinational Enterprise* (Ann Arbor: University of Michigan, 1982).

Holm, U. and Pedersen, T. (eds), *The Emergence and Impact of MNC Centers of Excellence* (London: Macmillan, 2000).

Hymer, S., *The International Operations of National Firms: a Study of Foreign Direct Investment* (Cambridge, MA: MIT Press, 1976).

Jarillo, J.C. and Martinez, J.I., 'Different roles for subsidiaries: the case of multinational corporations in Spain', *Strategic Management Journal*, 11(7) (1990) 501–12.

Jöreskog, K.G. and Sörbom, D., *LISREL 8: Structural Equation Modelling with SIMPLIS Command Language* (Chicago: Scientific Software International, 1993).

Kogut, B., 'Learning, or the importance of being inert: country imprinting and international competition', in Ghoshal, S. and Westney, D.E. (eds), *Organization Theory and the Multinational Corporation* (London: Macmillan Press Ltd, 1993).

Kuemmerle, W., 'The drivers of foreign direct investment into research and development: an empirical investigation', *Journal of International Business Studies*, 30(1) (1999) 1–24.

Malnight, T.W., 'The transition from decentralized to network-based MNC structures: an evolutionary perspective', *Journal of International Business Studies*, 27(1) (1996) 43–65.

Nohria, N. and Ghoshal, S., 'Differentiated fit and shared values: alternatives for managing headquarters–subsidiary relations', *Strategic Management Journal*, 15(6) (1994) 491–502.

Pfeffer, J. and Salancik, G.R., *The External Control of Organizations: A Resource Dependency Perspective* (New York: Harper and Row, 1978).

Porter, M.E., *The Competitive Advantage of Nations* (New York: Free Press, 1990).

Roth, K. and Morrison, A., 'Implementing global strategy: characteristics of global subsidiary mandates', *Journal of International Business Studies*, 23(4) (1992) 715–36.

White, R.E. and Poynter, T.A., 'Strategies for foreign-owned subsidiaries in Canada', *Business Quarterly*, 48(4) (1984) 59–69.

10

Subsidiaries, Knowledge Development and MNE Re-Investments

Ulf Andersson and Magnus Persson

10.1 Introduction

Scholars and policy-makers have long been interested in different methods of increasing the level of foreign direct investment (FDI). Many governments thus offer different incentives, explicit or implicit, to MNEs to invest and establish operations in their markets (Globerman and Shapiro, 1999). The primary goal of such policy-making is to increase the attractiveness of the market to potential investors. Although literature concerning this issue has mostly examined the establishment of new affiliates in host countries, the importance of enhancing the value adding role of already established subsidiaries has also received some attention as an important objective for policy-makers (Birkinshaw, 1996). Different subsidiaries can contribute to the competitiveness of nations to different degrees, depending on, for instance, their corporate strategic role and level of technology (O'Donnell and Blumentritt, 1999). As a result, understanding the process by which MNEs re-invest in host-country subsidiaries is important to policy-makers.

From a corporate perspective, the logic of FDI is that host countries provide opportunities not found in their home country. It is often argued especially that MNEs seek to tap local sources of knowledge to increase corporate knowledge assets (Bartlett and Ghoshal, 1986), and consequently invest in locations that provide good opportunities for learning and acquiring knowledge. Although knowledge-creating activities in subsidiaries are inter-linked with the conditions of the local environment (Birkinshaw and Hood, 1998; Frost, 2001), recent findings indicate that it is not the environment per se that induces knowledge development in already established host-country subsidiaries; rather, it provides *opportunities* for knowledge acquisition and creation. The extent to which the subsidiary can actually realize these opportunities depends on the past and present activities of the unit. Researchers have argued that it is foremost through linkages between subsidiaries and the environment that they acquire and integrate external knowledge (Almeida and Phene, 2004). In a similar vein Holm et al. (2005)

showed that the competitive conditions in the local market only influenced knowledge creation in subsidiaries through their local embeddedness in a network of relationships. It is through these linkages to other actors that the subsidiary can access and acquire valuable knowledge. Therefore, the network surrounding the subsidiary is an important factor influencing its ability to acquire capabilities (McEvily and Zaheer, 1999), create value, and achieve its economic goals (Echols and Tsai, 2005).

This chapter examines how such linkages (relationships) between the subsidiary and the local environment influences MNE re-investments in host-country subsidiaries. We focus on two different mediating mechanisms between subsidiary local linkages and MNE re-investments: 1) its strategic role as a provider of knowledge to other units, and 2) the performance of the subsidiary. Both these factors are in the second stage of the model hypothesized to induce re-investments into the subsidiary.

The chapter may potentially contribute to our current understanding of MNE investment behavior in three main ways. First, despite the reasonable assumption that re-investments in already established operations are at least as important as new investments for policy-makers, research examining MNE re-investments is relatively sparse. This is somewhat surprising since re-investments are a major source of resource inflow into nations; for instance, Forsgren (1989) found that about 81 per cent of all foreign investment into Sweden in 1975–82 were in fact re-investments into already established operations rather than *de novo* investments. Second, while there is a large and diverse body of literature aiming to explain the closely related topic of development and evolution of subsidiary roles and charter (c.f. Gupta and Govindarajan, 1991; Roth and Morrison, 1992; Birkinshaw, 1996; Birkinshaw and Hood, 1998) this research has seldom explicated the specific mediating mechanisms between subsidiary activities and subsequent investments in the unit, but rather examined the breadth of responsibilities and intra-organizational role as dependent variables. Third, by highlighting two different subsidiary characteristics, performance and knowledge provision that may influence MNE re-investments, anyone interested in how to stimulate investments in already established host-country subsidiaries through the usage of different public policies may draw tentative conclusions from our model.

10.2 Theoretical background

Earlier views of the MNE and FDI assumed that unfamiliarity with local circumstances of the host country incurred costs in subsidiary operations that needed to be compensated for. This 'liability of foreignness' was overcome by owning specific advantages developed in its home-country and subsequently exploited in the host-country location (Zaheer, 1995). In that sense, the rationale of internationalization was one of exploitation rather than exploration. This 'knowledge diffusion' model of the MNE (Almeida and Phene, 2004) reigned in early theoretical depicting of the multinational firm and its activities (c.f. Vernon, 1966; Buckley and Casson, 1976). The approach still has its merits, but the last decades have seen a fundamental shift in theorizing on the MNE. Research has instead come to recognize the critical role of subsidiaries as important sources of knowledge. The processes of acquisition, creation, and diffusion of knowledge at the subsidiary level have become central themes in recent literature (Frost, 2001; Schulz, 2001).

Consequently, learning from host-country locations may be an important source of value creation for MNEs. By engaging in their own entrepreneurial initiatives and local development processes, subsidiaries can come to hold knowledge not found elsewhere in the organization. This also entails knowledge being heterogeneously distributed inside the MNE. Some units may develop and accumulate knowledge to a considerable degree and thus become more important to the organization than others (Forsgren et al., 2000; Frost et al., 2002).

In terms of the local environment and its importance for subsidiary knowledge development, most research has focused on institutional level factors. This includes the existence of clusters and scientific institutions (Malmberg et al., 1996; Davis and Meyer, 2004), existence of knowledge spillovers (Cantwell and Piscitello, 2002; Fritsch and Franke, 2004), and market competition (Porter, 1990).

Despite the high merits of this research, it is largely silent on exactly how these factors influence subsidiary-level knowledge creation, and seems often to assume homogeneity in performance among firm subsidiaries located within the same region or nation. However, subsidiaries are differently linked to other actors within its environment. The network of relationships surrounding the subsidiary is its main source of information and knowledge (Andersson, 2003) and how subsidiaries are embedded in this network is therefore an important source of variation in their acquisition of new capabilities (McEvily and Zaheer, 1999). Therefore, MNE re-investments, which by definition occur in already established host-country subsidiaries, may be more influenced by the past and present activities of the subsidiaries than by institutional level factors.

In the following we argue that MNE re-investments are influenced by the activities of the host-country subsidiary (here represented by its network embeddedness), and explicate two subsidiary characteristics that are possible mediating factors; subsidiary performance and knowledge provision to other organizational units. We begin by examining the relation between subsidiary embeddedness and its characteristics, and then turn towards how these characteristics may influence MNE re-investments.

10.3 Embeddedness and subsidiary characteristics

Social network theory suggests that firms that are part of the same social structure can share knowledge and information more easily with each other than those that are not (c.f. Uzzi, 1997; Hansen, 1999; Gulati et al., 2000). Literature has focused on different facets of network embeddedness; some emphasize the structure of the network as an important determinant to its conduciveness in accessing knowledge and information (Burt, 1992; Ahuja, 2000; Gargiulo and Benassi, 2000), while others have focused more on the attributes of the constituent relationships (Hansen, 1999; Andersson et al., 2002). Here, the focus is on subsidiaries embedded in exchange relationships through which they access resources, that is the latter (relational) view (Dyer and Singh, 1998; Gulati et al., 2000; Andersson, 2003). The attributes of the relationships in the network that surround the subsidiary, such as their strengths or weaknesses (Granovetter, 1973), are believed to influence the subsidiary's ability to create and acquire knowledge.

Network embeddedness can be viewed as a strategic resource influencing subsidiaries' capability development (Andersson et al., 2002). But locally developed knowledge is of limited value if kept within a single subsidiary. Thus, we can assume that if the headquarters are informed about the existence of potentially valuable competence at subsidiary level, they will attempt to disseminate this to other units. By transferring knowledge from the subsidiary to other units in the firm the competence of the firm is upgraded. This may be done by formally assigning especially competent units strategic roles which include the creation and dissemination of knowledge, for example, 'Centers of Excellence' (Forsgren et al., 2000; Frost et al., 2002; Holm and Pedersen, 2002). Accordingly, we hypothesize that:

H1. Local embeddedness has a positive impact on subsidiary importance for other units' technical development.

Several studies have indicated that embeddedness in a set of relationships is conducive to organizational value creation and achievement of economic goals (Echols and Tsai, 2005). An important implication of network embeddedness is that organizations may develop a higher degree of trust towards each other, something that can limit expectations of opportunistic behavior (Gulati et al., 2000) and as such reduce governance cost. Inter-organizational trust has also been shown to influence the supplier performance in exchange relationships (Zaheer et al., 1998).

While direct relationships may be costly to maintain, they can also reduce the probability for information distortion (Hansen, 2002) something that can be vital when aiming to identify and exploit entrepreneurial opportunities, for instance, in innovation. Studying patent citation data, Almeida and Phene (2004) found that subsidiary linkages with the local environment positively influenced innovation. Tsai and Ghoshal (1998) found that social capital, as a 'relational resource', positively influenced resource exchange and combination, which in turn positively influenced innovativeness. A close relationship with local counterparts allows the subsidiary to gain better and more accurate knowledge about their operations and potential problems, which in turn can improve the efficiency of the exchange relation and the performance of the subsidiary. Accordingly, we hypothesize the following:

H2. Local embeddedness has a positive impact on subsidiary expected performance.

10.4 Subsidiary characteristics and MNE re-investments

The MNE has been described as an inter-organizational network in which subsidiaries have their own interests and agendas that might not coincide with those of headquarters (HQ) and of other units in the MNE group (Ghoshal and Bartlett, 1990). While earlier views of the MNE largely presumed that subsidiaries merely implemented strategies crafted at the HQ location and as such were the elongated arm of top-management, recent studies have indicated that subsidiaries are at the cross-section of two institutional realms (Kostova and Roth, 2002) or simultaneously embedded in two knowledge contexts (Almeida and Phene, 2004). Consequently, these units may be subject to the influence of stakeholders with conflicting interests.

Studies examining the HQ–subsidiary relationship have long recognized that this seldom is problem-free and characterized by mutual understanding and congruent interests. Factors such as information asymmetries, differing experiences between HQ and subsidiary managers, and decreasing subsidiary dependence on HQ may lead to dissonance and perception gaps in the relationship (Birkinshaw et al., 2000). For instance, increasing subsidiary embeddedness has been shown to negatively influence the feeling of being controlled by HQ with the subsidiary (Andersson and Forsgren, 1996). This finding is consistent with agency theory, which suggests that subsidiary control will be increasingly difficult as information asymmetries grow and that specialized competence at subsidiary level may give rise to difficulties in the HQ–subsidiary relationship (Roth and O'Donnell, 1996).

Research has also indicated that subsidiaries may actively be searching for independence and in some instances for influence inside the MNE. The 'political view' of the firm (Forsgren, 1989, 1990) suggests that power is distributed in the MNE organization according to the importance of controlled resources, that is, following the logic of resource dependence theory (Pfeffer and Salancik, 1978). As home-country advantages become less important for the foreign subsidiary – a situation that is amplified if the MNE originally invested in the market on a 'knowledge-seeking' rather than exploitative basis – the parent organization will decrease in importance as a resource provider to the subsidiary. Subsidiary units that have developed and acquired local resources vital to the operations of the firm may become powerful players in the MNE network, and use their bargaining power to attract resources and appropriate rents (c.f. Coff, 1999). We thus hypothesize that subsidiary embeddedness positively influences MNE re-investments.

H3. Subsidiary local embeddedness has a positive impact on re-investments into the subsidiary.

Literature on how subsidiaries evolve over time, in terms of charter or role, is rich and diverse. There is certainly no shortage of different theoretical archetypes of subsidiaries; however, less effort has been devoted to understanding the process by which these roles are given or taken inside the organization. Especially within the 'subsidiary choice' perspective on subsidiary evolution (Birkinshaw and Hood, 1998), the logic of the argument is that the subsidiary role changes as a result of, or in anticipation of, changes in its capabilities (Ståhl, 2004). Examining subsidiary mandates, Birkinshaw (1996) found that one important class of mandate extensions occurred when subsidiary development of 'distinctive capabilities' led to subsequent investments in the specific area.

Forsgren and Pedersen (2000) argued that knowledge creation only strengthens the role of subsidiaries inside the organization if other units can access and use this knowledge. By providing other units with important knowledge the subsidiary may establish itself as a central actor in the knowledge processes inside the MNE organization. Given that wholly-owned subsidiaries are legally controlled by parents, subsidiary power is most likely to be underpinned by intangible assets due to the difficulty in defining and enforcing property rights (Mudambi and Navarra, 2004). Basically, subsidiary control over important intangible resources makes the opportunity cost for other units, if not cooperating with the focal subsidiary, very high (Mudambi and Navarra, 2004). Especially, if FDI occurs not only as a means to reach sales of existing products and services, but also as a way to acquire new intangible resources, then host-country subsidiaries that are able to do so have a very important function in the organization, and are likely to attain more corporate resources to continue and expand their role as knowledge providers in the organization.

H4. Subsidiary importance for other unit's technical development has a positive impact on re-investments into the subsidiary.

The extent to which the subsidiary is important for other units' technical development is, however, not the only rationale for increasing investments in this unit. Some subsidiaries are not established in order to fulfill any such objective. Depending on the current strategy of the MNE, subsidiaries may be viewed more from a portfolio perspective. In such cases the subsidiary might not be called to diffuse any local knowledge or other developments, but instead focus on creating profitable business confined to its host-country market. Its main goal is to create rents in the local market, and not contribute to others' strengths. In such cases, the relative performance of the subsidiary – in terms of financial and market performance – rather than its importance for other subsidiaries, will be an important determinant of the distribution of resources within the organization.

H5. Subsidiary performance has a positive impact on re-investments into the subsidiary.

10.5 Data and methods

The sample incorporates MNEs involved in an array of manufacturing industries, such as petrochemicals, telecommunications, hard materials, paper, retailing, power systems, and transportation. The firms were approached at the business area/divisional HQ level rather than at the firm level for three reasons. First, business area HQ is the locus of control regarding the allocation of investments within the business areas. Second, the business area level of the firm is closer to the subsidiary operations and the business area HQ will have a direct managerial relationship with the subsidiaries. Third, knowledge about subsidiary activities is primarily an intra-business area matter since the divisionalization of the MNE separates different businesses from each other (Egelhoff, 1988).

We collected data from 20 such MNE business areas, in all 97 subsidiaries located in Europe (92) and North America (5). All business area HQ except one was located in the home country of the MNE. On average, five subsidiaries in each of the business areas have been studied. All the business areas were highly international and three-quarters of them had more than 50 per cent of their employees outside the business area HQ's home country.

The business areas' HQ assisted in the selection of subsidiaries that were representative of each chosen business area's business activities, with the intention of increasing the possibility of drawing general conclusions. On average, the subsidiaries in the sample accounted for over 50 per cent of the business areas' combined operations measured in terms of the number of employees. The number of employees in the subsidiaries varied from 50 to more than 5000. The investigated subsidiaries carry out both production and sales activities; thus, both product and process development are important activities in all subsidiaries studied.

To collect the data, interviews were carried out face-to-face with managers both at the subsidiary and at the HQ levels. First, the assessments of the subsidiaries' relationships with customers and suppliers were made with the sales manager and the manager responsible for purchasing in each subsidiary. A critical issue was the choice of which local business relationships to study because we had to limit the number of relationships to be investigated, as collecting information about all kinds of relationships that a subsidiary has would be an overwhelming task. Further, for both practical and analytical reasons, we restricted our analysis of subsidiary embeddedness to the subsidiaries' most important field of business, and involved relatively few of each subsidiary's customers and suppliers. The subsidiary managers interviewed were asked to describe and assess a total of six relationships – three with customers and three with suppliers – that they considered important in a broad sense. In the personal interviews with the managers we were very cautious not to make them select business relationships that were important for a pre-specified reason, but rather, for any reason as long as it was important to their business activities. In a second step, specific indicators of technical embeddedness were measured by asking the managers to characterize the relationships using a standardized questionnaire. A number of specific variables relating to subsidiary embeddedness in specific activities were measured, such as the number of functional areas involved in direct contacts between the subsidiary and its respective counterparts. Other measures concerned the degree of adaptation between the subsidiary and its partners in certain technical respects, measured on a five-point Likert-type scale for each relationship. At this relationship level the data show a normal distribution, although mean values differ somewhat.

Altogether the managers chose 514 relationships to customers and suppliers as important; of these 405 related to local external business partners, that is, 79 per cent. This means that, on average, the sample consists of 97 subsidiaries and their four, on average, most important local external business partners.

After interviewing subsidiary managers in one business area, we went back to the business area HQ and conducted a personal interview with the business area manager, using the same type of standardized questionnaire. Through these interviews, we gathered information about the HQ manager's view of the subsidiaries considered in terms of: their respective importance for the development of their sister units' products and production development; their expected market performance; and the re-investments made in each subsidiary. The personal interviews lasted for about two hours, during which any problems involving concepts and interpretations in the questionnaire were discussed and explained.

10.6 Construct analysis

The hypothesized model shown in Figure 10.1 was empirically tested using linear structural equation modeling with the LISREL program. The validity of LISREL models is estimated by considering the validity of the entire model; that is, by assessing the nomological validity (see, for example, Anderson and Gerbing, 1988; Bollen, 1989; Bollen and Long, 1993; Jöreskog and Sörbom, 1993). Before estimating the nomological validity of the model, with the causal relations specified, it is important to judge the convergent validity of the items and constructs. That is, the researcher must examine the

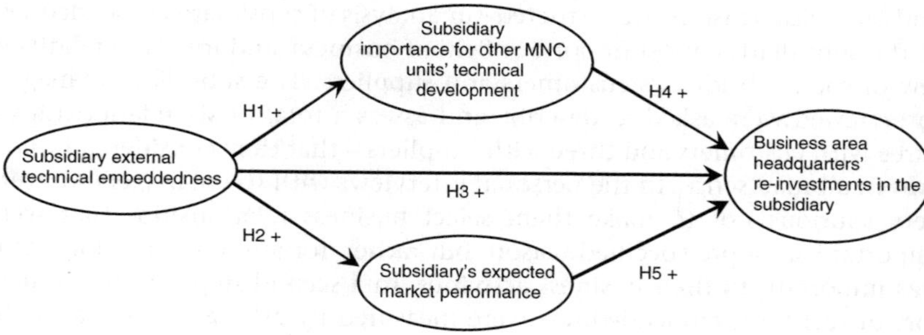

Figure 10.1 The hypothesized model

homogeneity of the constructs included in the model and, also, the discriminant validity. The latter validity measure being the extent to which the constructs are independent. First, we will describe the operationalization of the constructs included, and then we will evaluate the different forms of validity.

10.6.1 Subsidiary external technical embeddedness

Technical embeddedness ought to reproduce the value of a business relationship in terms of the ability of the subsidiary at issue to absorb new technology through a relationship, in this case, a local external business relationship. In some research it has been argued that the development of a company's technology is mirrored in its development of new products and/or production processes (see, for example, Mansfield, 1968). Thus, we need reliable indicators of how embedded the subsidiary is in its local relationships in terms of how it has adapted its development processes to those of its local business partners, and how extensive is its contact pattern with those business partners. We started by asking the subsidiary's sales and purchasing managers to evaluate to what extent a specific relationship with, respectively, a local external customer or supplier had caused the subsidiary's product, and production development to be adapted, using a five-point Likert-type scale from 1 (not at all) to 5 (very much) for each indicator. Secondly, we also asked sales and purchasing managers to quantify the number of different functional areas involved in direct contacts with customers and suppliers. The functional areas that they could select are the chief executives, the administration, the purchasing department, the sales department, the production department (that is, the technical staff) and the R&D department. The more functional areas were directly involved in the relationship, the greater the investment in that particular relationship and, also, the better the prospect of absorbing useful technical knowledge. Adding the scores of each of the subsidiary's local external relationships and then dividing the score obtained by the number of such relationships identified for each subsidiary gave the subsidiary's average local external technical embeddedness. As can be seen in Table 10.1, the mean values for the indicators were 2.22 and 2.65 for the measures of adaptation and 3.52 for the number of functional areas having direct contacts. The standard deviations were between 0.90 and 0.95.

To increase the relevance of our indicators, we have been very careful to relate all questions about business relationships, adaptations, product development and production development to a specific product or product group, rather than to the subsidiary's overall activity.

10.6.2 Subsidiary importance for other MNE units' technical development

This construct was judged by managers at business area HQ and measured by two indicators, the focal subsidiary's importance for sister units' 1) product- and 2) production development. The reason for conducting the assessment in this way was that the importance of a specific subsidiary should be evaluated relative to other units in the same business area. Our assumption is that the HQ of each studied business area is in a better position to make such an assessment compared to the individual subsidiaries, as HQ naturally has a more holistic view of the business area and is in a position to better compare different subsidiaries. A further benefit of using the business area headquarters managers' responses to these questions is that it ensures separation from 'technical embeddedness', where the respondents were subsidiary sales and purchasing managers, thereby avoiding the problem of common method bias. A five-point Likert-type scale from 1 (very small) to 5 (very high) was used. The mean values for the subsidiaries importance to other MNE units' product development were 2.82 (S.D. 1.25) and 2.18 for production development (S.D. 1.70).

10.6.3 Subsidiary expected market performance

There is a long tradition in the literature for measuring market performance. However, as firms are unwilling to make available information about their transfer pricing practices, tax considerations and other financial transactions inside the MNE, the conventional financial measures seem dubious and unsuitable when it comes to measuring subsidiary performance. Instead, measures like profitability, sales volume, and market share expansion seem more suitable as measures of a subsidiary's market performance.

How to measure performance has been meticulously discussed in earlier research (Chakravarthy, 1986; Chowdhury, 1992; March and Sutton, 1997). While some scholars (see for example, Wexley et al., 1980) imply that perceived performance in manager–subordinate dyads does not converge with more objective measures, other scholars have found that perceptual measures and more objective ones tend to have a high correlation (Dess and Robinson, 1984; Geringer and Hebert, 1991). In the literature on

Table 10.1 Descriptive statistics for the items used

	Subsidiary expected performance			Subsidiary importance for other MNC units' technical development		Subsidiary technical embeddedness			HQ re-investments in subsidiary
Statistics	**Increase in market share**	**Increase in sales volume**	**Increase in profitability**	**Product development**	**Production development**	**Adaptation of product development by subsidiary**	**Adaptation of production development by subsidiary**	**No. of functional areas in direct contact with counterpart**	**HQ's re-investments in the subsidiary the last five years**
Mean	3.30	3.58	3.34	2.82	2.18	2.65	2.22	3.52	2.65
Std. Dev.	0.86	0.90	1.02	1.25	1.70	0.91	0.90	0.95	2.59
Minimum	1.00	1.00	1.00	1.00	1.00	1.00	1.00	1.00	1.00
Maximum	5.00	5.00	5.00	5.00	5.00	4.50	4.40	6.00	5.00

JV performance, several researchers have turned to perceptual measures of the performance because of the concerns over the ability of objective measures to capture the performance (for example, Blanchot and Mayrhofer (1997) lists 24 studies that apply perceptual measures of JV performance).

To assess the perception of the subsidiary's expected performance, managers of the business area HQ were asked to estimate the expected increase in profitability, sales volume, and market share for each subsidiary. Apart from the advantage of separating the responses relating to technical embeddedness from those concerning performance, the measure also has the virtue of giving the business area HQ the possibility of making comparisons across subsidiaries and countries. A five-point Likert-type scale (1 = very small to 5 = very high) has been used to separate the answers. The indicators of this construct have an average ranging from 3.30 to 3.58, which is quite a low difference between the indicators. The three indicators are also fairly similar considering standard deviations, 0.86 to 1.02 (see Table 10.1).

10.6.4 Business area headquarters' re-investments in the subsidiary

This construct is measured by a single question put to headquarters. The manager was asked to estimate if the investments in the subsidiary had been smaller or larger considering its size the last five-year period. A five-point Likert-type scale (1 = much smaller to 5 = much higher) was used.

10.7 The testing

A measurement model was created in order to assess the convergent and discriminant validity. In Table 10.2, the convergent validity is judged by: the R^2-values measuring the strength of the linear relationships, the *t*-values, a significance test of each relationship in the model, and the factor loading for each indicator (Jöreskog and Sörbom, 1993). The constructs in this model all have good convergent validity, that is, they are homogeneous constructs. As can be seen in Table 10.2, the strength of the linearity of the relations between constructs and indicators is relatively strong in most cases. From Table 10.2 we can also conclude that the *t*-values for all indicators are highly significant (lowest *t*-value 4.59) and that their factor loadings are strong (all are above 0.47). Our set of latent constructs has good discriminant validity as key statistical estimates show an absence of uni-dimensionality.

In the next step we tested single causal relations with *t*-values and factor loadings between the constructs in the model. We assessed the entire model using Chi-squares (χ^2), degrees of freedom, and a probability estimate (p-value), which is a test of the non-significant distance between the data and the model, that is, the nomological validity (Jöreskog and Sörbom, 1993).

Table 10.2 Constructs and indicators

Constructs and indicators	Factor loading	*t*-value	R^2-value
Subsidiary performance			
How does HQ judge this subsidiary's future (3–5 years) potential:			
for increasing its sales volume?	0.95	9.41	0.90
for increasing its market share?	0.78	7.67	0.61
for increasing its profitability?	0.47	4.59	0.22
Subsidiary importance for other MNC units' technical development			
To what extent is this subsidiary important for other business area units':			
product development?	0.75	6.13	0.57
production development?	0.74	6.03	0.54
Subsidiary technical embeddedness			
To what extent has the relationship with this counterpart brought about adaptation to the subsidiary concerning its:			
product development?	0.95	9.02	0.89
production development?	0.69	6.60	0.47
number of functional areas involved in direct contact with people from this counterpart?	0.49	4.67	0.24
HQs re-investments in subsidiary			
Considering the subsidiary's size, have the last five years investment in this subsidiary been larger or smaller than would be the case if investments were made proportionately in the business area?	1.00		1.00

10.8 Results

The purpose of the LISREL analysis is to arrive at and confirm a model consisting of specified causal relations. When a specific relation cannot be verified, it is omitted from the subsequent estimation of the model (Bollen, 1989; Jöreskog and Sörbom, 1993). Thus, in our test, we generated a structural model that contained significant relationships corresponding to the stipulated hypotheses.

The first step of our analysis was to test all hypothesized causal relationships simultaneously. The overall model was significant ($\chi^2_{(df23)} = 26.97, p = 0.26$), and the results revealed that three of the five hypothesized relations were significant. The insignificant relations were the one between 'external technical embeddedness and re-investments', Hypothesis 3 (factor loading = 0.04, *t*-value = 0.32) and the one between 'subsidiary expected market performance and re-investments' Hypothesis 5 (factor loading = 0.14, *t*-value = 1.28). In the next step, therefore, we omitted the most insignificant relation, that is, Hypothesis 3, and ran the model with Hypotheses 1, 2, 4 and 5. This time the resulting model was again significant ($\chi^2_{(df24)}$ = 26.88, $p = 0.31$), although the relation between 'subsidiary expected market performance and re-investments', Hypothesis 5 (factor loading = 0.15, *t*-value = 1.40) was still insignificant. Hence, this relation was omitted and the model was once again run, but now with Hypotheses 1, 2 and 4. The resulting model was significant ($\chi^2_{(df25)}$ = 28.31, $p = 0.29$) and the remaining relations between the constructs were all significant, as shown in Figure 10.2. The statistics also indicate that no further development of the model is needed since the RMSEA measure (root mean square error of approximation) is 0.037, indicating a good fit (Browne and Cudeck, 1989).

The resulting model was based on three hypothesized relations. The first is a significant positive relation from 'subsidiary's external technical embeddedness' to 'subsidiary importance for other MNE units' technical development', supporting Hypothesis 1 (factor loading = 0.45, *t*-value = 3.48). Thus the data indicates that subsidiaries that have deep and intense relationships with local business partners become important to other sister units within the MNE. Secondly, Hypothesis 2 is supported, as there is a significant and positive relation between 'subsidiary's external technical embeddedness' and 'subsidiary's expected market performance' (factor loading = 0.27, *t*-value = 2.38). The data thus indicates that a subsidiary's external embeddedness not only develop knowledge of use for other internal units but also is of importance for the subsidiary's business performance on the host market. Hypothesis 4 is also supported as the model shows a significant, positive relation between 'subsidiary importance for other MNE units' technical development' and 'headquarters re-investments' in the subsidiary (factor loading = 0.35, *t*-value = 2.74). The result supports the view that the local environment represents an important source of knowledge for the subsidiary and, more specifically, that close relationships with network partners are conducive for subsidiary importance within the MNE. Together, Hypotheses 1 and 4 indicate that headquarters' re-investment in a subsidiary is mediated through the subsidiary's importance for others. Further, although the subsidiary's external technical embeddedness is a significant predictor of a subsidiary's expected market performance (Hypothesis 2), a subsidiary's expected market performance does not seem to influence the headquarters propensity to re-invest in that subsidiary.

10.9 Concluding remarks

For policy-makers in a region or a country, influencing domestic companies to participate in knowledge development projects together with foreign-owned sub-units can potentially pay off if these foreign owned units also become important for the development processes within their MNEs. The

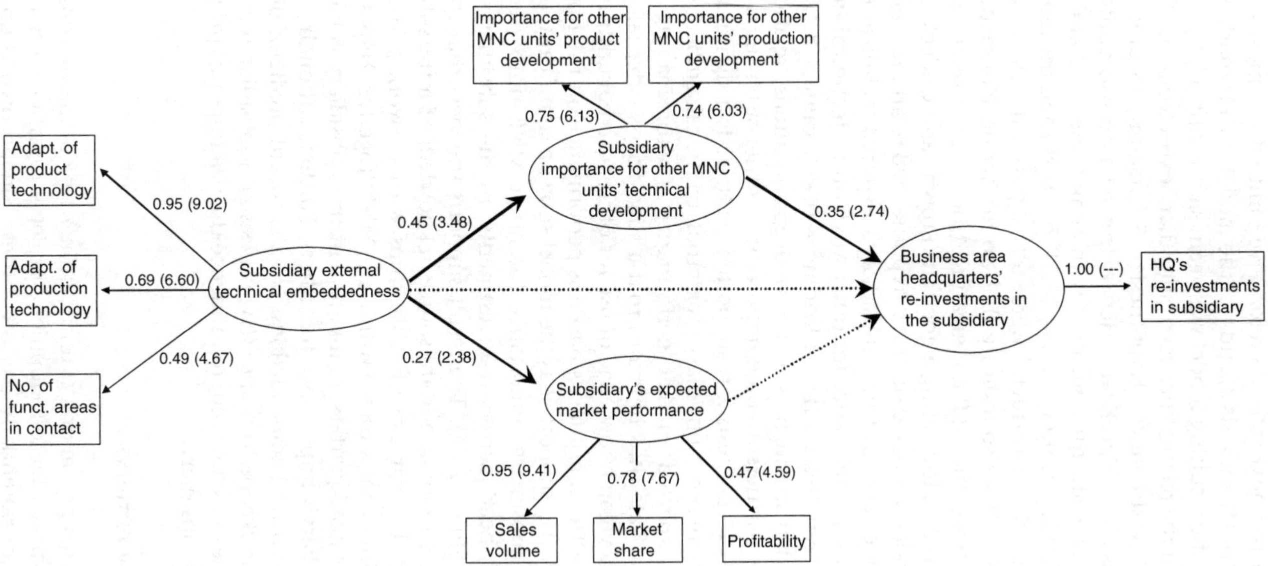

Figure 10.2 The final model

Notes: Model chi-square is 28.31 with 25 degrees of freedom at a probability of 0.293. The figures given are the factor loadings of causal relations with *t*-values written in parentheses.

finding that Hypothesis 4 was supported while Hypotheses 3 and 5 were not indicates that HQ are inclined to concentrate above normal re-investments into those subsidiaries that contribute to and become important for other MNE units technical development. Or, expressed differently, HQ in MNEs seem to value knowledge development that can be used by other units within the MNE as more important than high performance when it comes to investments in subsidiaries. Although a high degree of external embeddedness is important both for a subsidiary's market performance and its position within the MNE, embeddedness as such does not trigger re-investments. This can be due to the fact that subsidiaries always are more or less embedded externally and that the headquarters would need to put in a lot of resources to determine which subsidiaries have the most promising external contacts. How important the subsidiary's external network is for the MNE as a whole reveals itself in the form of subsidiary importance to other sub-units' technical development, and when that happens it is much easier for MNE headquarters to decide upon investments than would have been the case judging only from the subsidiary's degree of embeddedness.

10.10 Limitations and future research

One limitation of this study is the subjective measurement of the dependent variable 'Business area HQ re-investment in the subsidiary'. An ideal measurement would have been the actual investments done in euros or dollars together with the size of the subsidiary to determine if the subsidiary has attracted larger investments compared to its size. On the other hand, our measure has the benefit of taking care of the relativity problem concerning size, at the same time as the respondent had the opportunity to compare this subsidiary with all the other subsidiaries in the business area when deciding whether this subsidiary has had larger or smaller investments over the last five years. Another limitation regarding the dependent variable is that, due to the way we measure the variable, we cannot distinguish different types of re-investments. We do not for a fact know the origin of the investment, for example. It may very well be that the investment is not originating from headquarters but from the subsidiary itself as it has gained lower profit targets, and so on, from the HQ in order to increase the possibility of investments in that particular subsidiary. Further, we cannot know the purpose of the re-investment, that is, if it is in technology or some thing else. Future research should be more careful in characterizing the studied re-investments by using more indicators than has been done in this study.

Further limitations are connected to the fact that we have not controlled for aspects such as host-country development (for instance, GNP growth in the subsidiary's country), MNE strategy (for example, charters granted to the specific subsidiary), taxation levels, subsidies, and so on. Such things probably will have an impact on the propensity to re-invest into a subsidiary.

References

Ahuja, G., 'Collaboration networks, structural holes, and innovation: a longitudinal study', *Administrative Science Quarterly*, 45 (2000) 425–55.

Almeida, P. and Phene, A., 'Subsidiaries and knowledge creation: the influence of the MNC and host country', *Strategic Management Journal*, 25 (2004) 847–64.

Anderson, J.C. and Gerbing, D.W., 'Structural equation modelling in practice: A review and recommended two-step approach', *Psychological Bulletin*, 103(3) (1988) 411–23.

Andersson, U., 'Managing the transfer of capabilities within multinational corporations: the dual role of the subsidiary', *Scandinavian Journal of Management*, 19 (2003) 425–42.

Andersson, U. and Forsgren, M., 'Subsidiary embeddedness and control in the multinational corporation', *International Business Review*, 5 (1996) 487–508.

Andersson, U., Forsgren, M. and Holm, U., 'The strategic impact of external networks: subsidiary performance and competence development in the multinational corporation', *Strategic Management Journal*, 23 (2002) 979–96.

Bartlett, C.A. and Ghoshal, S., 'Tap your subsidiaries for global reach', *Harvard Business Review*, 64(6) (1986) 87–94.

Birkinshaw, J., 'How multinational subsidiary mandates are gained and lost', *Journal of International Business Studies*, 27 (1996) 467–95.

Birkinshaw, J. and Hood, N., 'Multinational subsidiary evolution: capability and charter change in foreign-owned subsidiary companies', *Academy of Management Review*, 23 (1998) 773–95.

Birkinshaw, J., Holm, U., Thilenius, P. and Arvidsson, N., 'Consequences of perception gaps in the headquarters–subsidiary relationship', *International Business Review*, 9 (2000) 321–44.

Blanchot, F. and Mayrhofer, U., 'Empirical literature on joint venture success: A review of performance measures and factors affecting longevity', Paper presented at the EIBA-conference in Stuttgart, December (1997).

Bollen, K.A. *Structural Equations with Latent Variables* (New York: Wiley 1989).

Bollen, K.A. and Long, J.S., *Testing Structural Equation Models* (Newbury Park, CA: Sage, 1993).

Browne, M.W. and Cudeck, R., 'Single sample cross validation rules for covariance structures', *Multivariate Behavioral Research*, 24 (1989) 445–55.

Buckley, P.J. and Casson, M.C., *The Future of the Multinational Enterprise* (London: Macmillan, 1976).

Burt, R.S., *Structural Holes: The Social Structure of Competition* (Cambridge, MA: Harvard University Press, 1992).

Cantwell, J. and Piscitello, L., 'The location of technological activities of MNCs in European regions: the role of spillovers and local competencies', *Journal of International Management*, 8 (2002) 69–96.

Chakravarthy, B.S., 'Measuring strategic performance', *Strategic Management Journal*, 7 (1986) 437–58.

Chowdury, J., 'Performance of international joint ventures and wholly owned foreign subsidiaries: a comparative perspective', *Management International Review*, 32(2) (1992) 115–33.

Coff, R.W., 'When competitive advantage doesn't lead to performance: the resource-based view and stakeholder bargaining power', *Organization Science*, 10 (1999) 119–33.

Davis, L.N. and Meyer, K.E., 'Subsidiary research and development, and the local environment', *International Business Review*, 13 (2004) 359–82.

Dess, G.C. and Robinson, R.B. Jr., 'Measuring organizational performance in the absence of objective measures: the case of the privately held firm and conglomerate business unit', *Strategic Management Journal*, 5 (1984) 265–73.

Dyer, J.H. and Singh, H., 'The relational view: cooperative strategy and sources of interorganizational competitive advantage', *Academy of Management Review*, 23 (1998) 660–79.

Echols, A. and Tsai, W., 'Niche and performance: the moderating role of network embeddedness', *Strategic Management Journal*, 26 (2005) 219–38.

Egelhoff, W.G., *Organizing the Multinational Enterprise: An Information-processing Perspective* (Cambridge, MA: Ballinger, 1988).

Forsgren, M., *Managing the Internationalization Process: The Swedish Case* (London: Routledge, 1989).

Forsgren, M., 'Managing the international multi-centre firm: case studies from Sweden', *European Management Journal*, 8 (1990) 261–7.

Forsgren, M. and Pedersen, T., 'Subsidiary influence and corporate learning: centres of excellence in Danish foreign-owned firms', in Holm, U. and Pedersen, T. (eds), *op. cit.* (2000) 68–78.

Forsgren, M., Johanson, J. and Sharma, D., 'Development of MNC centers of excellence', in Holm, U. and Pedersen, T. (eds), *op. cit.* (2000) 45–67.

Fritsch, M. and Franke, G., 'Innovation, regional knowledge spillovers and R&D cooperation', *Research Policy*, 33 (2004) 245–55.

Frost, T.S., 'The geographic sources of foreign subsidiaries' innovations', *Strategic Management Journal*, 22 (2001) 101–23.

Frost, T.S., Birkinshaw, J. and Ensign, P.C., 'Centers of excellence in multinational corporations', *Strategic Management Journal*, 23 (2002) 997–1018.

Gargiulo, M. and Benassi, M., 'Trapped in your own net? Network cohesion, structural holes, and the adaptation of social capital', *Organization Science*, 11 (2000) 183–96.

Geringer, J.M. and Hebert, L., 'Measuring performance of international joint ventures', *Journal of International Business Studies*, 22(2) (1991) 249–63.

Ghoshal, S. and Bartlett, C.A., 'The multinational corporation as an interorganizational network', *Academy of Management Review*, 15 (1990) 603–25.

Globerman, S. and Shapiro, D.M., 'The impact of government policies on foreign direct investment: the Canadian experience', *Journal of International Business Studies*, 30 (1999) 513–32.

Granovetter, M.S., 'The strength of weak ties', *American Journal of Sociology*, 78 (1973) 1360–80.

Gulati, R., Nohria, N. and Zaheer, A., 'Strategic networks', *Strategic Management Journal*, 21 (2000) 203–15.

Gupta, A.K. and Govindarajan, V., 'Knowledge flows and the structure of control within multinational corporations', *Academy of Management Review*, 16 (1991) 768–92.

Hansen, M.T., 'The search–transfer problem: the role of weak ties in sharing knowledge across organization subunits', *Administrative Science Quarterly*, 44 (1999) 82–111.

Hansen, M.T., 'Knowledge networks: explaining effective knowledge sharing in multiunit companies', *Organization Science*, 13 (2002) 232–48.

Holm, U., Holmström, C. and Sharma, D., 'Competence development through business relationships or competitive environment? Subsidiary impact on MNC competitive advantage', *Management International Review*, 45(2) (2005) 197–218.

Holm, U. and Pedersen, T., *The Emergence and Impact of MNC Centres of Excellence. A Subsidiary Perspective* (London: Palgrave Macmillan, 2002).
Kostova, T. and Roth, K., 'Adoption of an organizational practice by subsidiaries of multinational corporations: institutional and relational effects', *Academy of Management Journal*, 45 (2002) 215–33.
Jöreskog, K.G. and Sörbom, D., *LISREL 8: Structural Equation Modeling with the SIMPLIS Command Language* (Hillsdale, NJ: Lawrence Erlbaum, 1993).
McEvily, B. and Zaheer, A., 'Bridging ties: a source of firm heterogeneity in competitive capabilities', *Strategic Management Journal*, 20 (1999) 1133–56.
Malmberg, A., Solvell, Ö. and Zander, I., 'Spatial clustering, local accumulation of knowledge and firm competitiveness', *Geografiska Annaler. Series B. Human Geography*, 78 (1996) 85–97.
Mansfield, E., *Industrial Research and Technological Innovation* (New York: W.W. Norton, 1968).
March, J.M. and Sutton, R.I., 'Organizational performance as a dependent variable', *Organization Science*, 8(6) (1997) 698–706.
Mudambi, R. and Navarra, P., 'Is knowledge power? Knowledge flows, subsidiary power and rent-seeking within MNCs', *Journal of International Business Studies*, 35 (2004) 385–406.
O'Donnell, S. and Blumentritt, T., 'The contribution of foreign subsidiaries to host country national competitiveness', *Journal of International Management*, 5 (1999) 187–206.
Pfeffer, J. and Salancik, G.R., *The External Control of Organizations: A Resource Dependence Perspective* (New York: Harper and Row, 1978).
Porter, M.E. *The Competitive Advantage of Nations* (New York: Free Press, 1990).
Roth, K. and Morrison, A.J., 'Implementing global strategy: characteristics of global subsidiary mandates', *Journal of International Business Studies*, 23 (1992) 715–35.
Roth, K. and O'Donnell, S., 'Foreign subsidiary compensation strategy: an agency theory perspective', *Academy of Management Journal*, 39 (1996) 678–703.
Schulz, M., 'The uncertain relevance of newness: organizational learning and knowledge flows', *Academy of Management Journal*, 44 (2001) 661–81.
Ståhl, B., *Innovation and Evolution in the Multinational Enterprise* (Uppsala: Department of Business Studies, Uppsala University, 2004).
Tsai, W. and Ghoshal, S., 'Social capital and value creation: the role of intrafirm networks', *Academy of Management Journal*, 41 (1998) 464–76.
Uzzi, B., 'Social structure and competition in interfirm networks: the paradox of embeddedness', *Administrative Science Quarterly*, 42 (1997) 35–67.
Vernon, R., 'International investment and international trade in the product cycle', *Quarterly Journal of Economics*, 80 (1966) 190–207.
Wexley, K.N., Alexander, R.A., Greenawalt, J.P. and Couch, M.A., 'Attitudinal congruence and similarity as related to interpersonal evaluation in manager–subordinate dyads', *Academy of Management Journal*, 23(2) (1980) 320–30.
Zaheer, A., McEvily, B. and Perrone, V., 'Does trust matter? Exploring the effects of interorganizational and interpersonal trust on performance', *Organization Science*, 9 (1998) 141–59.
Zaheer, S., 'Overcoming the liability of foreignness', *Academy of Management Journal*, 38 (1995) 341–63.

11
Is Government Support Really Worth It? Developing Local Supply Linkages in Malaysia

Axèle L.A. Giroud

11.1 Introduction

This chapter focuses on the linkages established by multinational enterprises (MNEs) in host economies, with a focus on supply relationships and supply networks in the Malaysian electronics industry.

Since the mid-1980s, the Malaysian government has adapted its policies in order first to attract multinationals within its national boundaries, and second, to develop and deepen inter-firm relationships by encouraging MNEs to establish supply relationships with locally-owned suppliers on the one hand, and by trying to enhance local firms' capabilities on the other.

MNEs' backward linkages (or supply linkages) arise when multinational firms establish supply relationships with locally-owned or locally-based multinational firms in a host economy. MNEs tend to have complex supply networks, involving supplies that may be manufactured in-house by the company itself in the host economy or by sister companies established elsewhere in the world, and supplies that are purchased externally either from abroad or from suppliers located in the host economy. We focus on the development of supply linkages with external suppliers in the host economy, so as to evaluate the extent to which multinational firms contribute to the development and sustainability of local firms in the same or related sectors of activities. This allows us to engage in the discussion of the impact of multinational firms on the development of the host countries' production capabilities and in terms of knowledge and technology transfer from MNEs to local suppliers.

The electrical and electronics sector is well suited for the analysis of supply linkages by MNEs in host economies, inasmuch as companies need to rely on a pre-established network of suppliers for numerous parts and components and, in the case of hi-tech products, MNEs are often heavily involved in the design and development of parts and components together with their suppliers. The electrical and electronics sector in Malaysia is reliant on the presence and activities of MNEs, and thus the understanding of their role in the local economy is essential, particularly in terms of facilitating the capabilities of local firms.

The central debate in this chapter is about analysing the background and specificities about multinationals' supply relations in the Malaysian electrical and electronics industry, and to assess the success of government policies to date in encouraging supply linkages with MNEs. The discussion conducted is based on a series of empirical studies conducted by the author in the country, and on a comparative analysis of similar studies conducted by other academics in the region.

The chapter begins with an overview of the investment situation in Malaysia, so as to show the importance of the electronics industry in the country. Key concepts related to buyer–supplier development in a developing country context are then drawn upon to highlight the concerns and strategies of MNEs, with a discussion of policies adopted by the Malaysian government first to attract foreign investment in the electronics sector, and secondly, to deepen the relationships between foreign firms and local suppliers. The final part of the chapter reviews existing empirical studies on the relationships between MNEs' subsidiaries in the Malaysian electronics sector. We conclude with an assessment of the impact of current policies on the development of supply linkages by MNEs in Malaysia, bringing together some useful recommendations on ways for the government to encourage the deepening of these linkages.

11.2 Foreign investment in Malaysia: focus on the electronics industry

11.2.1 FDI at the heart of the Malaysian economy

While on a global scale Malaysia's share in total foreign investment has decreased, this developing country remains an attractive location for foreign MNEs, with firms already located there reinvesting to consolidate their activities. Malaysia started attracting foreign investment in the late 1960s and early 1970s, but it was from the mid-1980s that the country truly became a major recipient for FDI. In the late 1980s and early 1990s, Malaysia was among the top 10 main developing country recipients of world FDI. This success was the result of rapid economic growth during the same period, modern infrastructural facilities, with a good road network, port facilities and airports, adequate water and electricity supplies. Government policies were a major attraction to foreign investors with, for instance, industrial and technology parks being successful in attracting foreign firms. Finally, Malaysia has undoubtedly benefited from its central position within Southeast Asia and from being a neighbor country to Singapore. On a regional perspective, consolidation is taking place among ASEAN countries, the ASEAN Free Trade Area (AFTA) is in process of implementation, facilitating regional flow of parts and components in the electronics sector, and other regional industrial policies such as AICO (Asian Industrial Cooperation Scheme) are being strengthened. Yet, one may question the long-term attractiveness of Malaysia; first, Malaysia faces competition from lower-cost locations such as Vietnam and China; secondly, the Asia crisis in 1997 has impacted upon investors' political risk perceptions in neighboring countries such as Indonesia. As a result, inward FDI in many Southeast Asian nations has not regained their pre-crisis levels (Athukorala, 2001; Brook and Hill, 2004).

FDI inflows to Malaysia have been uneven over the past 30 years. In the 1970s and early 1980s, Malaysia attracted little foreign investment. A drastic change in government policies in the mid-1980s influenced inflows with a sharp increase in the late 1980s and early 1990s. Inward flows slowed down in 1997 with the outburst of the Asian crisis. Although Malaysia suffered less economic turbulence than its neighbors, Thailand and Indonesia, inward flows reached an annual average of US$5816 million over the 1992–97 period, before decreasing to US$2714 million in 1998, with the lowest level in 2001 (US$554 million). Inflows picked up again in 2002 and were US$2474 million in 2003 (UNCTAD, 2004). These figures indicate the dramatic impact the Asian crisis has had on FDI inflows. A large share of investment conducted in Malaysia nowadays is in the form of re-investment. Hence it has become very important for the government to retain existing investors as well as to deepen relationships between foreign and local firms in an attempt to further develop local players, particularly in the electronics industry.

Inward investments are essential to the Malaysian economy. Inward FDI stocks represented 20.7 per cent of GDP in 1980. This share increased steadily over the past two decades, to 23.4 per cent in 1990, 32.3 per cent in 1995, 58.5 per cent in 2000 and 57.2 per cent in 2003. Such figures demonstrate the dependence of Malaysia on foreign investment.

Sources of foreign investment in Malaysia changed widely over the past three decades. Japan and neighboring New Industrializing Economies (NIEs) emerged as major investors in Malaysia in the 1980s and 1990s. In 1988, the key country/region investor in Malaysia was Japan with 25 per cent of total inward investment, followed by the NIEs with 24 per cent, Europe and the US. In the late 1990s, Singapore became the main foreign investor, accounting for 26 per cent of inflows during 1997–98. More recently, over the 1999–03 period, the US was the first investor, followed by Japan, Germany and Singapore. One noticeable trend is that European firms have increased their investment in the country in the aftermath of the Asian crisis, and the US has established itself as the major investor investing twice more than Japan.[1]

11.2.2 Focus on the manufacturing sector

Since the 1980s and early 1990s, two sectors have attracted the bulk of inward FDI: the manufacturing sector, with slightly less than half the gross annual FDI inflow, and the services sector, with just over a third. The manufacturing sector is central to the Malaysian economy (MIDA, 2005). In 2003, it represented 42.1 per cent of GDP, 32.4 per cent of employment, 80.6 per cent of total exports and 88.4 per cent of total imports.[2] In 2002, exports from the electronics industry amounted to 58.3 per cent of total manufactured exports, 36.8 per cent of the country's manufacturing workforce.[3] Overall, electrical and electronic products made up 52.9 per cent of total Malaysian exports in 2003, and products worth RM211.16 (US$55.57) billion were exported world-wide especially to the US, Singapore, Hong Kong SAR, Japan, China and Taiwan.

Approved manufacturing projects have increased steadily, both in terms of number of projects and in absolute terms. Over the 1999–03 period, 4215 manufacturing projects were approved with a called-up capital of RM19 994.9 and a potential employment of 385 416. Of these projects, the breakdown shows the strong reliance on FDI, which over the same period represented 63.4 per cent of total capital investment.[4]

11.2.3 Predominance of the electronics sector

The electronics sector is key to the Malaysian manufacturing industry.[5] Over 1999–03, Malaysia received a total number of 4390 applications for the establishment of manufacturing. A quarter of these applications (1070 in total) were received for electronics and electronic products; the second major group, machinery manufacturing, totaled only 453 applications. The concentration of foreign investment within the electrical and electronics sector demonstrates the high dependence of Malaysia on this sector. Malaysia has been successful at generating a solid electronics sector, but this sector is dominated by foreign players. Overall, there is an urgent need for diversification, and it is crucial to deepen the links between foreign firms and domestic suppliers so as to maximize the economic benefit of this foreign-dominated industry.

This section has highlighted the dominance of FDI in total investment in Malaysia, the key role played by the electronics industry in the Malaysian economy, and the predominance of foreign firms within this industry. We have shown how successful the Malaysian government has been at attracting foreign investment, particularly in the electronics sector, thus achieving its goal of developing this industry within its national borders. Questions remain as the industry is dominated by foreign firms, and a sound *domestic* industrial base has not developed (Ernst, 2004). Too few domestic players have developed. Some of these include I, MEC, Khind, and Pensonic, and those firms are making small inroads into export markets. In the components sub-sector, *domestic* companies can mostly be found in semiconductor packaging and assembly (with names such as Malaysian Pacific Industries, Unisem, Globetronics Technology, and AIC Corporation), but too few of these companies can provide key parts and components, which therefore leads foreign firms to primarily purchase from other foreign firms or from abroad.

11.3 MNEs and local supply linkages

MNEs have a substantial impact on host developing countries' economic development, including impacting on the country's balance of payments, the local investment rate and stock of capital, the employment level and local firms' competitiveness. For this reason, host governments design policies first to attract foreign firms within their national borders and, secondly, to try and maximize the positive impacts and spillovers from MNEs' presence in their country. MNEs may cover for market imperfections, such as bringing resources difficult to find in specific markets, transferring technology, management know-how and labor, providing access to international production networks and to major markets. Foreign firms have both a direct impact through their activities, and an indirect impact through activities that might be generated in the rest of the economy as a result of its operation. In this chapter, we focus on indirect impact and spillovers to local supplier firms, looking specifically at inter-firm technology transfer through backward linkages.

MNEs operating in host economies develop a portfolio of linkages with numerous companies such as suppliers, customers, competitors and the non-business infrastructure. Supply of goods can originate either from the intra-organizational linkages of the MNE or from the inter-organizational linkages. In some cases, MNEs generate close relationships with their suppliers, which involve information flows between the two partners. Relationships vary according to the succession of production processes and the business responsiveness adopted by the partners. Formal relationships can include subcontracting or partnership agreements, whereby the MNE maintains close links with key subcontractors and key suppliers in the inter-organizational linkages portfolio. On the other hand, MNEs also supply locally through informal relationships, such as a one-off, off-the-shelf exchange.

A firm located in a host economy has the choice between importing inputs, parts and materials going into the production process, or purchasing these inputs in the local market. Within the local economy, the foreign affiliate can purchase inputs internally (that is, the manufacturing of key inputs is performed either by the foreign affiliate itself, or by sister affiliates also located in the host economy) or externally, either from locally-owned firms or foreign firms established in the host economy. Tavares and Young (2006 forthcoming) list three key groups of factors conditioning the choice between local (host country) and foreign sourcing. These are related to the MNEs' overall strategy, the host country's policy and environment, and the industry characteristics and environment. Even if the foreign firm does purchase its inputs in the local market, as indicated earlier, in the case of developing countries and especially so in the non-resource sector, it occurs frequently that key inputs are purchased from other foreign-owned firms. One must therefore consider the local supplier industry, or the related and supporting industries. From the host country's perspective, the propensity to purchase locally will further depend upon the strategies of the seller, the capabilities of the buyer, host government policies and the nature of technology. In considering local purchase, one must look at the host country's economic climate. Economic growth and the investment climate are important (Lim and Pang, 1991), as well as the quality of infrastructure and the size of the local components industry (Belderbos et al., 2001).

From a MNE's perspective various factors have been attributed to the explanation of backward linkages in host economies. The level of economic development of the host explains the extent to which foreign firms may (or not) purchase local supplies, and their overall strategy. For example, in the case of the electronics and textiles industries, parent firms often transfer their older equipment to developing countries. This equipment continues to be operated by foreign affiliates, while new equipment can be purchased at home, moving to high-technology at home and still benefiting from older product lines. Other factors such as the sector in which the foreign firms operate will influence backward linkages creation. Within the electronics and electrical industry, the level of local purchasing may be limited in developing countries, but there is extensive scope for linkage creation between foreign affiliates and local suppliers (Halbach, 1989; Supapol, 1995; Giroud, 2003). Foreign affiliates making standardized products with mature, non-proprietary technologies tend to prefer externalized, arm's length procurement (UNCTAD, 2001, p. 137). Backward linkages are further dependent upon foreign firms' characteristics. Market-orientation is of key importance and has been found to lead to variation in level of local content and subsequent relations with host suppliers (Supapol, 1995). The selected entry mode leads to different purchasing and relationship-building behavior (Belderbos et al., 2001), and the age of the plant is a key factor in explaining embeddedness into the local economy (McAleese and McDonald, 1978; Chen et al., 2004). The greater the autonomy of the affiliate, the more likely it is to try and identify local suppliers and to create relationships with them (UNCTAD, 2001, p. 137). MNEs that operate through centralized purchasing systems, for instance, in the automotive and electronics industries, may adopt global purchasing strategies and allow less independence to subsidiaries. Finally the size of the affiliate may affect sourcing and linkages (UNCTAD, 2001, p. 138; Giroud, 2003).

Overall, MNEs contribute to local supplier capabilities' development by exchanging information and transferring technology and knowledge. The degree of transfer varies according to the entrepreneurial potential and the existing technological capabilities of the host country. There are a number of considerations to be taken into account by the MNE in the decision to transfer knowledge and in the implementation of information exchange. These include the host country's technological capabilities, the different levels of power and control applied by the parent firm on its subsidiary, the level of trust in the host environment, the age of the subsidiary, the investment and the purchase strategies of the firm.

Local governments try and encourage foreign firms to source locally, but rarely so by using local content requirements, which have not been found to be very effective. Instead, an increasing number of countries have adopted soft policies to promote not only local sourcing but also the deepening of relationships between foreign and local firms. Linkage promotion policies are highly context-specific and need to be adapted to the specific circumstances prevailing in each host economy (UNCTAD, 2001, p. xxiii). The role of host government is three-fold: it needs first to promote the competitive development of indigenous firms; secondly, promote information exchange and raise awareness of foreign firms' requirements to local firms and of local firms' existence to foreign firms; and, thirdly, address issues specific to foreign firms. In terms of government policies and their impact on supply linkages' development, government requirements do not cover for the poor state of local suppliers' technological capabilities. Thus, even when government policies are in place in terms of local content requirements, their efficiency is not shown. Hackett and Srinivasan (1998) study FDI in host economies and find that strict local content regulations impact on FDI, with a somewhat stronger negative effect on Japanese FDI than on US FDI. Finally, in a study of the determinants of backward vertical linkages by 272 Japanese electronics manufacturing affiliates in 24 countries, Belderbos et al. (2001) find that local content regulations have a modest positive impact on linkages and they do not stimulate procurement from locally-owned suppliers. This indicates that softer policies need to be adopted by governments in developing economies if they wish to encourage supply linkages.

To conclude, backward linkages may be beneficial not only thanks to the business generated for local firms, but also to the knowledge and technology flow from MNE systems to local firms, and governments are facing various challenges when providing incentives to further linkages between foreign firms and domestic suppliers (UNCTAD, 1995; Driffield and Mohd-Noor, 1999; Jomo, 2001; Scott-Kennel, 2004).

11.4 Government policies in Malaysia towards FDI

As indicated in the first part of this chapter, Malaysia has been very successful in attracting FDI up until the Asia crisis in the late 1990s, and has developed a competitive electronics industry, even though the key players are foreign firms. In this section, we discuss the policies adopted by the Malaysian government to attract FDI, to encourage supply linkages and diversify its industrial base.

Since its independence successive governments in Malaysia have followed a trade-oriented stance and kept open policies toward foreign investment in the productive sectors. Four stages appear in the Malaysian economic development pattern since its independence, broadly following the Malaysian Master Plans. As the second Master Plan comes to an end this year, this is a crucial time for the current government to devise some new objectives in order to strengthen the Malaysian economy, continue to attract foreign investors but, most importantly, to continue to strengthen its indigenous industrial base. In recent years, the Malaysian government has strongly emphasized the possibilities of spillover generation by foreign firms. Policies adopted have aimed at maximizing positive impacts of MNEs in various sectors, particularly by encouraging inter-firm linkages throughout the country and across sectors of activity, with particular emphasis on the manufacturing sector. The Malaysian government has had some success in developing local investment, the share of inward FDI flows in the country's GFCF has decreased from 18 per cent annually over 1992–97 to just 10.8 per cent in 2003 (UNCTAD, 2004, p. 396).

One key feature of the second Industrial Master Plan, which was launched in 1995 for the period 1996–05 has been to adopt a cluster-based approach, whereby strategies focus on the development of specific industry clusters. On other aspects, the Master Plan presented no major reversal in the government policy orientation, continuing the promotion of reinvestments, industrial linkages, export and training. One of the three key objectives in terms of clustering was related to international linkages generated amongst the electronics and electrical appliances, and textiles industries (<http://www.mida.gov.my>). The new cluster-based approach and a stronger concern for local linkage creation indicated the willingness of the government to go further than mere incentives to attract foreign firms. In line with this idea, the eighth Malaysia Plan, 2001–05, also aimed at deepening the knowledge content of the economy, and paid particular attention to the human resource enhancement, as well as the development of R&D facilities. As of February 2001, the government merged the cluster working groups into 10 key industry task forces, the first one being for the electrical and electronics sector. The aim of this task force is to further develop the industry cluster, enhance the cluster-based industrial development and manufacturing activities. This change highlights the importance paid by the Malaysian government to the development of the electronics industry.

As these plans are coming to an end some concerns remain, namely, that there is still a high degree of concentration of the activities, MNEs continue to dominate exports, and the electronics industry continues to be dominated by foreign companies, even though some local players have developed. The Malaysian technological capabilities are still insufficient, and the underlying supply of high-level technical skills is inadequate. Hence, while the government appears to address the key issues, changes are slow and many challenges subsist, particularly in terms of spreading the development throughout all areas in Malaysia (Jomo and Wee, 2003).

Some essential investment policies can explain the rise in inward foreign investment that took place over a decade ago. First, to support its export-oriented industrialization strategy, the government has to date established 14 Free Trade Zones (FTZs). The government also developed a solid infrastructure, with over 200 industrial estates throughout the country, some developed by government agencies, namely, State Economic Development Corporations (SEDCs), Regional Development Authorities (RDAs), port authorities and municipalities, and others by private developers in certain states. In addition, the government supported the creation of three specialized parks equipped with modern and up-to-date facilities for high-tech manufacturing activity and R&D companies, together with a Biotechnology Park. The latest attempt to facilitate high-tech development has been the conception of the Multimedia Super Corridor (MSC). The MSC is 15 kilometers wide and 50 kilometers long, including Cyberjaya, an intelligent city that aims to attract multimedia industries, R&D centers, a multimedia university and operational headquarters for multinationals (<http://www.mdc.com.my>). This ambitious project has not yet attracted a substantial number of high-tech companies, but it shows the dedication of the Malaysian government to modernize the country.

A list of investment incentives available to both local and foreign investments appear in the Promotion of Investment Act (PIA, launched in 1986). Three types of activities are promoted by the Malaysian government: the first list is general, the second targets high-technology companies, and the third small-scale companies. These incentives are offered generally to encourage investment, but also work in line with the industrial policy followed by the government, with, for example, an emphasis on the promotion of high-tech industries. The government also wishes to attract high value-added activities. For instance, it provides incentives for firms to set up their operational headquarters (OHQ), for investment in high-tech industries and the allocation of R&D facilities and headquarters within Malaysia.

In addition to attracting foreign investment and facilitating industrial activities through infrastructure development, the Malaysian government encourages foreign firms to develop local skills. This is shown by an investment tax allowance of 100 per cent for 10 years given to companies that establish technical or vocational training institutions. The government has launched in 1993 the Human Resource Development Fund (HRDF) to encourage direct private sector participation in skills development.

To complement other FDI policies, the latest master plan emphasized the need for inter-firm linkages and cluster development. Following this policy stance, soft incentives have been launched to try and encourage MNE–local supplier linkages. This section gives an account of the way the Malaysian government encourages MNEs to create partnerships with their local suppliers, in particular, soft incentives and existing initiatives.

All the policies discussed above highlight a targeted policy towards foreign investment and the wish to strengthen the cluster among electronics manufacturers. The Malaysian government has in parallel tried to promote the creation of backward linkages using soft 'voluntary' incentives rather than compulsory policies, often cumbersome for foreign investors. Malaysia has initiated special programs that work closely with both foreign investors and domestic suppliers in promoting mutually beneficial linkages between the two parties. Governments across Asia gradually abandon traditional types of import restrictions and content requirements in favor of more interactive policies, where the government intervenes in the upgrading process of the local supporting industries (Battat et al., 1996). In Malaysia, inter-firm linkages are encouraged by using market-friendly incentives (Meyanathan, 1994), such incentives include tax incentives, specific institutional arrangements, vendor development schemes, local content requirements, subcontracting exchange schemes and information provision and exchange.

The importance of the purchasing function of the firm is illustrated by the international procurement centers (IPCs) initiative. The Malaysian government has tried to attract procurement centers through special tax incentives, targeting firms that undertake procurement and sale of raw materials, components and finished products for its group of companies in Malaysia and abroad.

More specifically, the government wishes to foster linkages through its Industrial Linkages Program (ILP), encouraging large companies to undertake actions to enhance the quality of local vendors by the provision of income tax reduction. Small local vendors also receive incentives when they take part in the program. The ILP is supported by other initiatives such as the *umbrella strategy* (whereby an *umbrella* firm with financial resources and expertise helps to coordinate the production and marketing of another firm) and the *Global Supplier Program* (previously known as the *vendor development scheme* aiming at encouraging large firms to provide assistance to a local vendor).

11.5 Empirical results on backward linkages in the electronics industry

In the previous sections we demonstrated the role of foreign investment in the Malaysian economy, and discussed the importance of the electronics industry in the country as well as the predominance of foreign firms in this sector. In this section, we look at two studies conducted by the author in Malaysia. The first study was conducted in 1996, covering a total of 95 foreign affiliates in the electronics industry[6] (for a full report on the results, see Giroud, 2003). The sample is composed of 49 Japanese firms, 27 firms from the NIEs, nine firms from Europe, eight from Europe and two from Other origins. The second study was conducted by the author in 2001 and 2002, with interviews of 25 firms[7] in the industry (for the full results, see Mirza et al. 2003). The sample is composed of nine Japanese firms, seven firms from the US, five from Europe and three from the NIEs. Seven firms manufacture or assemble consumer electronics, 18 other electronics. The sole aim of the first study was to analyse supply strategies and relationships among foreign affiliates, while the second study was broader in nature, and the supply strategies were analysed in the context of foreign firms' linkages within the host economies. The key advantage of comparing these two surveys is to analyse the change in purchase behavior and technology transfer from MNEs to local supplier between 1996 and 2001–02.

11.5.1 Local purchasing

As shown on Figure 11.1, the 1996 study indicated how little direct purchasing was being carried out by foreign subsidiaries in 1996. Nearly 10 per cent of the firms purchased no direct supply locally. The overall share of internal supplies across all firms accounted for 38.6 per cent of total supplies. The level of local direct supplies was low, and supplies were often purchased from other foreign affiliates, rather than from locally-owned suppliers. Some key components were sourced locally but by foreign-owned firms. The picture is slightly more positive in the most recent study. Overall, companies interviewed purchased about a third of their key inputs from Malaysia. Of these inputs locally purchased, 27 per cent were purchased from affiliated companies located in Malaysia, against 73 per cent from non-affiliated companies and slightly less than half of these inputs purchased from non-affiliated companies were purchased from foreign firms located in Malaysia. In the 2001–02 study, half of the sample companies have a specific supply

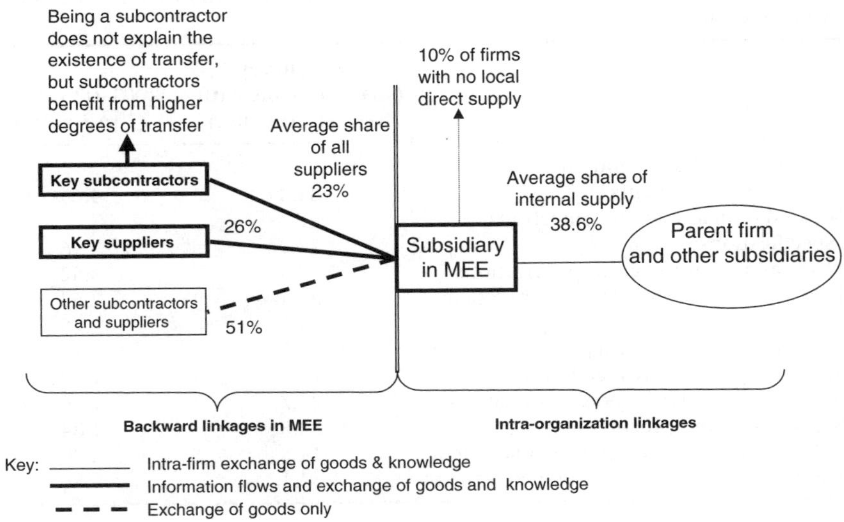

Figure 11.1 MNEs and backward linkages in the Malaysian Electronics and Electrical (MEE) industry

Source: Giroud (2003, p. 307).

partnership scheme developed for their local suppliers. Such schemes are a useful system of learning, whereby indigenous firms can benefit from knowledge transfer. 'The inputs these customers use must meet world standards; they are usually very specific products, of high quality, need to be produced within very narrow tolerances and involve not just "physical" technology, but also managerial, engineering and technical skills and techniques. It is in the interest of the purchasing companies to ensure that the inputs are produced efficiently and that the product is of a high order. Thus, considerable knowledge can flow down the supply chain, both explicitly and implicitly, and foreign companies are often willing to assist suppliers when problems are encountered' (Mirza et al., 2003, p. 110). Such local-supply partnerships were already used widely by firms in 1996.

11.5.2 Technology and knowledge transfer

No marked difference can be noticed in terms of the types of product and process technology transfer taking place between foreign firms and their suppliers between the two time periods. While the scales and samples are not perfectly comparable, the only difference arises from slightly less process technology transfer taking place in the 2001–02 survey. The type of knowledge most commonly transferred relates to specifications for standard materials and components (see Table 11.1). A bit further behind in terms of

Table 11.1 Product and process technology and knowledge transferred to locally-owned suppliers

	1996 survey (share of firms with frequent transfer)	2001–02 survey (1 to 5 scale)
Specification about standard materials/components	84.3	4.64
Method of manufacture, operational specifications	27.2	2.96
Physical or technological specifications	70.9	3.52
Joint design of materials and components	23.4	2.68
Joint design of operational requirements	19.1	2.28
Support in establishing a production plant	9.9	1.75
Support by supplying machinery	17.3	1.80
Support in supplying tools, materials, etc.	27.7	1.96
Support in input procurement	30.5	1.84
Support in supplier business management	14.8	2.00
Support in supplier technical management	34.1	2.56

Notes: The total includes one firm in Singapore; the 1–5 score range is 1 = never transferred; 5 = very frequently transferred.

Source: Adapted from Table 5.12 in Mirza et al. (2003, p. 112), and from Table 7.1 in Giroud (2003, p. 235).

likelihood of transfer are other specifications, such as operational requirements or technological aspects, followed – not uncommonly – by jointly designing materials or components. The main support offered to suppliers is in technical management (i.e. quality, production issues); occasionally, other types of support are also offered (e.g. supplying tools, machinery or materials), although such support seemed to have been more systematic in the 1996 study.

11.5.3 Supplier improvement

During the 2001–02 study, the researchers asked foreign affiliates the extent to which they thought local suppliers had succeeded in enhancing their capabilities by working jointly with foreign firms. Answers provided by managers appear in Table 11.2. Overall, suppliers in Malaysia have improved substantially, indicating the rise of specific skills/capabilities in the country. Key improvements made by local suppliers were noticeable in terms of delivery, quality, costs and service focus. This is in line with the key factors used by foreign firms to evaluate their suppliers. The strict requirements established by foreign firms encourage local suppliers to improve in this respect, but little improvements have been achieved in terms of design and innovation capabilities. Thus, while local suppliers undoubtedly benefit from support from foreign firms in terms of product and process technology, little impact occurs to enhance innovative capabilities.

Table 11.2 Supplier improvement as a result of working with foreign firms

Improved performance in terms of:	**Malaysia N = 24**
Cost	3.83
Quality	3.88
Delivery	3.96
Inventory control	2.88
Lead time performance	3.54
Continuous improvement	3.21
Technical skills	3.38
Design skills	2.58
Innovation skills	2.79
Safety	3.29
Business focus	3.35
Commercial awareness	3.42
Service focus	3.83
Professionalism	3.70

Notes: 1 = no improvement; 5 = very large improvement.

Source: Adapted from Table 5.15 in Mirza et al. (2003, p. 114).

The results discussed indicate that the indirect effects of FDI are not as great as might be expected and little difference can be noticed between the 1996 study and the 2001–02 study. Clearly, the purchase of inputs overseas and the existence of foreign-owned suppliers in Malaysia represent a major hindrance to the development of locally-owned suppliers but, for these backward linkages to develop fully, the Malaysian government needs to continue developing the technological capabilities of indigenous firms. If few good quality local suppliers are available, spillover from backward linkages will remain at a minimum. On a brighter note, spillovers *are* taking place and locally-owned suppliers are improving in their capabilities. In the future this should allow them to go some way toward supplanting foreign-owned suppliers.

11.6 Malaysian policies in perspective

11.6.1 Comparison with Singapore

In 1967, Singapore began its promotion of FDI, it developed industrial parks and one of the targeted sectors was the electronics sector. Policies adopted by Singapore have been successful, and today the city-state boasts some internationally competitive players in the industry. Singapore is the most attractive location in Southeast Asia. Even though it started its industrialization earlier than Malaysia, Singapore as a location and its investment policies have been more successful for three key reasons: it has maintained a high level of inward FDI even through the Asia crisis; it has succeeded in attracting investment in high value activities; and its own domestic companies in the electronics sector have grown into internationally competitive firms that now invest in neighboring countries and elsewhere in the world. Similarly to Malaysia, the electronics industry is key to the domestic economy, and electronics investments are the main contributors to the manufacturing value added per annum.[8] From a regional perspective, Malaysia has no doubt developed its electronics industry, but compared with its neighboring country Singapore, it has not established global manufacturing leadership and lags behind in terms of establishing itself as a trusted hub in FDI location for high-tech activities, such as R&D, innovation and design, headquarters and logistics and supply chain managements activities. Singapore, on the other hand, has.

11.6.2 Implications from the emergence of China

In the wake of China as the biggest recipient for FDI world-wide, Malaysia's challenge is two-fold. It needs to compete with China for labor-intensive activities or find niches of activity, and it needs to position itself to take advantage of the rise of China. Yusof (2003) recommends that Malaysia needs to work intensively with its partners in ASEAN, to form bilateral free trade agreements, to restructure its manufacturing industries and decrease its dependence on labor-intensive industries, and to improve the education and technological skills of its labor force. China does attract a large share of greenfield investment in labor-intensive industries, but Malaysia can draw some benefits, through increased trade with China (Lall, 2004) and by attracting investment coming from China. With lower levels of inward FDI, in part due to the emergence of China, the Malaysian government must face the challenge of nurturing its own industries and enhance their competitiveness in the international markets. If these challenges were to be met, the emergence of China would, over time, have brought in some beneficial changes to the Malaysian economic environment.

11.7 Discussion and conclusions

This chapter has addressed the issue of backward linkages development in the electronics industry in Malaysia, paying particular heed to the role played by the Malaysian government. The first section highlighted the substantial role of the electronics industry within the Malaysian economy and the heavy reliance on foreign firms. A careful analysis of government policies indicated the steps undertaken not only to attract investment in this particular industry, to develop key industrial zones and the efforts made towards the generation of a cluster in this crucial industry for the country's economy. The government has further adopted appropriate soft policies to encourage interfirm relationships and link larger foreign firms with small supplier firms. Finally, the last section highlighted that backward linkages remain limited and have not developed substantially between 1996 and 2001–02. Saying this, some marked improvements of locally-owned firms are noticeable. In the wake of China as the most attractive location world-wide, countries such as Malaysia that rely substantially on foreign firms for key industries – as is the case for the electronics industry – are faced with serious challenges. In the final section of this chapter, the Malaysian experience and policies are presented in perspective. First, the success of Singapore in developing its own domestic industrial base, and in attracting foreign investment in higher-value activities, contrasts with the experience of Malaysia, which still relies substantially on foreign firms. Second, given that China is attracting an increasing share of the world's foreign investment flows, Malaysian FDI policies are no longer sufficient to make it attractive. The government is faced with wider challenges, the most important of these being the restructuring and nurturing of its own industries. From a policy perspective, the market-friendly incentives offered by the government have only concerned a very small number of firms in Malaysia. They have produced few spillover effects, but selected domestic companies have succeeded in becoming competitive (Meyanathan, 1994; Battat et al., 1996; Jomo, 2001; UNCTAD, 2001; Giroud, 2003). Indeed, the few agreements between a foreign firm and a locally owned SME have led to positive links for the local firm. So, was government support really worth it? Undoubtedly, incentives have been successful in attracting FDI in the late 1980s and early 1990s, without which the electronics industry would not have developed. The policies aimed at attracting FDI were worth it; however, the picture is different regarding policies towards the development of supply linkages. They have not yet proved successful and one can question their effectiveness. The full benefits of such policies remain to be seen, but we believe there is scope for success as the programs gain importance. Malaysia is at a crossroads, with a new Master Plan due this year. It is imperative that further attention be paid to the deepening of inter-firm relationships.

Given the competitive dynamics taking place in today's world environment, efficient FDI policies are no longer sufficient to attract flows in Malaysia and, as we have shown in this chapter, the policies aimed at deepening supply linkages have not, as yet, brought sufficient beneficial spillovers for the development of a domestic supply base in the electronics industry. The Malaysian government's levy over foreign investment has decreased. Still, it must continue to provide attractive policies in an effort to attract FDI and, more importantly, it must consider the overall competitiveness of its domestic industry. Such policies will be viewed positively by investors and will stimulate spillovers from foreign firms with existing investment in the country.

Notes

1 Data extracted from <http://www.mida.gov.my> as of March 2005.
2 Data extracted from Table X.2, p. 305, *ASEAN Statistical Yearbook 2004*, <http://www.aseansec.org> as of March 2005.
3 Data extracted from <http://www.miti.gov.my> as of March 2005.
4 Data extracted and calculated from <http://www.mida.gov.my> as of March 2005.
5 The Malaysian government subdivides the electronics industry into three different categories. (1) Consumer Electronics: TV sets, audio visual products, VCD players, DVD players, home theaters and video and digital cameras. The capital investment under this sub-sector for the period 1998 till September 2003 amounted to RM2.9 billion (US$0.76 billion). (2) Electronic Components: semiconductor devices, passive components (capacitors, inductors, resistors) and display devices. Capital investment for the period 1998 till September 2003 amounted to RM24.9 billion (US$6.6 billion). The semiconductor industry is dominant. Exports of semiconductor devices accounted for RM72.5 billion (US$19.1 billion) or 35.1 per cent of total electrical and electronics exports in 2002 and RM57.9 billion (US$15.2 billion) or 42.9 per cent of total electrical and electronics export for the period January–September 2003. 3) Industrial Electronics: copier machines, fax machines, typewriters, automatic data processing machines, telecommunications equipment and industrial controllers. The capital investment under this sub-sector amounted to RM9.7 billion (US$2.6 billion) for the period 1998 to September 2003 (<http://www.miti.gov.my> as of March 2005).
6 The data discussed was collected through a mail survey among foreign affiliates in the electronics sector in Malaysia in 1996. A total of 95 questionnaires were

collected, and 13 face-to-face interviews were conducted by the author with selected foreign firms. Purchasing managers or managing directors were targeted in the survey.

7 The sample being discussed here is part of a larger sample. The study was conducted in four Southeast Asian countries, namely Cambodia, Malaysia, Thailand and Vietnam; and in two industries, the electrical and electronics and the textiles and garments industries. Data was collected by means of interviews, with a semi-structured questionnaire, conducted with either managing directors or/and other key top managers at multinational subsidiaries in each of the four countries mentioned. A total of 85 manufacturing foreign subsidiaries were interviewed, as well other key foreign affiliates and local companies.

8 <http://www.sedb.com/edbcorp/sg/en_uk/index/in_the_news/press_releases/2005/2004_investment_commitments.html>.

References

Athukorala, P.C., *Crisis and Recovery in Malaysia: The Role of Capital Controls* (Cheltenham: Edward Elgar, 1996).

Battat, J., Frank, I. and Chen, X., *Suppliers to Multinationals: Linkages Programs to Strengthen Local Companies in Developing Countries* (Washington DC: Foreign Investment Advisory Service, World Bank, 1996).

Belderbos, R., Capannelli, G. and Fukao, K., 'Backward vertical linkages of foreign manufacturing affiliates: evidence from Japanese multinationals', *World Development*, 29(1) (2001) 189–208.

Brook, D.H. and Hill, H., 'Divergent Asian views on FDI and its governance', *Asian Development Review*, 21(1) (2004) 1–37.

Chen, T.-J., Chen, H. and Ku, Y.-H., 'Foreign direct investment and local linkages', *Journal of International Business Studies*, 35(4) (2004) 220–33.

Driffield, N. and Mohd-Noor, A.H., 'Foreign direct investment and local input linkages in Malaysia', *Transnational Corporations*, 8(3) (1999) 1–24.

Ernst, D., 'Global production networks in East Asia's electronics industry and upgrading perspectives in Malaysia', in Yusuf, S., Altaf, A.M. and Nabeshima, K. (eds) *Global Production Networking and Technological Change in East Asia*, (Washington: World Bank, Oxford University Press, 2004).

Giroud, A., *Transnational Corporations, Technology and Economic Development: Backward Linkages and Knowledge Transfer in South East Asia* (Cheltenham, UK: Edward Elgar, 2003).

Hackett, S.C. and Srinivasan, K., 'Do supplier switching costs differ across Japanese and US multinational firms?' *Japan and the World Economy*, 10 (1998) 13–32.

Halbach, A.J, *Multinational Enterprise and Subcontracting in the Third World: A Study of Inter-industrial Linkages* (Geneva: ILO, International Labour Office Multinational Enterprises Programme, 1989).

Jomo, K.S., *Southeast Asia's Industrialization: Industrial Policy, Capabilities and Sustainability* (Basingstoke, UK and New York: Palgrave, 2001).

Jomo, K.S. and Wee, C.H., 'The political economy of Malaysian federalism: Economic development, public policy and conflict containment', *Journal of International Development*, 15(4) (2003) 441–56.

Lall, S., 'China's competitive performance: A threat to East Asian manufactured exports?', *World Development*, 32(9) (2004) 1441–66.

Lim, L.Y.C. and Pang, E.F., *Foreign Direct Investment and Industrialization in Malaysia, Taiwan and Thailand* (Paris: OECD Development Centre, 1991).
McAleese, D. and McDonald, D., 'Employment growth and the development of linkages in foreign-owned and domestic manufacturing enterprises', *Oxford Bulletin of Economic and Statistics*, 40 (1978) 321–40.
Meyanathan, S.D., *Industrial Structures and the Development of Small and Medium Enterprise Linkages: Examples from East Asia* (Economic Development Institute of the World Bank, 1994).
MIDA, *Report on the Performance of the Manufacturing Sector 2004* (Kuala Lumpur: Malaysian Industrial Development Authority, 2005).
Mirza, H., Giroud, A., Jalilian, H., Weiss, J., Freeman, N.J. and Than, M., *Regionalisation, Foreign Direct Investment and Poverty Reduction: The Case of ASEAN* (Bradford: University of Bradford, 2003).
Scott-Kennel, J., 'Foreign direct investment: A catalyst for local firm development?' *European Journal of Development Research*, 16(3) (2004) 624–52.
Supapol, A.B., *Transnational Corporations and Backward Linkages in Asian Electronics Industries* (New York: UNESCAP – United Nations Conference on Trade and Development, Economic and Social Commission for Asia and the Pacific. Monograph No. 5, 1995).
Tavares, A.T. and Young, S., 'Sourcing patterns of multinational subsidiaries in Europe: testing the determinants', in *Regional Studies* (2006) forthcoming.
UNCTAD, *World Investment Report 1995: Transnational Corporations and Competitiveness* (New York and Geneva: United Nations, 1995).
UNCTAD, *World Investment Report 2001: Promoting Linkages* (New York and Geneva: United Nations, 2001).
UNCTAD, *World Investment Report 2004: The Shift Towards Services* (New York, Geneva: United Nations, 2004).
Yusof, Z.A., 'Malaysia's response to the China challenge', *Asian Economic Papers*, 2(2) (2003), 46–73.

Part IV

Innovations and Linkages

12

Globalization of Innovative Capabilities in the Late-industrializing Context: Evidence from Local Firms and TNC Subsidiaries in Brazil

Paulo N. Figueiredo

12.1 Introduction

This chapter seeks to contribute to the debate on the issues of globalization of firm-level innovative technological capability and industrial strategy in the late-industrializing context. Although these issues have been examined since the early 1970s in the late-industrializing literature, it becomes increasingly important today as globalization and trade liberalization gather pace and some developing countries and areas are coping well, while others are being left further behind (Lall, 2004). Thus, it is important to know the extent to which some developing countries and/or areas are really responding to such phenomena. There are two additional reasons for addressing these issues here.

First, although it has been said that innovative technological capability has been spreading into developing countries (Lall, 2004), the main existing empirical literature related to globalization of technology and innovation comes short of examining this issue properly in the late-industrializing context. On the one hand, it is argued that globalization of capabilities hardly occurs – the *non-globalization perspective* (e.g. Vernon, 1966; Patel, 1995; Daniels, 1997, Patel and Pavitt, 1997); on the other hand, it is claimed that globalization occurs depending on the location and situation of the host and home countries of transnationals corporations (TNCs) and product types – the *'conditional' globalization perspective* (e.g. Mansfield et al., 1979; Mansfield and Romeo, 1984, Grandstrand et al., 1993; Dunning, 1994a, b; Cantwell, 1995, 1999; Zander, 1997). However, both perspectives have primarily been based on patenting, research and development (R&D) statistics and macroeconomic data. The problem with these arguments is that they are based on frameworks that are not in line with the technological characteristics of firms in developing countries. Additionally, they hardly examine capability building from the perspective of TNC subsidiaries located in developing countries and areas (Ariffin, 2000; Ariffin and Figueiredo, 2004).

Secondly, a set of assumptions has been spreading across Latin America arguing that the intensification of globalization and policy changes into an outward-oriented and deregulated pattern of industrial development – that began during the early 1990s in several developing countries – have generated negative implications for industrial technological capabilities, particularly in Southern Latin American countries such as Brazil (see, for instance, Cassiolato et al. 2001; Cimoli and Katz, 2003; Katz, 2004; Sutz and Arocena, 2004). However, such assumptions are based on aggregate analyses and on macro-level perspectives. Little is known about the extent to which such assumptions would hold from a standpoint based on fieldwork and firm-level investigation.

Additionally, such types of generalization emerge in relation to certain developing areas in Brazil. For instance, the Industrial Pole of Manaus has been considered as an area of very scarce innovative activities – a 'screw-drive' industrial site – and very low industrial competitiveness (see, for instance, *Forbes Brasil*, 2000; Fleury and Fleury, 2004). Such arguments, although interesting, still need support from first-hand empirical evidence.

This chapter is based on a firm-level empirical study concentrated on three sets of manufacturing firms – local firms and TNC subsidiaries – located in a developing area of a large late-industrializing country: the Industrial Pole of Manaus (Northern Brazil). Specifically, the study draws on first-hand empirical evidence from 46 manufacturing firms: 19 electro-electronics (EE) firms; nine motorcycle and bicycle firms (two-wheel – TW), and 18 supporting firms – key suppliers of both EE and TW firms. Globalization of innovative capabilities is measured here by the types and levels of technological capabilities that have been built up within firms. Indeed, the research underpinning this chapter sought to move a small step forward in relation to a previous research project that was implemented in the context of Manaus (Brazil), and in Penang and Klang Valley (Malaysia). That study examined the extent to which innovative technological capabilities had been spreading to the electronics industry in these two late-industrializing contexts. Drawing on a sample of 82 EE firms (29 in the Industrial Pole of Manaus and 53 in Penang and Klang Valley), the study contributed to shedding new light in this issue and also argued against some common generalizations relative to technological development in this industry in these areas (see Ariffin and Figueiredo, 2004).[1]

This chapter, although focusing only on the Industrial Pole of Manaus, examines the extent to which globalization of innovative technological capabilities has been spreading to such developing areas by investigating the extent to which firms' capabilities have progressed from basic operations levels into innovative levels not only within electro-electronics firms, but also in two-wheel firms and key suppliers. Additionally, the association between technological capabilities and export performance is examined. It should be noted, however, that even though this chapter is not about the globalization of research and development (R&D), it is also not about the globalization of production.[2] Thus, in the light of these reviews this chapter will address the following null hypotheses:

Null hypothesis 1. Local firms and TNCs subsidiaries have not built significant technological capabilities, and thus, the globalization of innovative capabilities has not spread to a developing area like Manaus.

Null hypothesis 1a. The technological capabilities of local firms and TNC subsidiaries in a developing area like Manaus have been confined to basic operations levels and are not moving into more innovative levels.

Null hypothesis 2. Firms located in Manaus are not competitive enough to engage in export markets. To put it differently, firms operate in a context of low market competition associated with low levels of technological capabilities.

We will test these null hypotheses as to whether they can be rejected and seek to explore some explanatory factors that are important in building up innovative capabilities. The next section presents the analytical framework for assessing types and levels of firms' technological capabilities. This is followed by a brief outline of the empirical setting of this study (Section 12.3) and the research design and methods (Section 12.4). The research findings are presented in Section 12.5. Finally, Section 12.6 discusses the chapter's conclusions and some policy implications.

12.2 A framework for examining firms' technological capabilities

Technological capability is defined here as the resources needed to generate and manage technological change, including skills, knowledge, experience and organizational systems (Bell and Pavitt, 1995; Figueiredo, 2001). To examine firms' capabilities, we make use of the framework developed in Figueiredo (2001), adapted from Lall (1992) and Bell and Pavitt (1995), because it uses a relatively fine disaggregation of different levels and types of technological capability. Following Bell and Pavitt (1995), Ariffin (2000) and Ariffin and Figueiredo (2004), this study distinguishes between 'routine' production capability and 'innovative' technological capability. 'Routine' production capability is the capability to produce goods at given levels of efficiency and given input requirements; it may be described as technology-using skills, knowledge and organizational arrangements. 'Innovative' technological capability is defined as the capability to create, change or improve products, processes and production organization, or equipment. It may be described as change-generating capability, consisting of technology-changing skills, knowledge, experiences and organizational arrangements.

In this chapter these frameworks have been tailored and calibrated to examine technological capabilities in the EE and bicycle and motorcycle (TW) industries and their related suppliers. The columns set out the technological capabilities by function; the rows, by level of difficulty. They are measured by the type of activity expressing the capability levels – the type of activity the firm is able to do on its own at different points in time.[3] The framework for capabilities in the EE was previously applied in Ariffin (2000) and Ariffin and Figueiredo (2004). Additionally, this framework has been adapted, tailored and validated to assess types and levels of technological capabilities in firms from the TW industry and their key suppliers. Both frameworks identify capabilities for three technological functions: process and production organization, product-centered, and equipment-related activities. Routine capabilities are classified from Levels 1 to 2 in both industries. Innovative capabilities range from Levels 3 to 6 (EE and TW firms).[4] Even though research-based levels, which involve activities at the tip of the iceberg, may be less applicable to firms in production-based EE and TW industries in a late-industrializing economy, they provide a perspective and link to total technological activity in the global EE and TW industries. This framework, thus, provides a basis for describing one of the two trajectories of technological development: progress from routine production capability to successively higher levels of creative and innovative technological capability. This trajectory should be distinguished from the other involving progress through increasingly complex and higher value products. That is, a firm's progression in technological activities (e.g. from minor product modifications in the mid-1980s to core design of its products in the early 2000s) should be distinguished from types of products or components it produces – for example, from simple analogue radios in the 1970s to the latest audio products of today.[5]

12.3 The empirical setting

The Industrial Pole of Manaus started up in 1967 under the Import Substitution Industrialization (ISI) policy and internal market protection regime in Brazil. The strategy of setting up a large industrial site in the heart of the Amazon Rain Forest sought not only to stimulate economic development in that region, but also to achieve more integration of that region to the rest of Brazil. While other international initiatives implemented at that time in other countries were based on export-oriented models (e.g. South Korea, Taiwan, India and Malaysia), the Industrial Pole of Manaus was designed to supply the domestic market – an inward-looking industrialization model. Since its creation there has been a specific tax incentive regime for the Industrial Pole of Manaus. This fiscal regime has been modified over time and today there are two types of tax incentives. At the federal level: (i) import tax: reduction up to 80 per cent over inputs to be manufactured; (ii) tax on manufactured goods: exempt; (iii) income tax: reduction of 75 per cent – based on net profit; (iv) social integration tax (Pis) and social security tax (Cofins): exempt for transactions within the Manaus Industrial Pole. At the state level: value added tax – compensation between 55 per cent and 100 per cent. In 2004 this fiscal regime of Manaus was extended up to 2023.

The Industrial Pole of Manaus consists of 17 different industrial sectors. The total annual revenue evolved from US$7.2 billion in 1999, to US$13.8 billion in 2004. In 1990 exports represented only 1.5 per cent of total revenue. Since the early 1990s, following the opening-up and deregulation of the Brazilian economy, the proportion of annual exports in total revenue evolved from 5.2 per cent in 1999, to 11.6 per cent in 2003. Seven manufacturing sectors respond for about 90 per cent of total revenue (by order of sales volume): EE, TW (motorcycle and bicycle), informatics goods, thermo-plastics, chemical, disposables, and watch-making. There are around 400 firms, of which 128 are of foreign capital. Main foreign investors are from the US, Japan, and South Korea, which altogether represent nearly 50 per cent of all foreign firms in Manaus. In 1999 there were 43 095 people directly employed by firms in the Industrial Pole of Manaus, whereas in December 2004 this number increased to around 66 000 (Suframa, 2005).

12.4 Research design and methods

12.4.1 Sampling

In order to address the questions and null hypotheses mentioned in Section 12.1, this chapter draws on a sample of 46 firms: 19 EE, nine motorcycle and bicycle, and 18 key suppliers. Following Hobday (1996), this chapter classifies the sampled firms in two groups: Group 1: *TNC subsidiaries* (EE and TW producers) of US, Japanese, European and South Korean origin or ownership; Group 2: *local EE and TW firms* (*local independent firms* – local firms that sell their products to a more general market, either domestic or export, and are fairly independent of specific TNC subsidiaries in Brazil for sales) and *suppliers* in the supporting sector that are mainly dependent on sales for Group 1 and EE and TW firms of Group 2. Unlike Hobday, this group consists of both large and smaller firms.

12.4.2 Data collection and analysis

Data collection in Manaus was implemented over three phases: exploratory work, pilot study, and main fieldwork. In each of these phases, in-depth interviews, casual meetings, and direct-site observations were used as strategies for collecting first-hand empirical evidence. Firms' publications in the form of report, brochures, books, videos and other sources (firms' websites and press reports) were also collected as sources of complementary empirical evidence. Face-to-face interviews were conducted with senior management involving the managing director, plant or production manager and those involved with innovative activities: R&D, design, engineering, quality and maintenance department managers and technical staff. These were followed by observation tours across plant and production support units.

The original data were initially entered into an Access database since it allowed qualitative interview data to be inputted as memo and text data. To allow statistical analysis to be conducted more efficiently, most of the qualitative text and memo data were converted into quantitative data according to the statistical package SPSS 12.0 format. Previous qualitative data on firms' activities were quantified according to types and levels of technological capabilities. For qualitative analysis, the technological capabilities types and levels of all 46 sampled firms were identified and graphed. For the statistical analysis, non-parametric statistical tests were used since levels and types of technological capabilities consisted of ordinal measurement.

12.5 Findings

12.5.1 Types and levels of technological capabilities in the sampled firms

This section addresses the null hypotheses 1 and 1a. The section presents the types and levels of technological capabilities that were found within the sampled firms by the time of fieldwork. Table 12.1 summarizes the number of sampled firms (local firms and TNC subsidiaries) that have attained specific types and levels of technological capabilities. The evidence in Table 12.1 shows that all 46 sampled firms have mastered basic operations across the three technological functions examined: process and production organization, product-centered activities and equipment related activities. For process and production organization activities, 61 per cent of EE firms and 5 per cent of supplier firms have attained Level 5, while in the TW sample one firm (11 per cent) reached Level 6, although in an incomplete manner, as the deepest capability level. In terms of product-centered activities, 17 per cent of EE firms and 5 per cent of suppliers have reached Level 4 as their deepest capability level, whereas in the TW sample the deepest level for this function was Level 6, again, though incomplete.

In order to test whether there were significant differences between TNCs' subsidiaries and local firms in terms of technological capability levels, the non-parametric Kruskal–Wallis test was used. This test was chosen because of the small sample of firms and because of the ordinal scale of capability levels. Within the EE sample, there were no significant differences between TNCs' subsidiaries and local firms in terms of capabilities for process and production organization and equipment-related activities. However, there were significant

Table 12.1 Number of sampled firms that have reached specific levels of technological capability

Types and levels of technological capabilities by sector	Electro-electronics (EE)			Two-wheel (TW)			Suppliers		
	Process and production organization	Product-centered	Equipment-related activities	Process and production organization	Product-centered	Equipment-related activities	Process and production organization	Product-centered	Equipment-related activities
Mastery of basic operations Level 1	18 (100%)	18 (100%)	18 (100%)	9 (100%)	9 (100%)	9 (100%)	19 (100%)	19 (100%)	19 (100%)
Mastery of basic operations Level 2	18 (100%)	18 (100%)	18 (100%)	9 (100%)	8 (89%)	4 (44%)	18 (95%)	18 (95%)	15 (79%)
Basic innovation Level 3	18 (100%)	13 (72%)	9 (50%)	8 (89%)	3 (33%)	3 (33%)	14 (74%)	9 (47%)	5 (26%)
Intermediate innovation Level 4	14 (78%)	3 (17%)	2 (11%)	3 (33%)	1 (11%)	2 (22%)	7 (37%)	1 (5%)	1 (5%)
High-intermediate innovation Level 5	11 (61%)	0 (not attained)	0 (not attained)	2 (22%)	1 (11%)	1 (11%)	1 (5%)	0 (not attained)	0 (not attained)
Advanced innovation Level 6	0 (not attained)	0 (not attained)	0 (not attained)	1* (11%)	1* (11%)	1* (11%)	0 (not attained)	0 (not attained)	0 (not attained)

Note: (*) Incomplete accumulation of the capability level.

Source: Derived from the empirical study.

differences ($p < 0.05$) between TNCs' subsidiaries and local firms in terms of capability for product-related activities. To put it differently, levels of technological capabilities of local firms were higher than those of TNCs' subsidiaries for this function. In relation to the TW sample, there was no difference between local and TNCs' subsidiaries across all three types of technological capabilities. In contrast, among the sampled supplier firms the capability levels accumulated in TNCs' subsidiaries were deeper than those accumulated in local firms. In other words, there were significant differences between the capability levels attained in local firms and those attained in TNCs' subsidiaries across all three functions: process and production organization ($p < 0.01$), product-centered ($p < 0.05$), and equipment-related ($p < 0.01$).

Considering that the number of Japanese firms is the highest among the sampled foreign firms, we tested whether there were differences between Japanese subsidiaries and those from the US, Europe, and South Korea. Within the EE sample, the results show that there were no significant differences between capability levels and firm nationality in terms of whether they were from Japan or from other foreign country. However, in relation to the TW sample, the results show significant differences ($p < 0.05$), between Japanese and other sampled firms (including Brazilian firms) in terms of capability levels for all three examined functions. Specifically, the capability levels for process and production organization, product-centered and equipment-related activities accumulated in the sampled Japanese firms were deeper than those accumulated in other firms. As far as the sampled supplier firms were concerned, there were significant differences ($p < 0.05$) between Japanese firms and other sampled firms in terms of capabilities for process and product organization and equipment-related activities. For these two technological functions, the capability levels of Japanese suppliers were deeper than those of other firms.

12.5.2 Export performance

This section briefly examines null hypothesis 2, i.e., the extent to which there is any association between the sampled firms' types and levels of technological capabilities and export performance. As mentioned earlier, the Industrial Pole of Manaus was created during the late-1960s on the basis of an inward-looking industrialization strategy. During the early-1990s, following the opening-up of the Brazilian economy and the exposure of local firms to foreign competition, firms in the Industrial Pole of Manaus – as with several firms in Brazil – and in some other developing countries, had to engage in the reorganization of their existing capabilities to keep competing in the market. Also, from the mid-1990s, some firms in Manaus began to look for export market much more as a survival strategy rather than as a result of industrial policy.

Of the whole sample of 46 firms, 41.5 per cent are exporters while 58.5 per cent are not. Among the exporters the great majority (35 per cent) are firms from Group 1 (TNC subsidiaries), whereas 6.5 per cent are domestic firms (see Table 12.2). Indeed, by the early 1990s exports represented only 1 per cent of the total revenue of the Industrial Pole of Manaus. By 2003 this percentage had increased to 11 per cent. During the 1988–2004 period, while sales for the domestic market increased by 5.9 per cent annually on average, the export sales increased at the rate of 19.8 per cent. Indeed, fastest rates of export growth were achieved from the mid-1990s. A similar picture was found in the EE and TW firms. During the 1988–2004 period while sales of EE products for the Brazilian market grew at 4.9 per cent annually, on average, exports sales increased at the rate of 32.7 per cent. Indeed, the TW sector showed an impressive growth rate of 46 per cent annually during 1988–2004. Exports increased by 19 per cent during the 1988–2004 period, whereas sales for the Brazilian market increased by 10.7 per cent annually for the same period.

Such impressive growth rates suggest that firms responded positively to the policy shift from an inward-looking to a more outward-looking policy

Table 12.2 Number of sampled firms distributed on the basis of their market orientation

	Exporters			**Non-exporters**			
Type of sampled firms	**Group 1: TNCs' subsidiaries**	**Group 2: local firms**	**Subtotal**	**Group 1: TNCs subsidiaries**	**Group 2: local firms**	**Subtotal**	**Totals**
EE	11	2	13	2	3	5	18
	61%	11%	72%	11%	17%	28%	100%
TW	2	1	3	0	6	6	9
	16%	11%	34%		66%	66%	100%
Suppliers	3	0	3	5	11	15	19
	16%		16%	26%	58%	84%	100%
Totals	16	3	19	7	20	27	46
	35%	6.5%	41.5%	15%	43.5%	58.5%	100%

Source: Derived from the empirical study.

Table 12.3 Kruskal–Wallis test for technological capability levels and export performance

	Exporter firms vs. Non-exporter firms								
	EE			**TW**			**Suppliers**		
	Proc	**Prod**	**Equip**	**Proc**	**Prod**	**Equip**	**Proc**	**Prod**	**Equip**
Chi-square	2.914	0.284	4.471	4.655	4.655	7.200	5.486	4.933	7.222
Df	1	1	1	1	1	1	1	1	1
Asymptotic Significance	0.088	0.594	0.034*	0.031*	0.031*	0.007**	0.019*	0.026*	0.007**

Notes: (*) Association significant at the 0.05 level; (**) Association significant at the 0.01 level.

orientation. Indeed, when we tested percentage of export for each sample (within EE, TW and suppliers) the export factor is significant for the three samples (Table 12.3). Within the EE sample, export levels were highly significant to innovative capability, particularly for equipment-related activities ($p < 0.05$). With the TW and supplier samples export is highly significant to innovative capability for process and production organization ($p < 0.05$), products ($p < 0.05$) and equipment-related activities ($p < 0.01$).

12.6 Discussion, conclusions and policy implications

This chapter has addressed the issues of globalization of innovative capabilities and outward-looking industrialization and implications for technological development in the latecomer context. These issues have been examined in a sample of three sets of manufacturing firms in a developing area in Brazil (the Industrial Pole of Manaus). The study has drawn on first-hand empirical evidence collected through original and detailed fieldwork. The results show that the technological capability of most TNC subsidiaries and local firms in Manaus, far from being confined to very basic levels for long periods, has constantly been upgraded to carry out diverse types of innovative activity. Additionally, the evidence suggests that these capability efforts have generated positive benefits for export performance. Thus, all three null hypotheses are rejected.

More specifically, at least for this sample of firms, the findings and analyses do not support the generalizations and assumptions related to the *non-globalization* of innovative capabilities, the general deterioration of industrial technological capabilities, and the low competitiveness as a result of a change into an outward-looking pattern of industrial development. Indeed, the results show that the majority of the sampled firms have been responding to globalization by strengthening their capabilities. Additionally, outward-looking industrialization has stimulated firms to improve their export performance.

This research has applied a systematic and comprehensive framework to measure capabilities in the sampled EE, TW, and suppliers' firms in a late-industrializing context. This new classification system provides a framework for measuring latecomer firms' technological capabilities. Additionally, it provides an alternative to conventional measures such as patent statistics and R&D expenses. Since these proxy indicators tend to focus only on activities at the highest technological level, the rest of the bulk of technological activities tends to be ignored. Thus, for the Manaus industries that are dominated by production-based subsidiaries of global TNCs, they are not likely to attain the highest level of product development and R&D, as these activities are retained in corporate R&D centers or carried out in advanced and historically established R&D locations. The framework here explicitly identifies different types and levels of industrial technological capability. It is useful to draw attention to the extremely important types and levels of technological capabilities that are concerned with neither *basic routine production* nor 'R&D'. Such intermediate capability levels are a *pre-condition* for the accumulation of more advanced levels of innovative capabilities.

Additionally, this chapter adds new insights to the debate of internationalization of innovative technological capabilities by examining detailed empirical evidence that tends to be ignored in the existing literature. The framework and empirical analysis in this study challenge the relevance of the existing perspectives on the internationalization of innovative capability to the late-industrializing context. For instance, in relation to the *non-globalization* perspective (e.g. Vernon, 1966, 1979; Patel and Pavitt, 1992; Patel, 1995; Daniels, 1997) and the *conditional globalization* perspective (Mansfield et al., 1979; Mansfield and Romeo, 1984; Dunning, 1994a,b; Cantwell, 1995), a combination of R&D personnel and R&D expenditures and of international patenting capability – through US patents – and the percentage of qualified engineers have been used to assess technological innovative capability. However, there are situations in which some of these measures may have their own limitations and are less relevant because:

1. Results from studies have shown that R&D and international patenting capabilities are only prevalent in advanced industrialized countries (e.g. the US, Japan and selected European countries such as the UK and Germany), where firms have sufficiently advanced levels of innovative capabilities to conduct R&D and produce international patents. Thus, R&D and international patenting statistics would only be relevant to assess the extent of internationalization of technological capabilities in firms in these advanced industrialized countries. Applying the same proxy indicators to firms in some late-industrializing countries, which do not have adequate levels of innovative capabilities to conduct such R&D and produce international patents, in the first place, would be quite *irrelevant*.
2. In the case of TNC subsidiaries, it is generally known that TNC subsidiaries began operations from the early-1970s (of US-based) and mid-1980s (of Japan-based) in different developing locations such as Southeast Asia and Brazil (Manaus) as offshore production sites for labor-intensive and simple products. Thus, if the activities and capabilities of these subsidiaries have changed for the production of more complex products, using more capital intensive process, or involving more engineering and design work, these changes would not likely be reflected in the US patent statistics and R&D expenditures. This argument may also apply to local firms, i.e., in terms of analysing progressive *changes* in firms' innovative capabilities.
3. Thus, the possibility of the internationalization of *other* types of innovative activity should not be ignored, as they could even be more relevant to firms in those late-industrializing countries. These activities range from the most engineering-based (i.e. process, mechanical, electronics, electrical,

industrial engineering); design and development-based (i.e. prototyping, product improvements, cosmetic/aesthetic, design for manufacture, PCB design and circuit design); to precision tooling, mould and die design, and automation design.

Finally, this research sheds some light on some of the perspectives that underlie common approaches to policy. First, we have found that the relative small number of firms that have reached Level 6 capability could be interpreted as a gap in relation to other industrialized areas. The evidence here does not imply that those innovative firms will continue through subsequent phases of capability development. Neither can we suggest that other innovative firms (at Levels 2 to 4) will move into the development of other types and levels of innovative capability. However, firms' move into the accumulation of more advanced levels of innovative capability will involve different types of intra-firm efforts and different government policy measures.

Secondly, in line with Ariffin (2000) and Ariffin and Figueiredo (2004), some questions should carefully be addressed during policy design: (i) how to increase the number of firms that have built up innovative technological capability; (ii) how to make sure that local people are occupying (or will occupy) key managerial positions in TNC subsidiaries; (iii) how to assess the role of the local technological infrastructure (universities, research centers) in contributing to firms' technological development efforts.

Thirdly, in tackling these questions – and also in line with Ariffin (2000) and Ariffin and Figueiredo (2004) – it is important to distinguish between two fundamentally different dimensions of technological development: (i) movement through increasingly 'advanced' and complex products and processes, and (ii) movement through increasingly creative roles in connection with those product/process technologies (e.g. from their basic operation and use through various kinds of design and engineering to differing 'depths' of R&D). Progress along these two dimensions involves the creation of very different kinds of resources and the use of different learning mechanisms. Thus, while governments may have interests in accelerating both types of progress, different policy measures will be necessary in each case.

As a final recommendation, empirical studies of technological capability development, particularly in the latecomer context, should not draw solely on a macro-level design, based on surveys and aggregated data. Studies could combine an industry-level design with a firm-level approach. This would be beneficial to gather more detailed empirical evidence and build a more realistic understanding of what is taking place inside a certain industry in terms of technological development and innovation. This would also contribute to illuminating and improving the design and implementation of business and government strategies based on technological capability building.

Notes

1 The study was initially implemented from November 1999 to April 2001. The second phase was implemented from September 2001 to October 2002.
2 There is a vast literature on international business about the globalization of production (e.g. Henderson 1989; Clegg, 1992; Coffey, 1996; Drakakis-Smith, 1996).
3 There are other ways of assessing firms' technological capabilities, such as, R&D expenditure (Mansfield, 1979), individuals' qualifications (Pack, 1987; Jacobsson and Oskarsson, 1995), investments in R&D personnel (Wortmann, 1990), and patenting (Patel and Pavitt, 1997). However, most of these indicators, particularly those based on individuals' skills, do not take into account the organizational setting in which technological capability is developed and the technological characteristics of firms that operate in late-industrializing countries.
4 The methodological procedures to adapt this framework for the electronics industry are described in Ariffin (2000). An earlier adaptation and empirical application of the original framework, as developed in Lall (1992) and Bell and Pavitt (1995), has been undertaken in Figueiredo (2001). The adaptation of this type of framework for the TW industry was carried out during the early stage of the fieldwork for the study underlying this chapter.
5 There are other frameworks for technological capability in latecomer firms: the 'reversed product-cycle' (Hobday, 1995) and the 'acquisition-assimilation-improvement sequence' (Kim, 1997). However, these frameworks, despite their merits, are more focused on product capabilities; they do not cover other technological activities such as process and production organization, equipment and project engineering.

References

Ariffin, N. and Figueiredo, P. N., 'Internationalisation of innovative capabilities: counter-evidence from the electronics industry in Malaysia and Brasil', *Oxford Development Studies*, 32(4) (2004) 559–83.

Arifin, N., 'The internationalisation of innovative capabilities: the Malaysian electronics industry', Unpublished Ph.D. thesis (Brighton: SPRU, University of Sussex, 2000).

Bell, M. and Pavitt, K., 'The development of technological capabilities', in ul Haque, I. (ed.), *Trade, Technology and International Competitiveness* (World Bank: Washington, D.C., 1995).

Cantwell, J., 'The globalization of technology: what remains of the product cycle?', *Cambridge Journal of Economics*, 19 (1995) 155–74.

Cantwell, J., 'From the early internationalisation of corporate technology to global technology sourcing', *Transnational Corporations*, 8(2) (1999) 71–92.

Cassiolato, J., Szafiro, M., Lastres, H.M.M. and Vargas, M., 'Local systems of innovation in Brazil, development and transnational corporations: a preliminary assessment based on empirical results of a research project', *Conference Paper for DRUID's Nelson–Winter Conference*, Copenhagen, Denmark (June, 2001).

Cimoli, M. and Katz, J., 'Structural reforms, technological gaps and economic development: a Latin American perspective', *Industrial and Corporate Change*, 12(2) (2003) 387–411.

Clegg, J., 'Explaining foreign direct investment flows', in Buckley, P.J. and Casson, M. (eds) *Multinational Enterprises in the World Economy: Essays in Honour of John Dunning* (Cheltenham: Edward Elgar, 1992), pp. 54–73.

Coffey, W. J., 'The "newer" international division of labour', in Daniels, P.W. and Lever, W.F. *The Global Economy in Transition* (Harlow: Longman, 1996), pp. 40–61.

Daniels, P.L. 'National technology gaps and trade: an empirical study of the influence of globalization', *Research Policy*, 25 (1997) 1189–207.

Drakakis-Smith, D., 'Less developed economies and dependence', in Daniels, P.W. and Lever, W.F., *The Global Economy in Transition* (Harlow: Longman, 1996), pp. 215–38.

Dunning, J. H., 'Re-evaluating the benefits of foreign direct investment', *Transnational Corporations*, 3(1) (1994a) 23–51.

Dunning, J.H., 'Multinational enterprises and globalization of innovatory capacity', *Research Policy*, 23 (1994b) 67–88.

Figueiredo, P.N., *Technological Learning and Competitive Performance* (Cheltenham: Edward Elgar, 2001).

Fleury, A. and Fleury, M.T.L., 'Por uma política industrial desenhada a partir do tecido industrial', in Fleury, A. and Fleury, M.T.L. (eds), *Política Industrial 1* (São Paulo: Publifolha, 2004).

Forbes Brasil, 'A nova cara do free shop', 1(3) (25 October, 2000) 62–6.

Grandstrand, O., Hakanson, L. and Sjolander, S., 'Internationalisation of R&D: a survey of some recent research', *Research Policy*, 22 (1993) 413–30.

Henderson, J. *The Globalization of High Technology Production* (London and New York: Routledge, 1989).

Hobday, M., 'Innovation in South-East Asia: lessons for Europe', *Management Decision*, 34(9) (1996) 37–48.

Hobday, M., *Innovation in East Asia: the Challenge to Japan* (Aldershot, England: Edward Elgar, 1995).

Jacobsson, S and Oskarsson, C., 'Educational statistics as an indicator of technological activity', *Research Policy*, 24 (1995) 127–36.

Katz, J., 'The limits of the prevailing orthodoxy: technology and education as restrictions to productivity growth and international competitiveness in Latin America', *Proceedings of the DRUID Summer Conference 2004 on Industrial Dynamics, Innovation an Development*, Elsinore, Denmark (June 2004) pp. 14–16.

Kim, L., *Imitation to Innovation. The Dynamics of Korea's Technological Learning* (Boston, MA: Harvard Business School Press, 1997).

Lall, S., 'Technological capabilities and industrialization', *World Development*, 20(2), (1992) 165–86.

Lall, S., 'Reinventing industrial strategy: the role of government policy in building industrial competitiveness', *Trade and Industrial Policy Strategy (TIPS) Working Paper 9* (2004).

Mansfield, E. and Romeo, A., 'Reverse transfer of technology from overseas subsidiaries to American firms', *IEEE Transactions on Engineering Management*, 31(3) (198[illegible]) 122–7.

Mansfield, E., Teece, D. and Romeo, A., 'Overseas research and development by US-based firms', *Economica*, May, 46 (1979) 187–96.

Pack, H., *Productivity, Technology and Industrial Development. A Case Study in Textiles* (New York: Oxford University Press, 1987).

Patel, P. (1995), 'Localised production of technology for global markets', *Cambridge Journal of Economics*, 19 (1995) 141–53.

Patel, P. and Pavitt, K., 'Large firms in the production of the world's technology: an important case of non-globalization', in Granstrand, O., Hakanson, L. and Sjolander, S. (eds) *Technology Management and International Business: Internationalisation of R&D and Technology* (Chichester and New York: John Wiley & Sons, 1992).
Patel, P. and Pavitt, K., 'The technological competencies of the world's largest firms: complex and path-dependent, but not much variety', *Research Policy*, 26 (1997) 141–56.
Suframa (Superintendency of the Manaus Industrial Pole), *Report of Activities* (Manaus, Brazil, 2005).
Sutz, J. and Arocena, R., 'Emerging neoperipheral structures and gardening policies', *Proceedings of the DRUID Summer Conference*, Elsinore, Denmark (14–16 June, 2004).
Vernon, R., 'International investment and international trade in the product cycle', *Quarterly Journal of Economics*, 80(2) (1966) 190–207.
Vernon, R., 'The product-cycle hypothesis in a new international environment', *Oxford Bulletin of Economics and Statistics*, 41(4) (1979) 255–67.
Wortmann, M., 'Multinationals and the internationalisation of R&D: new developments in German companies', *Research Policy*, 19 (1990) 175–83.
Zander, I., 'Technological diversification in the multinational corporation: historical evolution and future prospects', *Research Policy*, 26 (1997) 209–27.

13

Assessing Value Creation and Backward Linkages in Foreign Investment: A Combination of Macro and Micro Tools

Teresa Morais da Costa, Alexandre Videira and Francisco Veloso

13.1 Foreign Direct Investment, value creation and public policy

13.1.1 On the importance of Foreign Direct Investment

Foreign Direct Investment (FDI) is considered an important pillar of economic development (Katz, 1969; Aitken and Harrison, 1991; Markusen and Venables, 1997; Sjöholm, 1999a,b). There are multiple avenues through which FDI can contribute to an economy (de Mello, 1997; Moran, 1998). The foreign investor can provide a product that the local companies are not capable of supplying, or it can stimulate the national economy by creating an additional demand for intermediate or primary goods, as well as for labor. It can also complement national savings, thus contributing to the accumulation of capital. In addition, FDI has the potential to supply knowledge and technology capable of generating externalities that lead to higher productivity and increase national output (Haddad and Harrison, 1993; Aitken et al., 1996), for example, by bringing technologies that increase the marginal productivity of existing ones (Romer, 1986), or training and developing new capabilities in the workers (Borensztein et al., 1995).

Successive studies going as far back as Hirschman (1958) have also suggested that the links between multinationals and companies in the same region located upstream in the value chain ('backward linkages') are as important for the development of the region as foreign investment itself (e.g., Morris, 1992; Dunning, 1993; Turok, 1993; Crone and Roper, 1999). The idea is that country productivity growth materializes not through FDI per se, but by its potential role in inducing the creation of local sectors or clusters with higher technological levels than those in the nation (Rodrik, 1996; Rodríguez-Clare, 1996; Markusen and Venables, 1999). The critical issue is that the emergence of such sectors depends on the existence of a relatively dense set of local suppliers to the foreign investor, in quantity and diversity, and with international competitiveness.

The importance of foreign direct investment for industrial development has been recognized by governments for several decades. Thus, there has been an array of policies to try to attract FDI and condition its activity, with the purpose of maximizing the impact of investment in the region (Kokko, 1994; Braunerhjelm and Svensson, 1996; Görg and Ruane, 2000; Mazolla and Bruni, 2000; Blomström and Kokko, 2003a). Traditionally, government policies would focus on restrictive measures (Dunning, 1993; Globerman and Shapiro, 1999). For instance, they would require investors to buy certain amounts from local firms or export particular quantities. However, the perspective and policies toward FDI have recently begun to change. Most countries have been lowering the conditions for investment, freely opening new sectors for entry by foreigners, and creating important incentives for investment (Blomström and Kokko, 2003b). Nowadays, policies directed at FDI are meant to address issues such as market failures, notably by focusing on promoting linkages to national companies, and on fostering education, training and R&D.

Given these concerns, it is not surprising to find a body of literature addressing the estimation of impacts of these projects in a host economy. One of the most popular methods is the use of Leontief input–output models (I–O models). This method allows us to quantify the total impacts (direct and indirect) on the production, value added and employment at the macroeconomic and sectoral level. It is quite a powerful approach for several reasons. First, it aggregates impacts up to common monetary units of an economy, enabling easy comparisons across projects and areas of policy; second, it enables an account of both direct and indirect effects of a particular investment; third, because of its wide diffusion, reliable and extensive data is often available simplifying the work of the policy analyst. Finally, there is considerable experience with the method, allowing an easy understanding of the results of an analysis by decision-makers and even by the broader public.

Because of its own structure, the use of this method of analysis is also related to the establishment of backward and forward linkages by inward investors. In fact, some of the most popular and useful areas where one can find examples for the application of these models is precisely the study of the relevance and impact of backward linkages in the local economy. For example, the works of Forsyth (1972), Lever (1974), and Stewart (1976) studied backward linkages between investors and the local supply networks in the United Kingdom and Ireland. Turok (1993) and McCann (1997) discuss the level of integration of the electronics industry in Scotland, looking at the backward linkages and the type of I–O multipliers to be considered.

Brand et al. (2000) studied the direct and indirect contributions of foreign industrial enterprises in the regional economies of Wales, Scotland, and the West Midlands again analysing backward linkages.

Despite the significant experience with I–O models, most work addresses *ex post* evaluations of particular FDI projects. In these assessments the objective is to look at an investment that has already taken place and evaluate the significance of its impact. However, there is also the need and opportunity to study *ex ante* or at least 'what-if' scenarios, i.e. how different configurations of an FDI project might impact the local economy, helping to guide the creation of investment policies. This work aims at addressing these shortcomings.

13.1.2 Research question, approach and application context

As noted above, results suggest that MNEs can be credited with positive contributions to economic development if they are able to build a robust level of backward linkages. To address these concerns this chapter presents a model that allows an analysis of the impact of FDI for different scenarios of backward linkages to the host economy. The critical research question that is meant to be answered is: How to account for different levels and scenarios of backward linkages created by the investor (henceforth designated as an OEM – Original Equipment Manufacturer) to the local industrial fabric, and what is the value generated by each context to the host economy?

To address this question, this chapter presents a tool that can be used to study what-if scenarios in FDI projects, looking particularly at the role of backward linkages. The baseline approach is the use of the established I–O models. Still, while existing models allow straightforward computation of how increases in the amount of investment or domestic purchases might impact overall value creation, established assessment methods overlook the fact that local purchases are far from homogeneous and have the potential for very different impacts in the domestic economy. This chapter contributes to the literature by looking at the black box of backward linkages in the context of a new investment. It extends traditional I–O value estimations by coupling it with a detailed account of the structure of existing and potential domestic purchases and an engineering-based cost estimation method called System Cost Modeling (SCM) (Veloso, 2001). This method,[1] which will be referred to as the 'combined model', allows a quantification of results associated with FDI and an evaluation of public policies related to incentives, and restrictive policies. As a result, it enables a more profound understanding of how domestic value creation by foreign MNEs increases the potential for creating wealth in the region. This combined approach is then tested in the context of a case study. The case chosen to illustrate the method is the Autoeuropa plant located in Portugal. As described below, this was a major FDI project that has all the desirable characteristics to test the proposed method.

13.2 Research methods

13.2.1 Leontief model

The Leontief model, also called I–O, is an 'open static model' that explores the purpose of economic activity, i.e. the satisfaction of final demand (Rasmussen, 1963; O'Connor and Henry, 1974; Miller and Blair 1985; Leontief, 1986). As Leontief (1986, p. 19) notes, 'Economic I–O analysis is a method to systematically quantify the interrelationship among various sectors of an economic system. The economic system may be as large as a nation or even the entire world, or as small as the economy of a metropolitan area'. The importance of the Leontief model lies in its flexibility and capacity to take into account the complex network of simultaneous interdependencies that characterize economic activity at many levels (Rasmussen, 1963; O'Connor and Henry, 1974; Miller and Blair, 1985). This set of relations can be an important source of information for supporting eventual development policies for the economy, in particular for FDI (Rasmussen, 1963; Miller and Blair, 1985; Lopes and Amaral, 2002). For example, it's possible to use the I–O analysis to guide the establishment of incentives to improve the local offer in certain sectors, allowing a decrease of the dependency on foreign supplies.

I–O matrices can be seen as a double entry accounting system that represents the factory of economy in each year (Leontief, 1986). The economic analysis is based in tracing intersector acquisitions and identifying the direct and indirect impacts of each sector in the economy. Movements by final demand (exogenous component) are translated into bigger or smaller production of goods and services (endogenous component) and these, in turn, pull the necessary primary resources inherent to the stimuli made by each sector of intermediate demand of the economy, in each year. The analysis is based on matrix algebra with three main tables that reflect the relations: a Transactions table, also called I–O matrix; a Technical Coefficient table and an Interdependency Coefficient table. The structure of each productive process for a given sector, i.e. its production function, is a vector of structural coefficients that describe the relations between consumed input and produced output in terms of quantities, i.e. a linear production function (O'Connor and Henry, 1974; Leontief, 1986; Hewings, 2003). Technical coefficients quantify the first order or direct effect that the increase of one monetary unit of final demand for a certain sector generates on the economy's output (O'Connor and Henry, 1974; Costa, 2005). The interdependency coefficients reflect the total economic effect (direct and indirect) generated by the increase of a monetary unit of final demand for a certain sector, while keeping everything else constant (Rasmussen, 1963; O'Connor and Henry, 1974; Miller and Blair, 1985; Lopes and Amaral, 2002; Costa, 2005). The interdependency coefficients are obtained using matrix algebra (Leontief, 1986).

The analysis of the impact of exogenous changes (final demand) on the economic activity of the sectors in each year, i.e. on the short term, relies on indicators which are called simple[2] I–O multipliers (O'Connor and Henry, 1974; Miller and Blair, 1985). The calculation of the multiplier relies on the relation between initial effects of an exogenous change (final demand) and the total effects of that change, resulting in an adimensional value. The initial output effect in the economy is defined as the initial amount of a given monetary unit of output of sector *j* required to meet the increase of one monetary unit of final demand for that same sector. This initial effect is also called the direct effect, while indirect effects, the first to *n* order effects, are the supply chain effects (Costa, 2005). Typically, multipliers establish these relations ignoring concerns with production capacities for each sector, as well as other macroeconomic variables.

To evaluate the economic impacts the most important multipliers include (Costa, 2005):

Production multipliers – quantify the increase in production of the economy to meet one more unit of final demand for production in a certain sector, *ceteris paribus*. The value of the multipliers for each sector reveals its level of dependence relative to, on the one hand the locally available products, and, on the other hand, the corresponding production structures. When local activity is not capable of supplying the necessary intermediate inputs, there is a need to import inputs (Lopes and Amaral, 2002).

Gross value added (GVA) multipliers – translate the total impact that an increase of a unit of final demand (initial effect is equivalent to the stimulus created by the final demand) has on the GVA of the economy. These multipliers allow a quantification of the efficiency effect for each sector of the economy (Lopes and Amaral, 2002).

Despite its attractiveness, the Leontief's I–O model also has important limitations in a number of aspects (Rasmussen, 1963), in particular if one would like to use it in an *ex ante* evaluation. In fact, when used as a projection tool, the I–O analysis 'doesn't provide a suitable projection prognostic, as it can only be thought to produce a conditional prognostic' (Rasmussen, 1963, p. 36). The main driver of this limitation is the fact that the model is static, with constant technical coefficients. Thus, it assumes that the amount each industry acquires from other industries, and the value-added amount, per unit of its own output or production, stays constant as demand increases or diminishes (Costa, 2005). To reduce these limitations this research proposes coupling the traditional Leontief model with a costing method for complex products, called System Cost Modeling (Veloso, 2001). The resulting combined model allows an adjustment of the technical coefficients of the I–O analysis to reflect structural changes in the industry as to mirror different scenarios of national incorporation of intermediate sectors.

13.2.2 System Cost Modeling and SCM-IO analysis

The System Cost Model (SCM) method, developed by Veloso (2001), allows an evaluation of the cost for complex systems composed by a high number of individual components and subsystems, such as a motor vehicle, by building bottom-up cost structures. This method is in the general category of *Process Based Cost Modeling*, which uses engineering principles and relies on detailed information about the transformation processes and the nature of the products themselves to estimate production cost (Boothroyd et al. 1994; Humphreys and Wellman, 1996; Clark et al., 1997; Kirchain and Clark, 2001). However, because traditional *Process Based Cost Models* are very thorough in the process and product information used, they have important limitations when applied to systems with a high number of individual components, such as in the case of a motor vehicle. Thus, to allow the analysis of a large number of components and processes, SCM uses engineering information associated with the manufacture and assembly of each component, but the inputs are very simplified. As a result, the SCM loses some rigor when calculating the cost of the individual component, when compared with the *Process Based Cost Model* (Veloso, 2001).

The cost estimation method for individual components and the approach to group these into subsystems relies on two essential aspects: i) the cost factors of the manufacturing processes, and ii) the processing time associated with the particular technology and the component (Veloso, 2001). The necessary information for applying the model and the calculation logic are presented in Figure 13.1. The metrics considered for each component in the cost calculation are: the weight, the material, the level of complexity, and the manufacturing process. These variables are inputs in a set of functional relations that combined with the exogenous variables associated with the place of production of the components (e.g. salaries) allow an estimation of the fixed and variable costs for each component.

The main aspect of this estimation process is a set of parametric functions that relates the four characteristics described above with each of the cost variables. These relations are established using engineering principles and information on the cost of the systems associated with each production process. For example, the clamping force of a machine to inject a plastic part can be statistically related with the weight and complexity of the latter. This force can in turn be statistically related to the cost of the equipment. A second dimension of the model is the grouping of individual components into subsystems, and the subsequent grouping of subsystems to the system level. This estimation is based on statistical information about handling and assembly time, as well as on associated equipment for joining individual parts with certain characteristics into subsystems (Veloso, 2001). The complete SCM estimation procedure allows us to analyse the cost of the whole equipment as well as the influence of the key parameters (such as input costs) on the cost function's behavior.

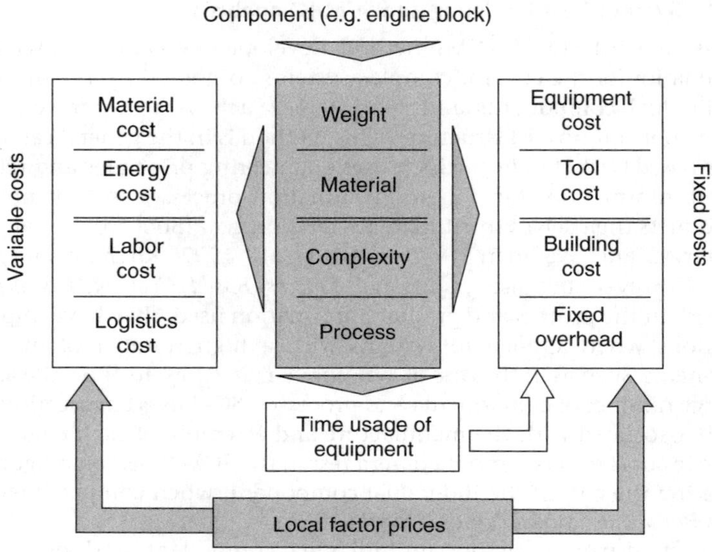

Figure 13.1 SCM estimation for a component production
Source: Veloso (2001).

The proposed analysis integrates the SCM method and Leontief's I–O model. SCM allows generating 'market basket' scenarios of components (individual, subsystems or systems) for the OEM, quantified from the perspective of the cost of acquisition. These scenarios take into account the variables inherent to the physical and technological characteristics of the components and the external variables of the country or region where they are manufactured. For any given scenario, it is possible to know the products and the corresponding manufacturing and assembly processes, which allows us to predict the sectors that will supply all the components required by the OEM. As a result, it is possible to breakdown national acquisitions for each sector of activity of the host country and then use traditional I–O evaluation. For instance, if we wish to study what happens if the car's interiors were imported from another country instead of being manufactured locally, the SCM is used to calculate the cost structure corresponding to both scenarios and to break down the impact of both scenarios in the different sectors, which then allows us to calculate the economic impact of both situations through I–O analysis. This approach has the advantage of providing information about the direct and indirect impacts of the intersector demand that result from changes on the technical structure of a sector (e.g. when more interior parts are being produced in the region). This overcomes some of the main limitations associated with an *ex ante* analysis using Leontief's model, and allows results for several scenarios of a particular FDI, especially in what concerns its level of backward linkages.

Specifically, the process of integrating both models to study project scenarios is as follows: first, possible 'market baskets' of OEM purchases are quantified based on the SCM in the perspective of acquisition cost. In this case study, the comparison between costs is done considering manufacturing either in Portugal or, alternatively, in a country like France and Germany. Then host country acquisitions for each sector of activity used in the I–O evaluation are broken down. Based on the distribution of acquisitions for OEM's intermediate inputs, a linear production function is then established. This function is equivalent to the technical coefficients column for each one of the scenarios. Simultaneously, for a given target-year scenario, the input–output matrices are created, breaking down OEM sector versus the rest of the sector. The technical coefficient matrix is obtained in this process. From the technical point of view, the SCM/I–O combination consists of introducing the data produced by the SCM, i.e. the different linear functions of the OEM's acquisitions of intermediate inputs in the national economy, into the technical coefficient matrix of the open, static Leontief I–O model, replacing the column corresponding to the OEM for each scenario. The technical structure of the economy (columns in the technical coefficients matrix) is a constant for all sectors, except the column corresponding to the OEM. As a result, all the economy has the same intersector acquisition relations, except for the OEM purchases where the different scenarios are being considered. Once the technical coefficient matrices have been obtained for each scenario, the total effects (direct and indirect), i.e. the technical coefficients, are calculated. Based on these the global impact on production, the GVA, the imports, the employment, and wages can be known, keeping everything else constant (Rasmussen, 1963; O'Connor and Henry, 1974; Miller and Blair, 1985; Lopes and Amaral, 2002).

13.3 Application of the model and results analysis

13.3.1 The Portuguese auto industry as an application

The auto industry has been an important driver of economic development in Portugal since the 1960s. However, it was in the 1980s, with the Renault investment project, that the foundations for a true national auto industry started. The project, now extinct, is considered a stepping stone for the development of the national components industry. For the first time in the history of the industry it enabled the development of components' supplier companies competitive at the international level. It also trained countless workers, induced the establishment of skilled employees, and generated managers who would have important roles in the development of the components industry when they moved on from Renault to suppliers.

At the end of the 1980s the government created an Industrial Development Program for Portugal (PEDIP), which included a series of important financial incentives for areas such as R&D, training, export promotion. PEDIP also included strong incentive packages for attracting foreign companies. Although there were no explicit target sectors, the Portuguese government considered that the establishment of an assembly unit by an important international car manufacturer would be a unique opportunity for the consolidation and further development of an emergent Portuguese auto industry, particularly at the level of components. This effort resulted in the Autoeuropa (AE) investment, a joint venture between Ford Motor Co. and Volkswagen AG (fully owned by VW since 1999) for the production of a minivan. This project, a 2.54 billion dollar investment, represents the second critical landmark in the development of the national auto industry and the Portuguese industrialization process in general.

In 2003, AE represented approximately 0.23 per cent of the national GDP, 6 per cent of total exports and 0.33 per cent of industrial employment. More importantly, the project induced a significant evolution in the component market, having subsequently attracted 22 additional foreign units as suppliers and establishing important contracts with existing local plants. As Figure 13.2 shows, the components sector saw its sales quadrupling between 1990 and 2000. Currently it represents approximately 1.35 per cent of GDP and 10 per cent of national exports.

This context justifies the choice of AE as a case study to apply the combined SCM/I–O model. The direct results of this project are generally seen as extremely positive. Moreover, existing studies suggest that this investment has played a critical role in the domestic economy, not only because of its magnitude, but also because it facilitated the introduction of advanced

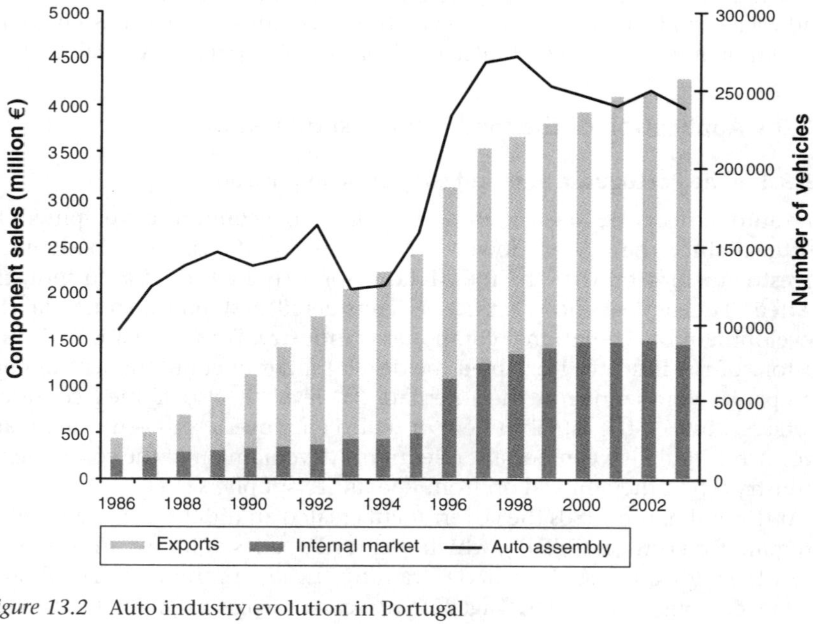

Figure 13.2 Auto industry evolution in Portugal

manufacturing technologies and modern organizational methods across the national supply chain (Guterres, 1996; Lima et al., 1996; Neves, 1996; Santos, 1996; Maia, 1999; Selada et al., 1999; Vale, 1999; Valente, 1999; Correia, 2000; Simões, 2000; Lobo and Melo, 2002). The problem is there is no work that allows us to analyse what the impact of this project would have been under different circumstances, and to understand alternative value generation configurations. This is precisely the objective of this chapter.

13.3.2 Backward linkages of the auto sector and AE

The data obtained from the national I–O matrices[3] for the automotive production function allows us to verify that the total production multiplier for the sector is relatively stable (Figure 13.3a). But, while the rest of the auto

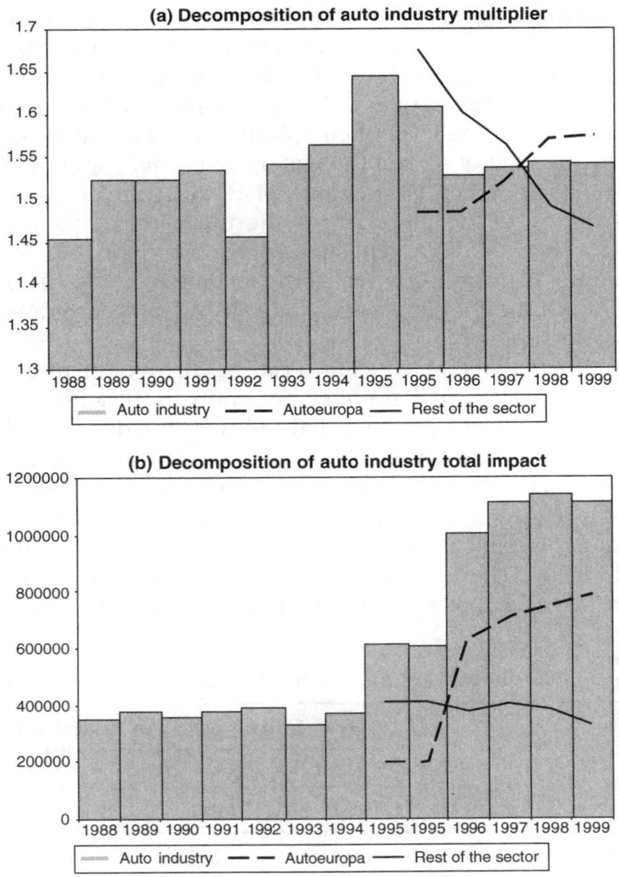

Figure 13.3 Impact evolution of auto industry production

Source: Costa (2005).

sector has suffered a progressive decline in its relations with the local supply chain, AE has slowly generated stronger bonds with the local supply network. Figure 13.3b shows that the total impact value, obtained by combining the production multiplier with the corresponding final demand, is growing and that it is the AE operation that drives the trend.

In addition to production, it is perhaps as relevant to analyse the direct and indirect components of the GVA multiplier over time. Table 13.1 shows that between 1995 and 1999 each monetary unit of the AE vehicle itself grows to generate approximately twice as much value as the other national assembly units combined. This means that the internal industrial activity of AE has much more value generation than other units. However, the indirect impact on GVA generated by the supply chain practically does not change during the same period. Moreover, while this value is clearly higher than that of the remaining assembly lines in 1999, this only happens because these other units have lowered their preference for incorporating local value. In fact, the indirect impact of the rest of the sector in 1996 was above that of AE in 1996, and the same applies to 1999. This suggests that there is not a significant difference in the level of complexity of the products manufactured internally by suppliers of AE and those of other assembly units in the country.

The combination of the GVA multiplier of AE with its trade volume generates an impact for AE of 1.5 per cent[4] of the total economy's GVA in 1999. Compared to the total GVA generated by the national auto sector in that year, this means that three-quarters of the income were directly or indirectly created by AE. Of the GVA impact generated by AE, 36 per cent[5] was indirect produced by its suppliers.

Based on this analysis, we conclude that the AE investment created significant value to the Portuguese economy, not only in the auto sector, but also in almost all the sectors supplying that industry. In addition, AE worked as an important lever in the level of GVA incorporation for each monetary unit of vehicle manufactured in the country.[6] Nevertheless, this lever was largely determined by a greater value added in AE itself. Its indirect effect in the rest of the economy is equivalent to the other assembly lines in terms of quality or intrinsic value. These results suggest that the arrival of the Ford–VW Group provided an important scale factor in the assembly lines sector and in

Table 13.1 Autoeuropa and the rest of auto industry GVA multiplier

	GVA impact per $1 of final demand			
	1995		**1999**	
	Indirect	**Direct**	**Indirect**	**Direct**
Autoeuropa	0.173	0.15	0.175	0.3
Rest of the sector	0.197	0.16	0.162	0.18

the purchases from its local supply chain, but it did not imply a substantial difference in value and diversity of first line supply offer in the national industrial fabric.

13.3.3 Application of the combined model and results analysis

Combined model applied to AE's minivan

While the calculations of the previous section were performed using standard *ex post* I–O calculations, the proposed combined method enables the evaluation of 'what-if scenarios'. The idea is to be able to study how departures from a proposed or existing investment and, in particular, the level of backward linkages, affect the economic impact and value generation of a particular investment. The existing information regarding AE supplies used in the previous sections to evaluate the national I–O matrices reflects only costs associated with the purchases of AE from suppliers located in Portugal. To be able to analyse the situation observed in 1999 using the SCM it was necessary to cross existing aggregate information with data about companies and their manufacturing volumes and competences and then estimate AE purchases by type of product and origin (national and foreign). This correspondence was then used as a basis for the study of scenarios that differed from what was observed in 1999 (Costa, 2005).

For the integrated evaluation, individual components and subgroups for AE's minivan were identified. These are the relevant level for the cost estimates using the SCM method. In the AE's minivan, the breakdown of individual components results in 4832. Subgroups are then generated according to sets of components that reflect typical sourcing decisions by the manufacturers, as noted in Table 13.2. It is then possible to further aggregate the subgroups for exposition purposes. Ensuing typical decisions of OEMs, seven groups identified in Table 13.2 were considered. The table also presents information on the correspondence between groups, subgroups and components.

Table 13.2 Minivan individual components and the respective groups

Group	Sub-assemblies	Components	Estimation costs but not by the model	Costs not modeler
Power train	36	860		
Chassis	45	770		
HVAC	12	322	Small parts	
Interior	37	1333	Electric & Electronic components	Body in white final assembly
Body	24	380		
Exterior	27	307		
Electr, Info, Control	25	859		

For each component there is detailed information regarding each of the four critical input dimensions for the SCM, i.e. process, material, complexity, and weight. There is also access to the variables of production and logistics of AE in 1999 (production volume, product and equipment life-cycle, interest rates, salaries, use of the assembly line). The cost of domestic supplies incorporates local manufacturing conditions. The remaining imported components are calculated in the same way, using SCM, but the manufacturing and logistics variables considered in the model are those typically found in foreign suppliers for the benchmark region (i.e. with exogenous factors, production efficiencies, etc. different from the Portuguese).

Based on the 1999 benchmark scenario, and according to the cost values estimated for the acquisition of imported and national products, the incorporation of national supplies verified in AE was approximately 43.3 per cent of total purchases[7] from first line suppliers. As can be seen from the estimates of Figure 13.4, the incorporation of national products in the AE assembly line was particularly strong in the areas of Chassis, Interiors, and Electric and Electronic groups.

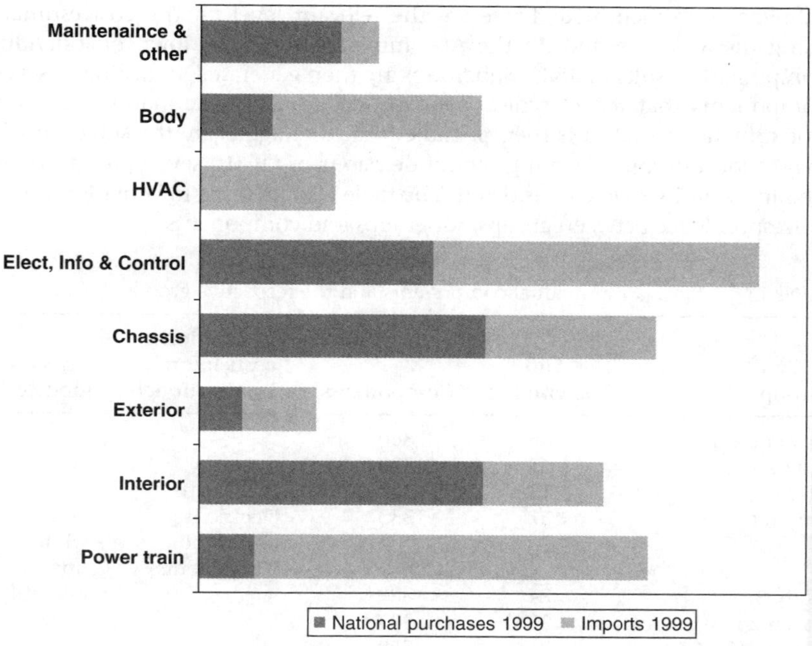

Figure 13.4 Local *versus* foreign content per each minivan group
Source: Costa (2005).

Optimistic and pessimistic scenarios for the SCM/I–O model

The next step is to apply the results of the cost estimate to the I–O method. For this, sector classifications must be given to the data resulting from the SCM so that, in each scenario, the 'market basket' of products manufactured in the host country can have a sector correspondence. For the 1999 baseline scenario, the sector correspondence for the intermediate inputs purchased by AE was established using SIC codes[8] of the supplying companies. The corresponding linear production functions for AE are then quantified, using as dependent variables the 15 sectors of the national I–O matrix, with the auto sector broken down (Costa, 2005).[9]

But replicating the I–O analysis using a micro-structure of costs calculated with SCM is an exercise that per se adds limited value. The SCM/I–O integrated model is particularly useful because it allows the study of the impact of backward linkage scenarios that are different from those observed, or the analysis of various options *ex ante*, before a particular investment actually takes place. In the present context, two scenarios of domestic purchases are considered: an optimistic and a pessimistic scenario. The method for generating the scenarios was as follows. First, information about the supply structure in 1999 and studies on firm capabilities done in the context of a large automotive project to characterize the national industry were gathered. Second, the potential cost advantages and disadvantages of manufacturing each subgroup in Portugal in comparison with the alternative of imports from suppliers in the Germany–France axis were estimated using SCM. This information was then used to generate a first approach to both scenarios. In a third stage, the initial version of both scenarios was discussed with specialists directly involved with AE to determine necessary adjustments based on their own perception. The changes they proposed were then integrated into the base scenarios, which resulted in the final scenarios.

The resulting contribution for each group is noted in Table 13.3, where a comparison to the reference year of 1999 is also provided. Under these

Table 13.3 Distribution of local content in each scenario

	Local content			
Group	**Pessimistic**	**Optimistic**	**1999**	**Total**
Power train	2	13	2	17
Exterior	1	4	2	5
Chassis	2	16	11	18
Electr&Electronic	3	11	9	22
Interior	2	14	11	16
Body	1	7	3	11
HVAC	0	5	0	5
Other	5	7	5	7
Total	17	77	43	100

Source: Costa (2005).

circumstances, the percentage of national incorporation in AE's purchases for the pessimistic scenario is 17 per cent, and for the optimistic scenario, 77 per cent. The two scenarios were then considered for evaluation using I–O analysis. For each one the column of the technical coefficients referring to AE is replaced, creating different I–O matrices for each scenario. Using these matrices, it is possible to estimate the impacts of each scenario on the GVA to the national economy in 1999 (assuming that the technical structure of the remaining sectors suffers no change).

Figure 13.5 presents the differentials of total impact in the production for each national economic sector. As one might expect, the auto components sector is the one absorbing the majority of the positive impact associated to the optimistic scenario. Nevertheless, other sectors, including metals and non-metals are also subject to important changes as a result of the two scenarios considered. Globally, the impact of both scenarios in the national flows of intermediate demand, quantified by the production multiplier means that, in the optimistic scenario, each monetary unit results in an increase of approximately 25 per cent of production in the domestic economy compared with 1999. In the pessimistic case, it represents a decrease of 21 per cent.

If one looks instead at the volume purchased in the whole economy, it can be shown that there is an increase of 83 per cent for the optimistic scenario, when compared with the total of intermediate demand of AE in 1999, and a decrease of 61 per cent for the pessimistic scenario. Figure 13.6 presents the

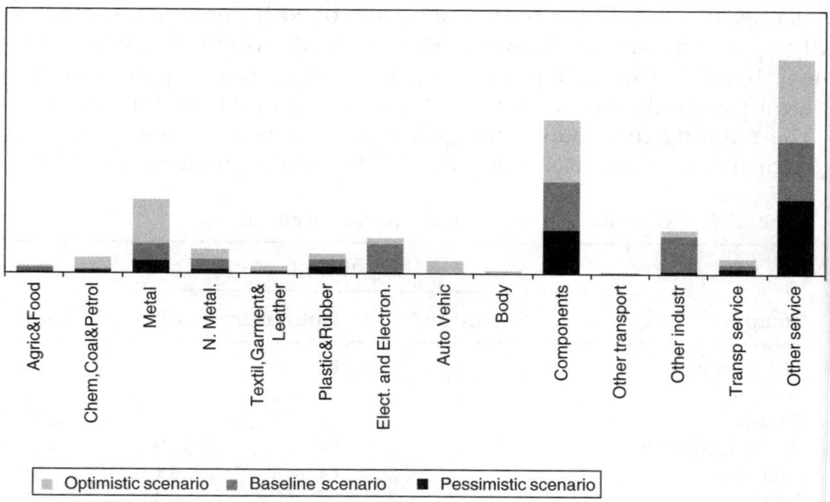

Figure 13.5 Autoeuropa induced production impact per scenario and per industry
Source: Costa (2005).

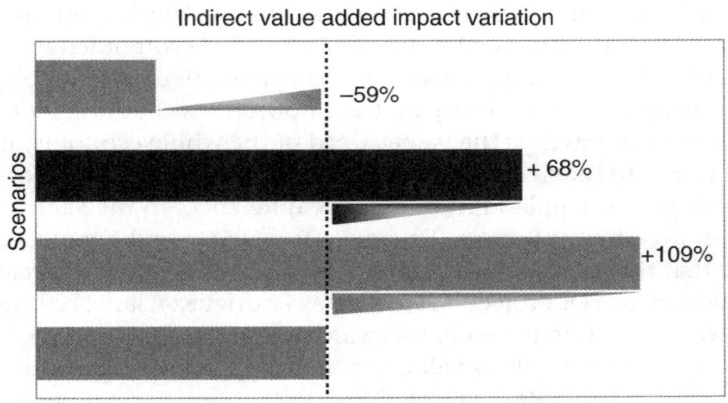

Figure 13.6 Comparative analysis of the impact on GVA per scenario
Source: Costa (2005).

variation of the indirect impact component measured in GVA. This change represents an increase of approximately 68 per cent for the optimistic scenario, when compared with the 1999 baseline, and a decrease of approximately 59 per cent for the pessimistic scenario.

13.4 Conclusions

This chapter looks at the black box of backward linkages. It shows how a combined model that accounts for the structure and costs of existing and potential domestic purchases can be combined with traditional I–O estimation to have a deeper understanding of domestic value creation by multinational investment. This method is used to analyse a particular auto industry foreign investment in Portugal, the AE mega-project. Results emphasize the distinct behavior of AE when compared with other assembly lines.

The AE investment created significant value to the Portuguese economy. Not only did it forge an operation that had much higher level of internal value creation than any existing assembly line in Portugal at the time, but it also grew to represent now 75 per cent of the income generated in the auto sector, while creating an important demand for almost all other sectors supplying this industry. Nonetheless, despite its magnitude, the economic lever of the investment was largely determined by greater value added in the unit itself. Its indirect effect in the rest of the economy is equivalent to the other assembly lines in terms of quality or intrinsic value. These results suggest that the arrival of the Ford–VW Group provided an important scale factor in the assembly lines sector and in the purchases from its local supply chain, but it did not imply a substantial difference in value and diversity of first-line supply offer in the national industrial fabric.

The results of this work reach beyond stating the importance of this particular FDI case. This chapter presents a SCM/I–O combined method that can be used to forecast and evaluate results of alternative scenarios of domestic content incorporation, quantifying the implications that different levels of local content can have for the value added of the whole economy. In fact, as the application to the AE investment shows, variations in the level of incorporation of domestic supplies mean dramatic differences in the value generated by a given investment project. The magnitude of these differences endorses the view that the levels of local content are a key driver of wealth creation in the context of an FDI project. Moreover, as Rodríguez-Clare (1996) suggests, certain density levels in the set of backward linkages might be needed to propel the industry to new levels of value creation. Even in the case of the AE project, where a reasonable level of domestic content was achieved, it was shown that the indirect value-added multiplier, a key indicator of the level of sophistication or intrinsic value of the local supply activity, did not grow significantly.

Achieving certain targets in the levels of domestic supplies for an FDI project is difficult. Not only the investor typically prefers freedom in this dimension, but there is also a scarcity of instruments that allow the policy analyst and the firm to have an objective discussion on how might a certain level of backward linkages induce value creation for the project and the local economy. The development of tools that combine the macroeconomic analysis typically found in the policy realms with engineering-based cost estimation more naturally used in the context of a firm may help to bridge this gap.

The SCM/I–O model presented here, combining I–O analysis with technical cost modeling, shows how such tools can help define profitable targets and perhaps reasonable incentive mechanisms that drive both investors and policy-makers to make informed decisions that are economically sound and politically acceptable. For example, it is possible to use the combined SCM/I–O analysis to guide the establishment of incentives to improve the local supply in certain sectors, allowing a decrease of the dependence on foreign supplies, or the increase of employment, allowing rather precise evaluations of the impact that particular changes might have.

Generically, from a public policy view-point, comparing output or GVA multipliers and how they might be affected by different supply chain scenarios, can show in which sectors any potential expense can generate greater impacts in terms of total output value of the economy. If the government's purpose is to maximize the output effects, the rational decision will be to invest in the development of a sector with the highest multiplier.

Notes

1 The approach is presented for the context of FDI. Nevertheless, the method is rather general and can be also used to study the economic impact of various backward linkage scenarios for national investments.

2 The effects may be divided into direct, indirect, and induced, if we consider Leontief's inverse matrix of Leontief's static model in which the private consumption variable (families) is endogenous (Miller and Blair, 1985). The multipliers based on this model are called total multipliers (Miller and Blair, 1985; O'Connor and Henry, 1974). This model was not analysed in this study.
3 The treatment of the information collected for the multi-sector characterization within the scope of this case study is described in Costa (2005).
4 The value corresponds to 239.2 billion Escudos.
5 Approximately 88 billion Escudos of indirect impact (GVA AE ≈ 151 billion Escudos).
6 Multiplier of sector 341 in the GVA of the economy = 0.44 in 1999 compared to 0.32 in 1988.
7 These values were considered without margins.
8 The SIC Code in Portugal is the CAE, Code of Economic Activity, established by the National Statistics to allow each company to be positioned in one national sector of activity, and in one of the groupings done according to the described SCM procedure.
9 The building of the national I–O matrix relies in the steps mentioned in Costa (2005), and it only excludes the differentiated deflation procedure of the prices of the total flow matrix.

References

Aitken, B. and Harrison, A., 'Are there spillovers from foreign direct investment? Evidence from panel data for Venezuela', *Mimeo* (Cambridge, MA and Washington, D.C.: MIT and World Bank, 1991).

Aitken, B.J., Harrison, A.E. and Lipsey, R.E., 'Wages and foreign ownership: a comparative study of Mexico, Venezuela, and the United States', *Journal of International Economics*, 40 (1996) 345–71.

Blomström, M. and Kokko, A., 'Human capital and inward FDI', *European Institute of Japanese Studies Working Paper*, 167 (January 2003a).

Blomström, M. and Kokko, A., 'The economics of foreign direct investment incentives', *Bundesbank Conference Working Paper* 168 (January 2003b).

Boothroyd, G., Dewhurst, P. and Knight, W., *Product Design for Manufacture and Assembly* (New York: M. Dekker, 1994).

Borensztein, E., De Gregorio, J. and Lee, J.W., 'How does foreign direct investment affect economic growth?', *NBER Working Paper* No. W5057 (1995).

Brand, S., Hill, S. and Munday, M., 'Assessing the impacts of foreign manufacturing on regional economies: The cases of Wales, Scotland and the West Midlands', *Regional Studies*, 34 (2000) 343–55.

Braunerhjelm, P. and Svensson, R., 'Host country characteristics and agglomeration in foreign direct investment', *Applied Economics*, 28(7) (1996) 833–40.

Clark, J.P., Roth, R. and Field III, F.R., *Techno-Economic Issues in Materials Selection. Materials Selection and Design* (Materials Park, OH: ASM International, 1997).

Correia, A.D., 'A AutoEuropa: um modelo de produção pós-fordista', *Análise Social*, 35 (2000) 739–82.

Costa, T., 'Análise do impacto do IDE do sector automóvel na economia portuguesa: aplicação de um modelo combinado ao caso de estudo Autoeuropa', Unpublished Masters thesis (Lisbon: Technical University of Lisbon, 2005).

Crone, M. and Roper, S., *Knowledge Transfers from Multinational Plants in Northern Ireland: Local Learning in the Supply Chain*, NIERC Report No. 15 (Belfast: Northern Ireland Economic Research Centre, 1999).

Dunning, J., *Multinational Enterprises and the Global Economy* (Reading: Addison-Wesley, 1993).

Forsyth, D., *U.S. Investment in Scotland* (New York: Praeger, 1972).

Globerman, S. and Shapiro, D.M., 'The impact of government policies on foreign direct investment: the Canadian experience,' *Journal of International Business Studies*, 30(3) (1999) 513–32.

Görg, H. and Ruane, F., 'An analysis of backward linkages in Irish electronics', *Economic and Social Review*, 31(3) (2000) 215–35.

Guo, D. and Hewings, G., *Comparative Analysis of China's Economic Structures Between 1987 and 1997: An Input–Output Prospective* (REAL 01-T-04, <http://www2.uiuc.edu/unit/real/d-paper/01-t-4.pdf> May, 2001).

Guterres, C., 'Desenvolvimento da indústria de componentes para automóvel em Portugal: estudo da interdependência entre as indústrias de fabricação automóvel e de componentes em Portugal', Unpublished Masters thesis (Lisbon: Technical University of Lisbon, 1996).

Haddad, M. and Harrison, A., 'Are there positive spillovers from direct foreign investment? Evidence from panel data for Morocco', *Journal of Development Economics*, 42 (1993) 51–74.

Hewings, G., 'New developments in input–output modelling lectures notes manual, short course', *5th Edition Summer School Instituto Superior de Economia e Gestão* (Lisbon, July 2003).

Hirschman, A., *The Strategy of Economic Development* (New Haven: Yale University Press, 1958).

Humphreys, K.K. and Wellman, P., *Basic Cost Engineering*, 3rd edn (New York: Mercel Dekker, Inc. 1996).

Katz, J.M., *Production Functions, Foreign Investment and Growth* (Amsterdam: North Holland, 1969).

Kirchain, R.E. and Clark, J.P., 'Process based cost modeling: understanding the economics of technical decisions', *Encyclopedia of Materials Science* (Oxford: Elsevier, 2001).

Kokko, A., 'Technology, market characteristics, and spillovers', *Journal of Development Economics*, 4 (1994) 279–93.

Leontief, W., *Input–Output Economics*, 2nd edn (Oxford University Press, 1986).

Lever W.F., 'Manufacturing linkages and the search for suppliers and markets', in Hamilton, F.E.I. (ed.), *Spatial Perspectives on Industrial Organisation and Decision Making* (London: Wiley, 1974) pp. 309–34.

Lima, M.P., Pires, M.L., Rodrigues M.E. and Duarte T., 'Organização da indústria automóvel na península de Setúbal', *Análise Social*, 31 (1996) 1117–82.

Lobo, A. and Melo, M., *O Automóvel: Um Cluster (Globalmente) Inovador* (Lisbon: ME/GEPE, 2002).

Lopes, J.C.D., and Amaral, J.F., 'External dependency and structural change: The Portuguese case', Paper prepared for the 14th International Conference on Input-Output Techniques (Montreal, Canada, UECE–ISEG–UTL, 10–15 October 2002).

Maia, J.N., 'Investimento directo estrangeiro e gestão industrial na periferia: as indústrias automóvel e siderúrgica no caso Português', Unpublished Ph.D. thesis (Lisbon: Instituto Superior de Ciências do Trabalho e da Empresa, Lisbon University, 1999).

Markusen, J. and Venables, A., 'Foreign direct investment as a catalyst for industrial development', *European Economic Review*, 43 (1999) 335–56.

Mazolla, F. and Bruni, S., 'The role of linkages in firm performance: evidence from southern Italy', *Journal of Economic Behavior & Organization*, 43 (2000) 199–221.

McCann, P., 'How deeply embedded is Silicon Glen? A cautionary note', *Regional Studies*, 31 (7) (1997) 697–705.

de Mello Jr., L.R., 'Foreign direct investment in developing countries and growth: a selective survey', *Journal of Development Studies*, 34(1) (1997) 1–34.
Miller, R.E. and Blair, P.D., *Input–Output Analysis: Foundations and Extensions* (Englewood Cliffs, N.J.: Prentice-Hall, 1985).
Moran, T.H., *Foreign Direct Investment and Development: The New Policy Agenda for Developing Countries and Economies in Transition* (Washington D.C.: Institute for International Economics, 1998).
Morris, J. 'Flexible internationalization in the electronics industry: implications for regional economies', *Environment and Planning C*, 10 (1992) 407–21.
Neves, A.O., 'Avaliação ex-ante do impacte de grandes projectos sobre o desenvolvimento local: um contributo metodológico a propósito do projecto Ford/VW', *Sociologia-Problemas e Práticas*, 22 (1996) 43–59.
O'Connor, R.E. and Henry, D.W., *Análise Input-Output e suas Aplicações* (Lisbon: Biblioteca de Estudos Económicos, 1974).
Rasmussen, P.N., *Relaciones Intersectorales* (Aguilar, 1963).
Rodríguez-Clare, A. 'Multinationals, linkages, and economic development', *American Economic Review*, 86(4) (1996) 852–73.
Rodrik, D., 'Coordination failures and government policy: a model with applications to East Asia and Eastern Europe', *Journal of International Economics*, 40(1–2) (1996) 1–22.
Romer, P., 'Increasing returns and long run growth', *Journal of Political Economy*, 94(5) (1986) 1002–37.
Santos, R.G., 'Os efeitos na economia portuguesa do investimento directo estrangeiro no sector automovel: do projecto Renault ao projecto Auto Europa', Unpublished Masters thesis (Lisbon: Technical University of Lisbon, 1996).
Selada, C., Felizardo, J.R. and Féria, L.P., *A Integração das Infra-Estruturas Tecnológicas na Rede de Excelência para o Desenvolvimento da Indústria Automóvel em Portugal: Uma Metodologia de Avaliação* (Lisbon : ME/GEPE, 1999).
Simões, V., *Efeitos do Investimento Directo Estrangeiro sobre a Modernização do Tecido Produtivo Nacional: O Caso da Indústria Automóvel*, (Lisboa: CISEP/ISEG Final Report, June 2000).
Sjöholm, F., 'Productivity growth in Indonesia: The role of regional characteristics and direct foreign investment', *Economic Development and Cultural Change*, 47(3) (1999a) 559–84.
Sjöholm, F., 'Technology gap, competition and spillovers from direct foreign investment: evidence from establishment data', *Journal of Development Studies*, 36(1) (1999b) 53–73.
Stewart, J.C., 'Linkages and foreign direct investment', *Regional Studies*, 10 (1976) 245–58.
Turok, I., 'Inward investment and local linkages: how deeply embedded is "Silicon Glen" ', *Regional Studies*, 27 (1993) 401–17.
Vale, M. 'Geografia da Indústria Automóvel num contexto de globalização – Imbricação Espacial do Sistema Autoeuropa', Unpublished Ph.D. thesis (Lisbon: Technical University of Lisbon, 1999).
Valente, F.M., 'A inovação tecnológica no sector de componentes para a indústria automóvel : estudos de casos', Unpublished Masters Thesis (Lisbon: Technical University of Lisbon, 1999).
Veloso, F.M., 'Local Content Requirements and Industrial Development Economic Analysis and Cost Modeling of the Automotive Supply Chain Policy and Management', Unpublished Ph.D. thesis (Cambridge, MA: MIT, 2001).

14

International Buyer–Supplier Relationships, Transfer of Knowledge and Local Suppliers' Capability

Celeste Amorim Varum

14.1 Introduction

One of the primary motivations for attracting foreign direct investment (FDI) is to obtain advanced technology and knowledge and then, based on this, to establish domestic innovation capability. As is well recognized in the literature, there are several important channels through which inward FDI can benefit capabilities of domestic firms in the host country. Of particular relevance are spillovers that may take place vertically from foreign firms to their local suppliers by means of technology transfer, staff training and so forth. These vertical spillovers can then enhance the capabilities of local suppliers. In a context of global production networks, they act as a catalyst for international knowledge diffusion, as mediators of local capability formation (Ernst and Kim, 2002).

The automotive industry in particular is often cited as an example in terms of globalization and buyer–supplier (B–S) relationships (Florida and Kenney, 1991; Dyer, 1996; Langfield-Smith and Greenwood, 1998; Okada, 2004). Rich B–S relationships have been found to benefit both the automotive buyer and the suppliers in improved product quality, productivity and cost reductions (Dyer, 1996) and key to promoting skills upgrading among domestic firms (Okada, 2004). Relationships with international automotive buyers represent a window of opportunity for domestic suppliers established in more peripheral economies. It may allow them to access the technologies and best practices of large international corporations renowned for their expertise, technology and techniques.

Virtually all recent literature shows that network forms of organization are conducive to inter-firm knowledge and skill development and transfer (Hakansson and Johansson, 2001). Yet, few studies examined which mechanisms within networks promote knowledge and skill development and transfer between firms (Okada, 2004). Research on informal knowledge transfer and on the importance of local capabilities is scarce (Ernst and Kim, 2002). Moreover, we need research that also explores the implications for suppliers, especially those from less favored regions (Ernst and Kim, 2002). The focus on component suppliers, for example, is unusual in the literature, which has mainly focused the perspective of auto assemblers.

Using the case of the automotive industry, this chapter examines B–S linkages as a key arrangement inducing innovation and skills development among domestic firms in the age of globalization. Instead of looking at the transfer of technology, the chapter looks at the development of capabilities at organizational and managerial level, and skills of workers in domestic firms, issues overlooked in the literature. The study involves firms located in Portugal and Spain.

The Portuguese and Spanish automotive industries provide an interesting focus for studying this topic. First, the automotive industry is often associated with a well-developed B–S relation on an international basis. Second, at least at a conceptual level, a number of techniques and practices, such as lean production and quality models, became highly popular (Walgenbach and Beck, 2000) in the automotive sector world-wide. Building on the ISO 9000 norms, international automotive assemblers developed their own industry-group specific quality systems (e.g. VDA 6.1, QS 9000), which are imposed on their direct and indirect suppliers world-wide. Third, the automotive industry in both economies gained increasing relevance in the last decades. Against this background, there is no systematic study on the role of international B–S relations for the upgrading of local suppliers' capabilities in these economies.

The next section analyses international B–S relationships and capability development of suppliers. The empirical part of the chapter provides evidence from the suppliers' perspective on the main transformations occurring at domestic supplier level, and how international buyers contributed to upgrade their techniques and practices. The last section summarizes the findings and concludes with policy implications.

14.2 B–S relations, upgrading capabilities of local suppliers

It is generally agreed that vertical networks, which involve close co-operation, promote incremental learning (von Hippel, 1988; Lam, 1997; Simonin, 1999). Another strand of the literature highlights the role of lead firms, often multinationals, in governing the supply chain by setting and enforcing parameters on suppliers in less favored regions and thereby upgrading their organizational capabilities (Humphrey and Schmitz, 2001).

Multinationals (MNEs) have an interest in domestic suppliers' practices and the power to influence them (Lincoln et al., 1998; Brennan and Turnbull, 1999; Javorcik, 2002). Regarding the automotive industry, buyers interfere with suppliers' practices by imposing, or strongly suggesting, a number of standards, lean production and quality techniques. Suppliers are further pressed to conform through auditing of operational routines (Okada, 2004).[1] Because the norms and buyers' requirements apply equally to all automotive suppliers, the process is likely to enforce standardization of practices across firms while promoting capacity upgrading in suppliers. In addition to stimulating change, buyers can also transfer knowledge to suppliers, technical and managerial skills, so that they can meet specifications to compete. Knowledge transfer is effective only when internalized and translated into capability of local suppliers. Much depends on the types of knowledge involved, the mechanisms used to transfer different types of knowledge, suppliers' knowledge base and efforts, and external contextual variables (Cummings and Teng, 2003).

Regarding managerial knowledge, Armbrüster and Kipping (2003) suggest distinguishing the following types of knowledge:

- *General knowledge*: concepts applicable to a wide and diverse range of organizations, such as total quality management (TQM), re-engineering;
- *Regulation-oriented*: more in-depth knowledge about particular management areas such as, IT, or a specific industry, such as automotive, telecommunications;
- *Change knowledge*: operational knowledge about how to carry out certain processes, e.g. how to analyse and change a company, how to implement a new quality system, an IT system, and so on.

While the first two types of knowledge are more explicit, capable of codification and therefore of communication, operational knowledge is mainly of experimental nature. Its transfer is slow and costly, its acquisition requiring careful interpersonal contact and extensive learning-by-doing. In line with this, Cowan and Foray (1997) argue that codified knowledge does not provide all the knowledge to implement an innovation. Companies need an operational type of knowledge, which is non-codifiable (Cohen and Levinthal, 1990). In terms of change knowledge, companies themselves obviously also have it, but this is 'routine' type knowledge; that is, it is carried out repeatedly and only evolves very slowly (Nelson and Winter, 1982). International buyers can provide such knowledge to a certain extent. Large international firms in particular developed processes which they apply to change the existing routines of suppliers through 'suppliers' support offices' or with interns who work temporarily with their suppliers (Dyer and Singh, 1998; Lincoln et al., 1998; Okada, 2004).

Companies may acquire knowledge externally, but learning is imperfect (Cohen and Levinthal, 1990). Firms are repositories of knowledge, which both determines what it makes, and the directions in which it searches (Rosenberg, 1982). Hence, firms with accumulated competencies and expertise related to the innovation in question are likely to show better ability in exploiting them. Firms are said to suffer from 'natural' inertia, mainly due to the predominance of their own 'routine' type of knowledge. The ability to overcome organizational inertia depends considerably on the resources available (capital and human), on management commitment and motivations (Nadler and Tushman, 1989; Kanter, 1993). Benson et al. (1991) and Rosett and Rosett (1999) for example, found these to be crucial on the adoption of TQM.

It is widely accepted that the national environment, which comprises political and legal ground rules, shapes the behavior of firms (Dosi and Malerba, 1996; Edquist, 1997; Lundvall and Borrás, 1997). National government programs have stimulated worker participation, for example, in Denmark, Sweden and Germany (Lundvall and Borrás, 1997; Benders et al., 1999), and also stimulated the diffusion of quality management practices in Japan (Cole, 1985). Firms' capabilities are also linked to the wider set of institutions in a country. They reside not only in firms' own know-how, but also in the strengths of its institutional relationships with customers, suppliers, or sources of know-how, a network of infrastructural organizations including knowledge producing and disseminating structures such as research and professional associations and consultants. An undeveloped supply of knowledge-based organizations may be a barrier for the diffusion of new practices in an economy.

Features of the labor market are also a determinant. Lorenz and Lazaric (2000) for example, found that owing to career and training specificities, Japanese multinationals experienced partial success on the transfer of Japanese employment practices to Britain and France. Strambach (2002) argued that differences in labor relations and labor market conditions partially explain the patterns and timings of diffusion of a number of organizational innovations in the UK and Germany in the 1990s.

The country effect is even more central when one considers the transfers of innovations created and developed in central countries into peripheral catching-up economies which face an increasing competitive environment, but also quite dissimilar institutional environments. Studies on the transfer of technologies into catching-up economies (Cantwell, 1992; Buesa and Molero, 1998) provide rich examples on how the import of technologies was constrained by the lack of supportive business services. More recently, results from the international Benders et al. (1999) program clearly documents international differences in Europe between northern core developed economies (Denmark, Sweden and Germany) and southern catching-up countries (Portugal, Spain, Italy). Differences in national traditions in training, industrial relations, labor market regulations, among others, are possible explanations for the difference (Lundvall and Borrás, 1997). Thus, we predict considerable country effects on the diffusion of practices and knowledge from international buyers to their suppliers located in catching-up economies.

14.3 Transfer of knowledge and capability development among Portuguese and Spanish automotive suppliers

14.3.1 Methodology

This chapter focuses on the upgrading of capabilities and managerial practices of automotive suppliers located in Portugal and Spain. The complexity of the issue is better examined with data of qualitative type. In these cases, and considering the intrinsic nature of the process of learning, the case study approach is a suitable methodology (Dosi, 1991). Instead of statistical representativeness, case studies offer depth and comprehensiveness for understanding the specific phenomenon, enabling inductive and rich description (Halinen and Törnroos, 2005). The data were analysed according to the practices recommended by Miles and Huberman (1994), with a variety of graphical and tabular formats for structuring the analyses. It uses a combination of interviews, internal documentation and business press to document the transformations at the companies.

We covered four case studies of automotive suppliers, two in each country. These companies have reported to have implemented a number of innovations in a broad questionnaire survey (Amorim, 2002). Table 14.1 provides additional information on the cases. All sites are small and medium firms producing automotive components. Firms differ at the level of the product share accounted for by this industry and their position within the value-chain. Automobile accounts for less than 50 per cent of C2, C1 turnover. Two firms (C2, C3) are second-tier suppliers.

14.3.2 Transformation in suppliers

This section analyses the changes at suppliers. Recent and rapid transformations in the automotive industry led suppliers to improve their production capabilities markedly, acutely aware of the need to raise quality. Changes in two areas deserve particular relevance: work organization, and skill development of workers.

Table 14.2 shows the adoption of new work organization, with emphasis on teamwork and quality consciousness. Suppliers introduced selected lean production techniques, such as QC circles, SPC, suggesting schemes, rotation, ISO certifications and other quality schemes.[2]

The transformations in the industry also led to changes in the patterns of workers' skill acquisition. All suppliers rely more on formal training, both standardized vocational training at specialized centers and formal training in-house. Training focuses mainly on quality improvement and problem-solving to improve production capabilities. One main motivation for introducing such training activities is firms' need to obtain ISO 9000. Since the mid-1990s they have conducted training programs specially designed for ISO certification. Occasionally firms use direct interactions and exchanges with their clients to provide their employees with on-site exposure to new practices.

Table 14.1 Main characteristics of the cases

Code	Number of interviews	Year of establishment	Ownership	Main product	Employees	Turnover €	Core clients within automotive	% sales to automotive	Supplier
C1	2	1943	Spanish	Metal parts	208	20 718.000	Seat, Renault, Ford, VW	Nearly 50	1st
C2	3	1926	Portuguese	Paint	452	70 112.160	Salvador Caetano–Toyota	<50	2nd
C3	2	1924	Spanish	Electric	200	22 620.000	Osram, Philips, GE	100	2nd
C4	3	1988	Portuguese	Seats	275	26 950.430	Keiper Recaro, TRW, Lear, Chrysler	100	1st

Source: Companies' annual reports and interviews.

Table 14.2 Main transformations at domestic suppliers

Practices introduced	**C1**	**C2**	**C3**	**C4**
TQM	1993	1996	1992	1998
Model of TQM	1998	1996(X)		
ISO9001	1997	1993	1999	1997
ISO9002		1992	1996	
QS9000	1998			2000
VDA 6.1				
EAQF	1995			
Q1 Ford	1983			
Relation with suppliers	✓	✓	✓	✓
Suppliers' evaluation index	1995	1992	1999	1994
Suppliers' assistance service	1995			
Joint development		1991		
Auditing				No
Measuring client level satisfaction	✓	✓		✓
Index	1997	1993		✓
Surveys	1998	X		
Meetings/Joint development				✓
Employees' involvement and motivation	✓	✓	✓	✓
Team work	✓	✓	✓	✓
Keizen		✓		1993
Project teams	✓		✓	1993
Problem-solving teams		✓	✓	
Multi-divisional teams	✓			
Suggestion box	1998			
Surveys	✓			
Performance-contingency payments	✓			
JIT		✓		
Kanban		X		X
Brainstorming	✓			
Benchmarking			1990	
Autocontrol	1986		1983	✓
Quality dev. function	✓			
Internal audits	✓	✓	✓	✓
Conducted by internals		✓		
With externals	✓	✓		✓
Taguchi		X		
Cause–effect diagrams	✓		✓	
Processes checklists /flowcharts	1999		1999	
FMEAs	✓		1999	1991/2
NQC		X		1994
SPC	✓		1983	✓
Pareto	✓		✓	1994
Control letters and plans			1999	✓
PPMs	✓		✓	✓
Capacity studies	✓		✓	

TQM: Total Quality Management; ISO: International Standard Organization; SPC: Statistical process control; VDA: Verband der Automobilindustrie, EAQF: Evaluation d'Aptitude Qualité Fournisseurs; JIT: Just-in-time; FMEAs: Failure mode and effect analysis; NQC: Non-quality costs; PPMs: Defective parts per million.

✓: Implemented. When available, the date of implementation is provided.

X: Used in the past but not currently.

Source: Case study analysis.

Behind this global picture, cross-firm differentiation exists: (i) different combination of practices; (ii) same practices but different meanings; and (iii) decoupling, that is, only formal adoption. It is possible to identify two main groups (see Figure 14.1). One is formed by firm C1 only, a well-established Spanish producer of metal parts. C1 developed and achieved a broad management program and presents the clearest evidence of having achieved higher infusion of lean production and total quality techniques. Its total quality program goes beyond implementation of simple compulsory QAS practices and adopted the EFQM model.[3]

The second group is formed by the other three firms. They achieved the quality certificates imposed by their clients, adopted a set of techniques and practices, but they do not have a broad framework within which the last are a part. As shown in Table 14.2, only C1 reported following a standardized broad approach to quality management. When asked for their background literature, guides and frameworks, only C1 using the EFQM model was able to provide something.

C2 is a paint manufacturer and distributor. In the early 1990s, it initiated a systematic program of quality management. As a result, in 1992 C2 received the ISO certificate, being the first Portuguese paint company to get the certificate. After the ISO 9002 process, the company moved into the implementation of the ISO 9001, achieving the certification in 1993. They further attempted, without success, to implement a continuous improvement model and to win a prize for excellence – the *Prémio de Excelência em Portugal* (PEX).

C3 is a well-established Spanish company. In the early 1990s it embarked on an ambitious process toward total quality. The final aim was the QS 9000 certification and the ISO certification was part of the process. Despite starting in 1992 they got the ISO-9002 certification only in 1996 and the ISO 9001 in 1999.

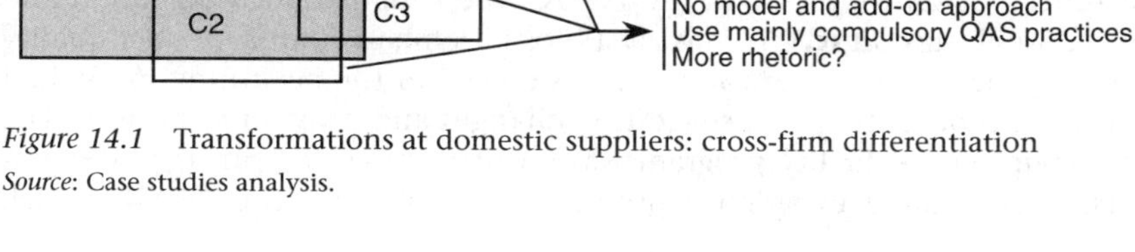

Figure 14.1 Transformations at domestic suppliers: cross-firm differentiation
Source: Case studies analysis.

C4 is a relatively young firm established in 1988. It soon gained a reputation in the market supplying diverse automotive assemblers. The development of its capabilities and the achievement of diverse quality certificates contributed to the success.

For firms in group 2, it is questionable to what extent the changes will actually lead to a modification of existing routines in the long-run. These firms often are revealed to have adopted a number of changes only formally, and when this was the case, the firms' processes and internal routines remained practically unchanged. For example, in the first instance C2 and C3 highlighted the major turnaround that had occurred in 'company culture'. However, later in the interview they also expressed considerable frustration and recognized that in fact 'the organizational practices changed very little. The quality system brought mainly formalization of the things we were already doing'. C4 manager made a similar remark. Managers at C1 started by arguing how things had changed little, but brought to light numerous changes that had occurred with the broadening and deepening of firm quality program.

The impact of the changes stays in a balance between the unclear impact over the company's costs and product quality and positive standardization benefits. One of the advantages of the quality systems refers to the upgrading of workers' skills and to the process of standardization of internal routines. Quality management programs helped to standardize practices, and to 'depersonalize' processes. Writing procedures helped to externalize tacit knowledge previously embedded in people and hierarchical positions. The certificates were also considered a protection against non-certified firms, helping to improve the image of the company.

14.3.3 B–S relations and demand-driven transformations

Mode of B–S relations

The four case studies indicate an industry characterized by a 'horizontal' structure, where motor vehicle manufacturers share the same suppliers (Helper and Sako, 1995). Firms adopted two main strategies: C1 and C4 of diversification, moving from a situation of reduced number of clients to a broad base of clients; C2 and C3 of client exclusivity.

Since the 1960s the C1 quality system developed following clients' own quality requirements. Seat and Ford were the main clients, and the main catalysts for change. Industry specific characteristics led Ford to increasing collaboration with its suppliers, reflected, for example, on an active 'suppliers' technical assistance office' in the 1990s. Ford supplied documentation and operational support to C1. In early 1990s C1 embarked in a broader quality (TQM) program. Ford contribution was crucial in the beginning, providing the TQM model, auditing, suggesting changes and providing training. The follow-up of the quality program was ensured by visits from Ford's engineers. Accumulated experience prompted C1 to adopt the EFQM model, implemented with the help of consultants. C1 also strengthened its presence in the French market through an International Alliance (IA) with a French firm offering complementary products. There is a joint 'Quality Committee' which discusses with the French customers (e.g. Renault) quality improvements plans yearly. Nowadays, a large base of clients is a good source of knowledge for further development. Exchanges occur through continuous and informal contacts, visits, and with the participation of foreign clients in internal development/quality teams.

As for C1, C4 quality system developed initially under the pressure and suggestions from its main client. In 1991 Keiper Recaro classified it with a C (with reserves), but in 1992, it was already classified already with an A by Lear. Since then, C4 has a preferential relation with Lear, collaborating on the development of new products. The knowledge acquired was translated into reputation, leading to broadening of the client base. C4 also had to implement several quality systems. Yet, there is no underlying frame bringing them all together. Similarly, there is no systematic mechanism to exchange knowledge with buyers. The process is mainly informal and document based.

C2 reflects the case of an exclusive relation with a large automotive buyer, SC, which holds a joint venture with Toyota. By early 1990s, SC initiated a 'supplier development program' to help a number of Portuguese suppliers to get the ISO 9000 certification. They worked together for one and a half years on designing new practices. As a result, C3 was indeed the first Portuguese paint company getting the ISO-9002. The contacts with SC are limited and more on an informal basis. Buyer support has been limited. The implementation and post-development of the system relied mainly on internal capabilities.

C3 reflects a different experience. The firm evolved from a diversified base to a preferential relation with one buyer. Buyers impacted mainly through quality manuals and post-implementation audits. Know-how sharing occurred also through visits to clients' plants. However, there has never been exchange of personnel nor training. In 1997–98 C3 became OSRAM 'preferential supplier'. In the past there were no major direct mechanisms to exchange information with OSRAM, but since then the flow of information with this firm has increased.

Demand-driven transformations and buyers as source of knowledge

In all cases, buyers played a leading role on the development of suppliers' practices. Buyers were a catalyst for change by imposing, or strongly encouraging, lean production techniques and practices, and quality systems. Suppliers were further pressed to conform through inspection by the buyer or accredited institution at the level of operational routines. Buyers had a further impact by transferring knowledge and less often by getting actively involved with the implementation of new practices.

Categories of knowledge and knowledge transfer mechanisms

The analysis revealed the role of international buyers in complementing the skills of suppliers (1) by familiarizing them with general quality concepts and ideas; (2) by providing them with additional know-how applied to quality programs in the automotive industry; and (3) with skills of the change type to carry-out the implementation.

The international customers provided suppliers with general knowledge and 'best practices' or 'benchmarks' based on collective knowledge of the automotive industry and its specific quality programs (Table 14.3). Buyers transferred knowledge to suppliers through various mechanisms. As shown in Table 14.3 and in Figure 14.2, demonstrations, seminars and detailed manuals, quality systems and improvement plans are largely used. Specific knowledge was also provided through training, but this has been limited. Three of the firms reported to have received such support. C1 mainly from Ford, C4 from Lear and C2 from a large Portuguese group working with Toyota. However, such knowledge is mainly of codified type.

Informal personnel contacts, namely between firms' quality managers remained an important mechanism for the transfer of more operational know-how. The transfer of tacit knowledge has been very limited (Figure 14.2). The horizontal structure, where motor vehicle manufacturers share the same suppliers, might explain reduced buyer involvement on the transfer of tacit knowledge. The benefits from supplier development will also be enjoyed by other buyers. In this regard, the Japanese 'vertical' model of B–S relations is said to create better conditions for knowledge transfer (Helper and Sako, 1995). The limited support at procedural level might explain firms' frequent use of external consultants on the implementation of quality innovations. On a few occasions firms used consultants that had previously worked for their buyers or for other MNEs operating in the same business as their clients.

Table 14.3 Knowledge transfer mechanisms

Knowledge		Transmission mechanism	
Category	**Level**	**Formal**	**Informal**
Explicit	General	Seminars; demonstrations	Personal contacts
	Specific	Manuals, seminars, demonstrations, training	
Tacit	Change	On-the-job; personnel exchanges	

Source: Case studies analysis.

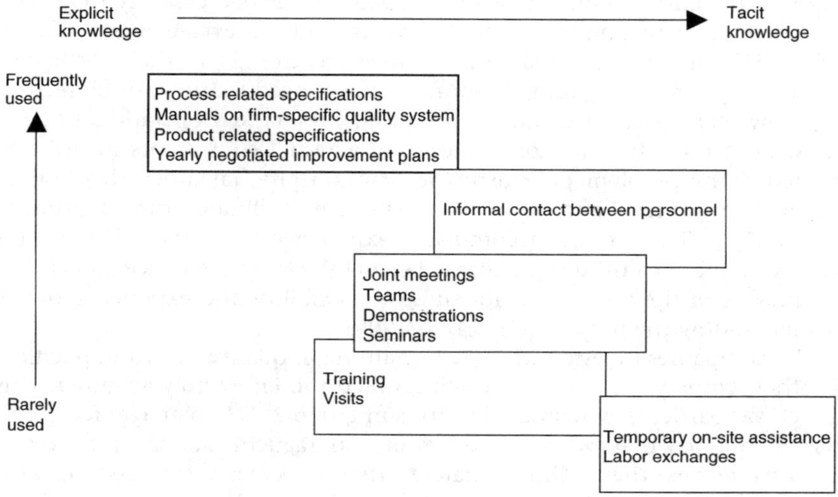

Figure 14.2 Main mechanisms to transfer knowledge from buyer to supplier
Source: Case studies analysis.

Despite being limited, the tacit knowledge received from the international buyers was highly regarded by suppliers. Overall, suppliers highlighted the specific and change-related knowledge received from buyers. Important also is the role of B–S relationship of having access to up-to-date and on-time information. The frameworks and requirements demanded by international buyers were, at least, considered as good starting points to build up supplier quality systems. Its main advantage consisted of being a mechanism to eliminating managers' uncertainty about which direction to take, and about buyers' own expectations.

For all suppliers, buyers were a source of external knowledge, but the management of the relation is complex and there seems to be limited inter-personal interaction. The mix of clients also creates problems for suppliers as they are forced to implement an overload of models and systems, sometimes overlapping, others confrontational. The existence of a broad quality program to frame all these models is an advantage (e.g. C1). Another set of problems emerged due to the internationality of the relationship: this has lead to language barriers (especially with shopfloor workers), and costly investments to maintain the link.

14.3.4 Supplier's absorptive capacity

Local suppliers can only effectively absorb knowledge and implement it if two elements are fulfilled: a knowledge base of competence, and intense effort and commitment. While some organizations had capacity to absorb, interpret, circulate and utilize the knowledge, others revealed great difficulties. All companies reported to have invested strongly in their absorptive capacity by directly sending personnel for advanced training, or by recruiting new managers, for instance. Yet, often they lacked operational know-how. Even though some companies used external consultants in order to overcome this problem, pre-existing relevant operational knowledge, both at the managerial level and at the employee level, facilitated the adoption of innovations. The internal accumulated experience made the difference for C1: 'we achieved a third stage characterized by active acceptance and enthusiasm: one of the secrets was for sure the flexibility and experience of the people leading the project' (C1, 14/06/1999).

The companies invested strongly in training in quality tools and practices for their employees, but these often ignored the know-how acquired. This effect was particularly relevant for firms in group 2. C2's manager for example, mentioned that 'we received training in Taguchi, Kanban practices ... but we rarely use them'. Unfortunately, other firms (e.g. C3) shared the same experience. Such problems were particularly strong when training had been given by external experts and it had been mainly of theoretical type. Employees continued lacking operational skills to take on the new practices. A more efficient alternative seems to have been followed by C1 through temporary exchanges of personnel, which gave its employees opportunity to learn with suppliers on-the-job.

The result confirms Nelson and Winter's (1982) argument that firms suffer from 'natural' inertia. Organizational inertia emerged not only because existing routines were difficult to change but also because organization members found it difficult to integrate them with existing practices.

This effect was more clearly identified in group 2 firms. C3, for example, began considering ISO by 1992 but it was not until 1996 that they received the ISO 9002 certificate. Three more years passed before achieving the ISO9001 certificate. In this regard, the quality manager and CEO highlighted that 'organizational culture' slowed down the standardization of the internal quality system (C3, 16/7/1999). C1 by contrast managed to overcome organizational inertia reaching deep organizational change. The interviewees at C1 recognized that before achieving the present system, and while working with external experts, the firm went through a major 'cultural revolution' subdivided in three stages: 'the first occurred when we introduced autocontrol per department and reduced the emphasis on control carried out by the quality department. The second occurred when we finally changed the concept of the quality department (already in the 1990s). Meanwhile, we had been convincing people that everybody was responsible for quality. However, there was some internal reaction because people did not want to change the way they were doing things. They used to argue that "quality is for the quality department". After, we went through an intermediate stage characterized by passive acceptance, and now we've reached a third stage – active acceptance and enthusiasm.'

Top-management commitment was determinant in overcoming inertia. C1's top managers were actively involved with the implementation of new quality practices, whereas in the other firms the lack of management's commitment to the implementation was obvious. Often, in the interviewees' statements it was implicit that new practices were seen as new ways for others to work. Top managers would define teams, set up their aims and reward systems but they would not participate in the implementation. C2's CEO's words clearly reveal his detachment toward quality related programs: 'ISO is a fashion ... the certification itself is bullshit. It consists in writing up what was already being done ...' (C2, 23/03/1999).

This behavior partly explains why C2 attempts to embark on a broad quality program did not go far. C4, by contrast, seems to have learned by its own mistakes. In 1994–95 the lack of management commitment seriously affected the progression of its quality program. As mentioned by the consultant working for C4 (and confirmed by C4's internal managers): 'I gave training but things would not be applied mainly because there was lack of involvement and commitment from the top managers.'

Despite the support received from buyers and consultants, C4's quality manager argued that 'I was too involved in other issues and let the time pass'. As a result, practices demanding higher commitment and enforcement were dropped. Since then things changed considerably, and the quality program gained a new boost. Administrators took the lead in the ISO process and nowadays they are actively involved in the teams. C4 decided to develop its program relying mainly on internal resources, which has been a slow, step-by-step process. C3 provides another example of how management positioning delayed the changing process. It took more than six years for C3 to achieve the ISO 9001. '[t]he company was growing at a fast pace (in five years it doubled the production and sales) and as a result there was no time to structure internal processes' (C3, 16/06/1999).

14.3.5 Country specificity

A detailed comparison across firms reveals that the transformations in suppliers were also shaped by policies and industry development reflected in terms of labor markets and development of business support industries.

Both governments have used external mechanisms through which certain practices became more attractive for managers. In this regard, and especially after the EC membership in 1986, Iberian governments have been using a plethora of incentives and grants in an attempt to increase firms' competitive advantages or/and to help firms' restructuring processes. The Portuguese and Spanish funding policies have now gone through two cycles. While in a first stage (1988–94) incentives were mainly directed to upgrading firms' equipment and production technology, in a second stage (1994–99) qualitative innovations were given preference (PEDIP, 1999). Human resources training, implementation of quality systems (ISO system), organizational issues, marketing and design, internationalization processes and environmental actions became the central focus of the industrial policy of both countries. In the Portuguese case, in the period 1994–99, more than 299.8 million euros were allocated to more than 1880 projects related to quality (PEDIP, 1999, p. 8). In the Spanish case, several programs with similar aims were set up. The program FORCEM, which funded training, was often mentioned by interviewees.

These mechanisms were an important catalyst for change, but also distorted the economic rationality behind the adoption of certain practices and interfered with the range of alternatives considered by decision makers. Our results are in line with Santos et al. (1999), which revealed that public funding and awards, clients' pressures, other incentives and, finally, the search for higher competitiveness were the most important factors explaining the acceleration of the quality movement in Portugal.

Other country specificities shaped the transformation of domestic suppliers. In the 1980–90s, Spanish automotive suppliers benefited to a greater extent from the existence of a pool of skilled workers and specialized business services, technological centers and professional associations. Portuguese companies by contrast could scarcely find local sources of specialized know-how, and often they had to search for it abroad. In this context, the support received from MNE clients became a crucial source of industry-specific know-how. The Portuguese established firm C4 highlighted that fact. This effect was also confirmed in other interviews with foreign automotive suppliers established in Portugal (Amorim, 2002).

In this regard, it is safe to argue that the automotive industry developed earlier and more fully in Spain than in Portugal. This might explain, to a certain point, the Portuguese 'backwardness' at the level of development of industry-specific business services. This result is in line with Kostova's (1997) findings that the Portuguese institutional profile was less supportive for the implementation of TQM practices than the Spanish one.

14.4 Conclusion

B–S relations are crucial to spur FDI spillovers. Moreover, most of these relationships occur within global production networks, creating new opportunities for international knowledge diffusion that suppliers should strive to exploit. We argue that B–S linkages motivated and forced domestic suppliers to develop their practices and capabilities. Three positive effects of the B–S relationship were clearly identified. Buyers acted as a catalyst in promoting learning among domestic suppliers by setting and enforcing strict performance standards. Suppliers increasingly must produce components based on standards provided by global assemblers. The need to improve products and processes has motivated suppliers to upgrade their workers' skills, confirming the important role of leading multinational corporations in shaping patterns of skills development and governing the performance of suppliers (e.g. Humphrey and Schmitz, 2001). Moreover, through certification schemes and other standardized procedures, it can be argued that modes of knowledge diffusion become more explicit, standardized and codified (Okada, 2004). This facilitates the process of accumulating and sharing common knowledge intra- and inter-firm.

The study also identified critical aspects of the B–S relationship. First, the transformation occurring at each firm is a complex multi-level process, thus contradicting predictions that management practices and tools are easily transferred between dissimilar environments. The implementation of new knowledge was often barred and, in some cases, reported changes appeared to 'have failed', i.e. did not actually lead to modification of existing routines. Most changes may be more rhetoric than real, as in a program that exhorts people to alter their behavior, but the ultimate interest is the achievement of a quality certificate. When investing in implementing buyer-specific requirements and models without a clear broader managerial strategy, suppliers might end up in a 'certification trap'. Not only are the internal benefits reduced, but also its external benefits are reducing as quality certificates become widely diffused. However, they are still a necessary condition to be in the business. Thus, it is up to the supplier to make the most of these circumstances. One choice is to develop broader and more efficient TQM frameworks as C1 did; another is to lobby at least for the harmonization of such procedures across different buyers. Second, the 'horizontal structure' identified creates co-ordination problems, both in terms of implementing different standards, and in terms of managing several relationships with different clients. Finally, and more critical, is the limited involvement of suppliers regarding the transfer of tacit knowledge. Buyers' approach to supplier development is a crucial factor, and suppliers must be more active and insist on getting more from them.

The study offers implications for policy in terms of upgrading industrial skills in the age of globalization. Increasing trade and industrial liberalization makes market forces assume greater roles. In this context, is there a role for policy-makers in this regard? How can policy-makers promote inter-firm learning linkages? One way is to work with assemblers to achieve high levels of local content. As we have seen, the development of the industry depends on the capacity of firms to meet international standards, which in turn depend on continuous investment on the upgrading of workers' capabilities. The development of such capabilities is determinant so that domestic suppliers remain competitive. Finally, there should be awareness of financial incentives as a means of promoting certain practices. As we discussed above, governments directed firms' innovative paths by using policies favoring innovations related to quality management. The evidence shows that these policies may suffer from important shortcomings that need to be addressed. Financial incentives must be complemented with a set of conditions facilitating the access to sources of know-how and firms' capacity to use and implement such knowledge. Within the latter we include, for example, the availability of complementary skills, information and intermediate inputs and capital goods. The lack of local specialized sources of know-how was more evident in Portugal. This is particularly important considering that domestic firms are the ones more affected by the national contextual variables. Secondly, there is the need to ensure that firms are actually changing routines and not adopting the practices merely formally, for the purposes of accessing to subsidies, and/or formal certificates. After-care policies and controlling post-subsidization are necessary to ensure that such 'innovations' go beyond formal adoption, and lead to effective change of routines and mentalities.

Notes

1 The automotive industry is just an example to demonstrate this trend. Organizations across many sectors are subject to growing external reviews of their quality practices, as well as increased incentives for participating in external evaluations (Powell, 1995).

2 The results seem to confirm the argument that from the original statistical ideas of Deming and Juran, TQM incorporated less technical elements with unclear meanings, and expanded into a diffuse, increasingly unclear concept (Hackman and Wageman, 1995).

3 EFQM model stands for the model created by the European Foundation for the Quality Management. In brief, the EFQM model it is based on the idea that customer satisfaction and impact on society are achieved through leadership, driving policy and strategy, people management, resources and processes. The EFQM is the base for the European Quality Award, which has been run annually since 1992.

References

Amorim, C., 'Diffusion of innovation in countries in the periphery-core transition', Unpublished Ph.D. thesis (Reading: University of Reading, 2002).

Armbrüster, T. and Kipping, M., 'Types of knowledge and the client-consultant interaction', in Sahlin-Andersson, K. and Engwall, L. (eds), *The Expansion of Management Knowledge: Carriers, Ideas and Circulation* (Stanford: Stanford University Press, 2003).

Benders, J., Huijgen, F., Pekruhl, U. and O'Kelly, K.P., *Useful but Unused – Group Work in Europe. Findings from the EPOC Survey* (Luxembourg: European Foundation for the Improvement of Living and Working Conditions, 1999).

Benson, P.G., Saraph, J.V. and Schroeder, R.G., 'The effects of organizational context on quality management: an empirical investigation', *Management Science,* 37(9) (1991) 1107–24.

Brennan, R. and Turnbull, P.W., 'Adaptive behavior in buyer–supplier relationships', *Industrial Marketing Management*, 28 (1999) 481–95.

Buesa, M. and Molero, J., *Economia Industrial de Espana: Organizacion, Tecnologia e Internacionalizacion* (Madrid: Biblioteca Civitas Economia y Empresa, Coleccion Economia, Editorial Civitas, 1998).

Cantwell, J. (ed.), *Multinational Investment in Modern Europe: Strategic Interaction in the Integrated Community* (Aldershot: Edward Elgar, 1992).

Cohen, W. and Levinthal, D., 'Absorptive capacity, a new perspective on learning and innovation', *Administrative Science Quarterly*, 35 (1990) 128–52.

Cole, R.E., 'The macropolitics of organisational change: a comparative analysis of the spread of small-group activities', *Administrative Science Quarterly*, 30(4) (1985) 560–85.

Cowan, R. and Foray, D., 'The economics of codification and the diffusion of knowledge', *Industrial and Corporate Change*, 6(3) (1997) 595–622.

Cummings, J.L. and Teng, B.-S., 'Transferring R&D knowledge: the key factors affecting knowledge transfer success', *Journal of Engineering and Technology Management*, 20 (2003) 39–68.

Dosi, G., 'The research of innovation diffusion: an assessment', in Nakicenovic, N. and Grübler, A. (eds), *Diffusion of Technologies and Social Behaviour* (Berlin, Heidelberg, NY: Springer-Verlag, 1991).

Dosi, G. and Malerba, F. (eds), *Organization and Strategy in the Evolution of the Enterprize* (London: Macmillan Press, 1996).

Dyer, J.H., 'Specialized supplier networks as a source of competitive advantages: evidence from the auto industry', *Strategic Management Journal*, 17 (1996) 271–91.

Dyer, J.H. and Singh, H., 'The relational view: cooperative strategy and sources of interorganizational competitive advantage', *Academy of Management Review*, 23(4) (1998) 660–79.

Edquist, C. (ed.), *Systems of Innovation. Technologies, Institutions and Organisations* (London: Pinter, 1997).

Ernst, D. and Kim, L., 'Global production networks, knowledge diffusion, and local capability formation', *Research Policy*, 31 (2002) 1417–29.

Florida, R. and Kenney, M., 'Transplanted organizations: the transfer of Japanese industrial organization to the U.S.', *American Sociological Review*, 56 (1991) 381–98.

Hackman, J.R. and Wageman, R., 'Total quality management: empirical, conceptual, and practical issues', *Administrative Science Quarterly*, 40 (1995) 309–42.

Hakansson, H. and Johansson, J., *Business Network Learning* (Pergamon Press: Netherlands, 2001).

Halinen, A. and Törnroos, J.-A., 'Using case study methods in the study of contemporary business networks', *Journal of Business Research*, 58 (2005) 1285–97.

Helper, S. and Sako, M., 'Supplier relations in Japan and in the United States: are they converging?', *Sloan Management Review*, Summer (1995) 15–28.

Humphrey, J. and Schmitz, H., 'Governance in global value chains', *IDS Bulletin*, 32(3) (2001) 19–29.

Javorcik, B.K., 'Spillovers of foreign direct investment through backward linkages: does technological gap matter?', *World Bank Working Paper*, 2002.

Kanter, R.M., *The Change Masters* (New York: Simon and Schuster, 1993).

Kostova, T., 'Country institutional profiles: concepts and measurements', Paper Presented at the *Academy of Management Proceedings*, 1997.

Lam, A., 'Embedded firms, embedded knowledge: problems of collaboration and knowledge transfer in global cooperative ventures', *Organization Studies*, 18(6) (1997) 973–96.

Langfield-Smith, K. and Greenwood, M.R., 'Developing co-operative buyer–supplier relationships: a case study of Toyota', *Journal of Management Studies*, 35(3) (1998) 331–53.

Lincoln, J.R., Ahmadjan, C.L. and Mason, E., 'Organizational learning and purchase–supply relations in Japan: Hitachi, Matshushita and Toyota compared', *California Management Review*, 40(3) (1998) 241–64.
Lorenz, E. and Lazaric, N., 'The transferability of business practices and problem-solving skills to Japanese firms in Britain and France', Paper presented at *DRUID Summer Conference*, Rebild, 2000.
Lundvall, B. and Borrás, S., *The Globalising Learning Economy: Implications for Innovation Policy* (Luxembourg: Office of the Official Publications of the EU, 1997).
Miles, M.B. and Huberman, M.A., *Qualitative Data Analysis: an Expanded Sourcebook*, 2nd edn (Thousand Oaks, California: Sage Publications, 1994).
Nadler, D.A. and Tushman, M.L., 'Leadership for organizational change', in Mohrman, S., Mohrman, G., Ledford, J., Cummings, T. and Lawler III & Associates (eds), *Large Scale Organizational Change* (San Francisco: Jossey-Bass, 1989).
Nelson, R.R. and Winter, S., *An Evolutionary Theory of Economic Change* (Cambridge, MA: Harvard University Press, 1982).
Okada, A., 'Skills development and interfirm learning linkages under globalization: lessons from the Indian automobile industry', *World Development*, 32(7) (2004) 1265–88.
PEDIP, 'O PEDIP II e a indústria Portuguesa', *Info-Pedip* (Ministério da Economia, Portugal, 1999).
Powell, T.C., 'Total quality management as competitive advantage: a review and empirical study', *Strategic Management Journal*, 16 (1995) 15–37.
Rosenberg, N., *Inside the Black Box* (Cambridge: Cambridge University Press, 1982).
Rosett, J.G. and Rosett, R.N., 'Characteristics of TQM: evidence from the RIT/USA today quality cup competition', *NBER Working Paper* Series 7241, 1999.
Santos, G.M., Pais, M.S. and Machado, V.C., 'Efeitos potenciadores da certificação dos sistemas da qualidade nas empresas Portuguesas', *Ingenium – Engenharia e Finanças*, January (1999) 92–4.
Simonin, B.L., 'Transfer of marketing know-how in international strategic alliances: an empirical investigation of the role and antecedents of knowledge ambiguity', *Journal of International Business Studies*, 30(3) (1999) 463–90.
Strambach, S., 'Organisational innovation in different systems of innovation – the case of Germany and UK', in Schätz, L. and Revilla-Diez, J. (eds), *Technological Change and Regional Development in Europe* (Heidelberg, NY: Springer Physica Verlag, 2002).
von Hippel, E., *The Sources of Innovation* (Oxford: Oxford University Press, 1988).
Walgenbach, P. and Beck, N., 'From statistical quality control, quality systems to TQM: The institutionalization of a new management approach', Paper presented at 16th EGOS Colloquium, 2–4 July, Helsinki, 2000.

Part V

Policy Issues

15

Multinational Subsidiaries and Innovation Policy in Central and Eastern Europe: Alignment and Autonomy

Katrin Männik and Nick von Tunzelmann

15.1 Introduction

The countries of Central and Eastern Europe (CEE) are in the process of emerging from a difficult process of transition out of a rather inward-looking socialism, toward fuller integration into the global economy. The countries considered in this study all became members of the EU during 2004. This implies new sets of responsibilities for autonomous development, alongside and in parallel with the extent of harmonization with EU activities (Hannula et al., 2006). The transition process was, in innovation terms, largely headed by an inrush of multinational companies (MNCs), but the question arises of how far these can be relied upon to continue being the main drivers of innovation into the new stage of development. In order to stand on their own feet, and develop autonomous and indigenous mainsprings for growth, CEE countries have to absorb technologies generated elsewhere into a wide range of their productive practices, and to begin to inaugurate (or relaunch) up-to-date technologies of their own. This carries strong implications for the autonomy of the subsidiaries of the MNCs, which are the main focus of this chapter. The study is therefore concerned with the interaction between autonomy at the national/regional/local level and autonomy between MNC parent companies and their subsidiaries. This issue is central to the perspective of 'alignment', which we have developed in other work on these and other countries (summarized in McGowan et al., 2004). These perspectives of reconciling autonomy with alignment constitute the basis for the policy implications drawn in the final section of the chapter.

15.2 Transition and the role of MNCs

Central planning before the 1990s involved top-down, centralized structures of decision-making, in which the autonomy offered to enterprises was very limited in terms of the technologies they could develop, the marketing they could undertake, the regional coherence of production, and so forth. Although the centralized model operated more weakly in some of these countries, like Slovenia (in former Yugoslavia) and Hungary, the transition process thus involved dramatic change not just in ownership but in resource, functional, and spatial terms.

The responses to the new conditions have had to come from both internal sources and from abroad via MNCs. The governments for the most part saw their task as a financial one, of privatizing the productive establishments (and those for other functions), and hence devolving all responsibilities to them. They believed they had enough to do with coping with new levels of administrative autonomy. As the enormity of changes required began to become more evident, foreign partners were encouraged to enter and then were increasingly welcomed. Initially seen as tapping into new sources of finance in the privatization process, the extent to which foreign models could supply both frameworks and resources for technology, organization, marketing and management became steadily more apparent. Thus began a new phase of contestation, between the new-found desire for local and regional independence and the realities of superior knowledge resting in the MNCs and in their home countries. In this context, the power of global market forces led to seemingly inexorable drives to replace the early shoots of indigenous operation with corporations experienced in what it took to survive and prosper in such a world.

The MNCs supplied ways of integrating functional, resource and spatial elements into a more cohesive whole that were clearly far in advance of what domestic capabilities, and still more histories of central planning, could offer. Bringing in organizational systems, technology, managerial and other skills as well as deep pockets of finance from their home countries, the MNCs often felt compelled to bypass local set-ups in a variety of ways. This process did not however come without cost. As the literatures on international business and on innovation studies have shown extensively (e.g. Zanfei, 2000), MNCs are themselves caught between centrifugal and centripetal tendencies. Although the problem is similar to that of the central planners of old, they have been much more successful in reconciling the conflicting needs, owing partly to their vastly stronger global market orientation, and partly to the more integrated character of their functional systems.

Initially, the motivations of the MNCs entering CEE markets tended to be either cost-seeking (taking advantage of supposedly cheap labor) or market-seeking (though many of these new national or regional markets were quite small). Neither offered great security to these countries. Cost-seeking could easily turn itself into a 'race to the bottom', where countries would bid against one another to attract MNCs through cheaper labor costs together with attractive (and often expensive) state support outlays, both of which delayed the gain in living standards, perhaps indefinitely. In the case of market-seeking, the smallness of the markets put indigenous activities at risk.

In the last analysis, what the countries needed was upgrading, in the sense of Ernst (1998) and other scholars, rather than simply cost-cutting or market-monopolizing. In the longer term this could be to the advantage of the MNC as well, though the costs of upgrading were likely to discourage immediate attempts to bring this about. Upgrading was attained by bringing in management, skills and technology from the western headquarters, but to the detriment of local resources in these spheres, as already noted. In the area of technology, which we have especially investigated, many MNCs dismissed out of hand the capacities of local research institutes or other bodies to equip them with technologies geared to local market conditions that could have been integrated into an upgrading process (Yoruk, 2004). No doubt in many circumstances this dismissal of local capacities was justified, in view of the inheritance of central planning and the lack of market orientation or even productive application that went with such a system; yet there were many local R&D establishments that – at least in conditions of market- rather than cost-seeking MNCs – could have been much better deployed. The MNCs needed to adjust to local product and factor markets on a centrifugal basis, and this could have been an effective means for so doing.

15.3 Network alignment and autonomy

From the side of the host country the limited extent of the integrating–upgrading process entailed much more than keeping some of its research workers employed. Upgrading was a basic condition for longer-term viability of development processes in the countries concerned. While a revolutionary overthrow of existing practices is possible, progress normally takes the form of blending between existing strengths with new strengths, and using the latter as well as possible to overcome existing weaknesses.

The MNC subsidiary is almost by definition interposed between two (or more) different spatial levels: global markets (for technology, etc., as well as products) on the one side, and local/regional/national resources and constraints on the other. This provides us with the context for assessing the extent of 'autonomy' in the actions of the subsidiary vis-à-vis its headquarters (HQ) in a more developed country abroad. On one side are the centripetal forces deriving from the global strata and world frontier levels of technology, productive and organizational efficiency, marketing, finance facilities, and so on. From this viewpoint, stronger ties to HQ and weaker links with the local environment facilitate catching-up to international best-practice. The less the subsidiary engages with its parent, the more it deprives itself of these possibilities. On the other side lie the centrifugal forces of interaction with the local conditions of supplies of goods and most of the above-noted factors of production, together with local markets (especially in the case of market-seeking MNC motivations). On this score, the tie to the parent is something of a disadvantage, while commitment to local circumstances involves a larger exploitation of the possibilities from such an environment.

If we put these two conflicting considerations together we might draw the conclusion that there is a nonlinear relationship between effectiveness and autonomy. Low autonomy maximizes the global possibilities, high autonomy maximizes the local potentials, and hence some kind of intermediate position might be expected to maximize (or possibly minimize) a combination of the two. This is entirely in line with the celebrated papers of Granovetter (1973, 1985) about the 'strength of weak ties', which adds in the capacity for response of the agent caught in the middle – here the MNC subsidiary – which is fostered by its partial detachment from both sides. An associated position is that of Evans (1995) on 'embedded autonomy'; while Evans's issue was rather different, about the role of bureaucracies, it also seems applicable here, in terms of the latter point. Applied to the MNC subsidiary, it needs to be embedded in both international and local/regional networks but also partially autonomous from both. The position of the host country in terms of the benefits to be derived is then a reflection of the position of the MNC subsidiary, together with all other companies and productive units in the country. As we have recounted elsewhere, this is an issue of 'network realignment'.

15.4 Autonomy and MNC subsidiaries in CEE countries

15.4.1 Method of analysis and variables

This section, which aims to analyse the above conundrum of how to balance centripetal and centrifugal forces through trading off effectiveness and autonomy, draws mostly on empirical investigations by one of us (Männik), using a database of such subsidiaries from five CEE countries: Estonia, Hungary, Poland, Slovakia and Slovenia.[1] The completed questionnaire covers 433 firms in these countries, spread over a range of manufacturing sectors (see detailed information about the sample in Männik et al., 2004). A major objective of the broader study from which this chapter is derived is to assess the degree of control of the foreign owner over the subsidiary located in CEE. The autonomy of multinational subsidiaries in relation to subsidiary embeddedness (local, international) and performance is examined across country-, industry- and firm-specific features.

Previous studies have attempted to explain variations in subsidiary autonomy, which can be divided into MNC characteristics, subsidiary characteristics and environmental factors (see Björkman, 2003). Usually, studies of MNC characteristics look at the size of the MNC and the effect of parent nationality on the subsidiary. The results for the impact of MNC characteristics have been mixed and there is no clear consensus. On the other hand, the studies of subsidiary characteristics are richer and show a little more consistency than those of MNC characteristics. The most recent literature overview and discussion about gaps in research in this area has been by Young and Tavares (2004). Much less has been analysed concerning the impact of environmental factors on autonomy, especially the host-country role in providing opportunities for the subsidiary to develop external networks and increase autonomy through capability building. The present analysis draws attention to the multi-dimensionality of the subsidiary business functions and their distinctive influences on subsidiary performance. In addition, country-, industry- and firm-specific contexts are involved in the context of the five CEE economies.

Table 15.1 below computes the productivity (value-added per employee) levels by the five countries and four industry groups in focus. First, the results clearly point to the presence in terms of productivity of two leading countries, namely Hungary (total $US 18 993 in 2001) and Slovenia ($US 18 753) – the countries with relatively greater autonomy pre-1990. Hungarian and Slovenian firms were roughly 1.4 times more productive than firms in Poland, and compared with Estonia and to Slovakia their advantage was respectively 2.3 and 2.5 times. However, they differ from the others by surprisingly bigger dispersions of productivity levels (see the differences between low, medium-low and medium-high-tech industries). Examining the industry distinctions across all countries, the results show higher productivity among medium-high industries compared to high-tech industries. Poland is the only country where the high-tech sectors have the highest productivity ($US 20 508 in 2000).

The extent to which each function is wholly, partly or little under the sway of the multinational (scale between 1 and 0) is included in the study.

Table 15.1 Value-added per employee in the manufacturing industries of five accession countries (in $US, annually)

Industry group*	**Slovenia (2001)**	**Hungary (2001)**	**Slovakia (1999)**	**Estonia (2001)**	**Poland (2000)**
High-tech	18 849	14 750	5290	6897	20 508
Medium-high-tech	23 485	30 446	8395	10 198	13 360
Medium-low-tech	18 210	18 383	8029	9746	14 954
Low-tech	15 870	10 128	6970	7334	12 063
Total	18 993	18 753	7687	8263	13 451

Note: * according to OECD classification high-tech sectors involve the following NACE industries: 24.4, 30, 32, 33, 35.3; medium-high-tech: 24.0–24.3, 24.5–24.7, 29, 31, 34, 35.2, 35.4–35.5; medium-low-tech: 23, 25, 26, 27, 28, 35.0–35.1; low-tech: 15, 16, 17, 18, 19, 20, 21, 22, 36, 37.

Source: Authors' calculations based on UNIDO, Statistical Database 2004, <http://www.unido.org>; Statistical Office of the Republic of Slovenia 2004, Business Subjects, <http://www.stat.si/eng/tema_ekonomsko_poslovni.asp>; Eurostat 2004, Structural Indicators, <http://europa.eu.int/comm/ eurostat/newcronos/reference/display.do?screen=welcomeref&open=/&product= STRIND_ECOBAC&language=en&depth=2>.

The survey asked about the decision-making processes between the local unit and the parent company. The question asked was: Which business functions are being undertaken: a) on your own only; (b) mainly on your own; (c) mainly by your foreign owner; or (d) by your foreign owner only? Answers were received about 13 business functions (all originally considered functions are presented in Table 15.2). Responses range between 0 and 1 of which the first denotes complete subsidiary autonomy and the latter one absolute dependence on HQ.

In the latter case, 13 different functions are too many to distinguish in subsequent analysis so they are collapsed into four summary functions using principal components analysis. This provides the factors for foreign determination of sales and marketing – a new variable called FACTMARK; production and technology – FACTTECH; management – FACTMAN; and finance – FACTFIN (see Table 15.2). The variable FACTMARK predominantly integrates 6 out of 13 business functions such as determining product price, market research, distribution and sales, after-sales services, and advertising marketing. It is followed by FACTTECH, which mainly combines the functions of product development, process engineering, supply and logistics. FACTMAN primarily consists of operational and strategic management, and finally the variable FACTFIN accounting, financial operations and investment finance. The inclusion of all variables in new factor components was also justified by the correlation matrix of the 13 variables across five countries.

In the present chapter the multidimensionality of the subsidiary autonomy (given by the summary factor variables FACTTECH, FACTMARK,

Table 15.2 Rotated component matrix*

Business functions	F1 FACTMARK	F2 FACTTECH	F3 FACTMAN	F4 FACTFIN
Product development	0.381	0.769	0.165	0.003
Process engineering	0.003	0.865	0.220	0.115
Determining product price	0.657	0.395	0.243	0.179
Supply and logistics	0.381	0.518	0.153	0.400
Accounting and financial operations	0.136	0.008	0.009	0.903
Investment finance	0.234	0.168	0.383	0.545
Market research	0.877	0.169	0.138	0.005
Distribution, sales	0.868	0.118	0.007	0.187
After-sales services	0.836	0.120	0.008	0.138
Advertising	0.875	0.152	0.215	0.144
Marketing	0.866	0.153	0.237	0.142
Operational management	0.007	0.248	0.794	0.259
Strategic management or planning	0.382	0.187	0.783	0.006

Note: * The method of extraction of the Rotated Matrix was Varimax.

FACTMAN, FACTFIN) is interrelated with some performance indicators characterizing the technology upgrading of subsidiaries and exports as a final output. Because the data involve limited dependent variables, ordered regression methods are then used to scale those dependent variables, which include change in productivity, in technology level of production equipment, in level of product quality, and in export share from output since the registration of a firm as a foreign investment company.[2] Answers of the firms are scaled between 0 and 1 (considerable reduction ... considerable increase). Expecting nonlinearity in relations between the subsidiary autonomy and performance we have also included squared values for four types of autonomy ($FACTTECH^2$, $FACTMARK^2$, $FACTMAN^2$, $FACTFIN^2$) in a second model of the regression analysis.

Dummy variables are entered for the individual country and for sector, as well as for ownership, size, age of firm and type of product. Industries are grouped on the basis of the OECD classification at the 3-digit NACE level (see note to Table 15.1). By size, firms are divided into small and medium-size enterprises (below 250 employees) and large enterprises (250 and more). Foreign involvement is distinguished according to whether the basis is minority (below 50 per cent) or majority (equal to or above 50 per cent). The year of establishment of the subsidiary is seen as having taken place either before or during 1990, between 1991–95, or from 1996 onwards. Firms are seen as producing only intermediate goods, only final products, or both intermediate and final products. In the ordered regression analysis the determination of so-called benchmarking variables is needed. Currently, Estonia, the low-tech industry group, and the firms with majority foreign ownership, large, established after 1996 and producing both intermediate and final products are taken as benchmarking variables against the others (see further analysis and Appendix).

15.4.2 Results of ordered regression models

We build on the hypothesis established in section 15.3 that too little autonomy (high control by the foreign owner) and too little control (high autonomy of the local subsidiary) are both likely to be 'inferior' outcomes, and the relationship with the performance variables may be posited to be of an inverted U-shape. This relationship may however vary from one function to another, and moreover a proper U-shape is possible, implying that one or both extremes are 'better' (full autonomy and/or full control may give superior results to a 'halfway house' in certain circumstances).

We run four ordered regression models separately for each performance indicator as dependent variables (see table with regression results in the Appendix). In the second regression model there are included four new dependent variables with squared values for all four types of subsidiary autonomy. The distinctions in the relationships between autonomy and performance are represented by different dimensions of autonomy. Finally the country-, industry- and firm-specific characteristics will be discussed.

We find *technology autonomy* (FACTTECH) to have the expected relationship with the technology performance of multinational subsidiaries. If we take into account the country-, industry- and firm-specific characteristics then support can be seen for the hypothesis assuming the presence of a non-linear relationship between effectiveness and autonomy of the subsidiary. Technology autonomy has a certain intermediate value for achieving a maximum level of performance in a combination of both local and international corporate-specific knowledge. Statistically significant results for the autonomy in technology issue are present in Model 2, which upholds the existence of an inverse U-shape relationship between autonomy and changes in productivity, technology level and product quality. This does mean that neither high nor low technology autonomy of subsidiaries represents the best solution for technology upgrading in the observed transition countries. Neither the local technology potential nor the global possibilities are maximized, but rather some intermediate position has been developed to maximize the subsidiaries' performance. Calculating these middle values for technology autonomy we get three different results: 0.423 is the level of the autonomy for achieving the best performance in productivity, 0.529 for changes in technology production, and 0.510 for increasing the product quality, where complete autonomy equals 0 and no autonomy equals 1.[3] Thus, despite the low expectations in terms of technology and innovation potential in catching-up economies, foreign firms have still been using the local sources of knowledge, while in the meantime opening up corporate knowledge and practices to the local units.

Similarly to technology autonomy, *marketing autonomy* (FACTMARK) has a strong influence on the performance of multinational subsidiaries, either from estimating changes in technology upgrading or from export shares over roughly the last two decades in CEE economies. High marketing autonomy negatively affects the technological upgrading of the company (implementation of new technology, quality improvements) and also hinders exports. For example, the results from Model 1 in the Appendix indicate that for an increase of the FACTMARK by one unit (which means a reduction of marketing autonomy), the technology level grows by 1.160 units. Hence there is a positive impact of lower marketing autonomy on the level of technology of production equipment. Low marketing autonomy is also associated with positive changes in product quality and export share of the turnover. This seems to imply that subsidiaries with high marketing autonomy were not supported by their parent firms to create firm-specific competencies, especially technological competencies. Only for autonomy in marketing has its high level hindered the growth in the technology level of subsidiaries. Furthermore, subsidiaries with high marketing power were mainly oriented toward the domestic market, which in turn constrained the development of links with foreign countries and impeded additional access to international knowledge flows. Such might be the case of foreign subsidiaries in Poland

and Hungary where domestic markets are much bigger in comparison with countries like Estonia or Slovenia.[4] Too much domestic marketing orientation in the case of transition countries does not seem the best solution for the catching-up process.

The opposite results are received for *financing autonomy* (FACTFIN), particularly examining the relationships with changes in level of product quality and export shares. We can reasonably conclude that, the greater the autonomy in financing, the greater the improvement experienced in the quality level of products. This indicates that higher financial autonomy reflects a subsidiary that has already obtained a solid level of production technology combined with strong improvements in the quality of products. It is inevitable that marketing and financial autonomy are somewhat reciprocal to each other. Low marketing autonomy induces positive changes in the technology competency-building of foreign subsidiaries. Subsidiaries with high marketing autonomy have limited resources for technology upgrading and they are presumably involved in the production of relatively low-tech products. From another side, high autonomy in financing is an advantage for capacity growth, both through increasing product quality and from more active export performance.

Moreover, the degree of control is related to other determining variables, such as the level of per capita GDP of the country (as a proxy for the stage of development). Subsidiaries from the more developed CEE countries, Slovenia and Hungary, had the highest scores for autonomy.[5]

Country differences do not affect the results obtained in the case of productivity and production equipment. But in Slovenia and Hungary improvement in the quality of products was significantly weaker compared to Estonia (e.g. estimates of −1.239 in Model 1, −1.305 in Model 2 for Slovenia in the Appendix). This difference between Slovenia and Estonia as well as between Hungary and Estonia could be explained by differences in the development level of the countries reflected in the different starting position of subsidiaries as foreign-owned firms. In Slovenia and Hungary the general economic development level and also productivity level were much higher than in Estonia in the early and mid-1990s. There is also a smaller change in export orientation in subsidiaries in Poland and in Hungary in comparison with Estonia. The latter result supports the idea of different influences on subsidiary performance depending on the size of domestic market and FDI orientation. Relative to the Polish and Hungarian subsidiaries, Estonian rivals are really more foreign than domestically oriented.

What of the analysis of *sector differences*? Subsidiaries in the high-tech sector have a significantly smaller contribution to the growth of productivity (estimate −1.054 in Model 1, −1.276 in Model 2), technology level (−0.846, −0.970) and level of product quality (−0.831, −0.866) compared to low-tech industries as the benchmark sector of analysis. It can already be seen from the previous analysis that high-tech industries in CEE countries are less productive (measured by value-added, see Table 15.1) compared with medium-high, medium-low or even low-tech industries (except in Poland). The result from the ordered regression model supports this fact.

On the contrary, compared to low-tech or even other medium-tech sectors, subsidiaries in medium-low-tech sectors have experienced higher improvement in production technology (estimates 0.501 in Model 1, 0.536 in Model 2) and in share of exports (0.581, 0.592). The result is surprising from the perspective of an expected greater technological intensity in high-tech sectors in comparison with low-tech sectors. Again, this could be explained by the much lower level of production technology in the low and medium-high-tech industries, which could therefore achieve much more rapid growth in the technology of production over time. But it may also indicate that foreign firms have used much more new technology in the production of low and middle-tech products, reaping benefits of factor cost differentials in labor costs. Or external links of subsidiaries in the technology areas could be relatively weak. The inflow of technological knowledge from the mother company is weak because it is taken to be a misuse of resources to invest in these subsidiaries in the environment of under-developed national innovation systems.

The relationship between autonomy and *firm size* differed across the five countries. Large firms seem to have superior productivity capabilities, as well as *majority-owned foreign subsidiaries, those firms registering as a foreign investment enterprise after 1996*, and *producing only intermediate or only final products* (see precise results in Appendix). Therefore, from the perspective of the host economy, the establishment of new foreign investment enterprises should be stimulated, as well as those larger in their numbers of employees and foreign involvement. According to expectations, large foreign subsidiaries have also contributed more to the level of production technology and product quality improvements as well as to the growth in export orientation compared to small and medium-size firms.

Following Rugman and Verbeke (2001), the aim of subsidiary development is to build subsidiary-specific advantages which may involve production-related assets like technological, productive or marketing know-how, or be associated with the organizational capability to coordinate and control efficiently the MNC's asset base. It is important for the multinational to determine the proper combination of centralization and autonomy under which foreign subsidiaries could maximize their value-creating roles (Hewett et al., 2003). The stronger the embeddedness of the subsidiary within its external relationships via local demand, sourcing and links with the local system of innovation, the lower is supposed to be the control by the parent firm, and vice versa for stronger corporate links (Andersson and Forsgren, 1996). As already observed in earlier sections, neither too much subsidiary autonomy nor too little autonomy seems likely to be efficient in the case of transition countries. The main aim is to use the capacity of foreign subsidiaries in an optimal way, taking into account the context where the subsidiary is located. On the basis of the present analysis we can confirm that the relationship between the subsidiary autonomy and the performance in fact differs across business functions depending on the capacities provided by countries and/or foreign/domestic firms.

15.5 Conclusions and policy implications

The driving forces of innovation for economic development in CEE countries have changed radically as a result of the transition from socialism. The old state-based apparatus for innovation has been bypassed or closed down, while foreign MNCs have swept in to take over much of the responsibility for introducing new technologies.

Though the arrival of western (and some Asian) multinationals has led to a steep improvement in productivity and performance in many areas, it has rather inadvertently led to a new form of imbalance. The socialist pattern was one of a national system that was strong in terms of its power, but weakly connected to international systems and only moderately connected to local and regional innovation systems within the countries concerned. The transition pattern has been instead one of powerful links to the international system, modest but growing connections to local systems, and the crumbling of the national systems that lie between the international and the local systems. The national level has become what we have referred to elsewhere as the 'weakest link' in CEE innovation systems (McGowan et al., 2004). This comes at a cost to their prospects for continuing economic progress. In contrast with the general pattern in the 'Asian tigers' of harnessing international forces via MNCs to indigenous development efforts, in the CEE countries they have tended to become dominant elements in the structure.

In our previous work, we have outlined the need for a new alignment of the overall innovation systems in the CEE countries, in place of the lurch from one misaligned system based on the centrally planned state to a new misaligned system dependent on foreign multinationals. In this chapter we have studied in detail the role of subsidiaries of foreign MNCs in CEE countries at the 'functional' level. For our empirical work we used a database drawn from a questionnaire circulated in five CEE countries: Estonia, Hungary, Poland, Slovakia and Slovenia. The completed questionnaire covers 433 firms in the five countries, spread over a range of manufacturing sectors.

The results support the basic idea concerning the heterogeneity of autonomy. The relationship between the subsidiary autonomy and its efficiency differs across functions. Autonomy in marketing and financing appear as reciprocal processes that influence technology upgrading and export performance in CEE countries. The low level of marketing autonomy and on the other side the high level of financing autonomy give positive impulses for technology improvement and export orientation. High marketing autonomy is therefore signaling that the subsidiary is not in a strong position to create subsidiary-specific competencies, nor to benefit from access to complementary foreign knowledge transferred through exports. Larger domestic markets (Poland, Hungary) might become rather disadvantageous for international knowledge flows. Autonomy in technology issues appears to have different 'rules' in terms of subsidiary performance compared with other types of functional autonomy. Having neither too much autonomy from nor too much dependence on the parent company positively affects the increase in productivity, the level of technology and product quality. Some intermediate position of technological autonomy satisfies the parent company for achieving the maximum performance in the subsidiary in such countries. We demonstrate that the precise trade-off depends on which objective is under consideration (maximum productivity growth, increase in product quality, etc.), although the outcomes are similar for all the objectives discussed. The subsidiary is using both the local potential and international sources for technology upgrading. From the perspective of international knowledge flows the subsidiary locating in technologically under-developed transition countries maintains deep corporate links and this does imply the presence of an additional knowledge inflow. From another side the subsidiary exploits the local knowledge sources, which in turn are expected to develop local innovation potential.

Moreover, the degree of control by the parent company is also country-, industry- and firm-dependent. In general, majority-owned, large subsidiaries in medium-low-tech sectors, which were established later than 1996 and producing only intermediate or only final products have achieved more extensive positive effects in regard to the performance of subsidiaries. FDI in these economies is largely oriented to medium-tech sectors, which also provide the majority of value-added today. In fact, this might indicate that foreign firms have used more new technology in the production of low- and medium-tech products, or that parent companies develop more sophisticated technology in their home or in other advanced economies. In terms of country distinctions Estonia, which is one of the smallest by local market, one of the less developed among the examined CEE countries and one receiving foreign investments later than others, has achieved more significant effects in export orientation compared with Poland and Hungary, and in the improvement of the level of product quality compared with Slovenia and Hungary.

At the subsidiary level, it is the role of managers to combine different areas of autonomy to gain maximally from the relation with the parent company located in a more advanced country. In the case of transition countries it might be appropriate to have lower autonomy in fields of shortages of specific knowledge (for example, in technology). Excessive dependence on the parent company might impede the development of its own absorptive capacity while excessive independence might leave the local unit in a circle of 'internationally uncompetitive' knowledge.

In terms of innovation policy, we had expected to find that the level of autonomy granted to the CEE subsidiaries is too low to be 'optimal', even though too much autonomy may also be undesirable (for instance, too little engagement with foreign technologies). Host nations have to consider the interactions among the firms that constitute them. Just as firms have to promote dynamic capabilities in production and suppliers in technology, so policy-makers at regional or national level have to promote dynamic capabilities in respect of the institutional and economic environment in which the firms can best flourish.

The transition economy evidently has to 'buy' high technology, advanced processes and mission-oriented science to a large degree, in the sense of acquiring them externally. However blending in external advances will be very much internally inspired, even if external agents are brought in to try to smooth the process. This raises the issue of the absorptive capacity of the region in question. The seminal studies of absorptive capacity by Cohen and Levinthal (1989, 1990) showed that its extent is highly dependent on activity carried out by the actor in question. Thus, as they demonstrated, much R&D carried out by firms is not intended to extend the technological frontier so much as to increase the firm's awareness of the vicinity below the frontier – a vicinity that other firms may be occupying but where the firm in question is still feeling its way. The same applies to a (regenerating) region, implying that it will need to carry out a fair amount of research just to absorb developments originating in other places. That is, even before the blending process proper begins, the region must commit itself to expenditures that represent 'make' in the domain of technology. The pay-off is that without these expenditures the blending is likely to be much more problematic and perhaps undoable.

The converse, it may be pointed out, applies to an external agent seeking to export its technology (processes, etc.) to the nation in question. The MNC will have undertaken the necessary expenditures already to develop the technology in its headquarter country; it then faces the reverse type of blending by which it seeks to adapt its accumulated expertise to a new spatial context. It is usually less awkward to change the new environment to suit the accumulated knowledge of the MNC than the opposite. The MNC is therefore likely to carry out modest local research – often in the form of product development and modification – while maintaining its core knowledge base as is. These adjustments may or may not be in the best interests of the region involved – the problem of reconciling autonomy with alignment.

Notes

1 The questionnaire was carried out under the EU 5th Framework Program project HPSE-CT-2001–00065 'EU Integration and the Prospects for Catch-Up Development in Central and Eastern European countries: The Determinants of the Productivity Gap' (see website <http://www.iwh-halle.de/projects/productivity-gap/default.htm>).

The analysis was prepared with financial support received from the Estonian Science Foundation, Grant 6493.

2 E.g. productivity growth or improvement in technology production is ranked as follows: 0 (considerable decline), 0.25 (decline), 0.5 (no change), 0.75 (growth), 1 (high growth).

3 These values are calculated as the roots of a quadratic equation formulated from the coefficients of each regression with squared terms.

4 See Männik et al. (2004) for a thorough analysis of the distinctions about interrelations between country-, industry- and firm-specific variables with the degree of autonomy, where multivariate methods (ANOVA, MANOVA) were used.

5 See Männik et al. (2004). The results for control variables presently in the Appendix are given in the form of a benchmark indicator (for example, Estonia compared to the other four countries).

References

Andersson, U. and Forsgren, M., 'Subsidiary embeddedness and control in the multinational corporation', *International Business Review*, 5(5) (1996) 487–508.

Björkman, A., 'Subsidiary power and autonomy', 29th Annual EIBA Conference (Copenhagen, December 2003).

Cohen, W.M. and Levinthal, D.A., 'Innovation and learning: the two faces of R&D', *Economic Journal*, 99 (September) (1989) 569–96.

Cohen, W.M. and Levinthal, D.A., 'Absorptive capacity: a new perspective on learning and innovation', *Administrative Science Quarterly*, 35 (1990) 128–52.

Ernst, D., 'Catching-up, crisis and truncated upgrading: evolutionary aspects of technological learning in East Asia's electronic industry', *DRUID Working Paper* no. 98–16 (1998).

Eurostat 2004, Structural Indicators, <http://europa.eu.int/comm/eurostat/newcronos/reference/display.do?screen=welcomeref&open=/&product=STRIND_ECOBAC&language=en&depth=2>.

Evans, P.B., *Embedded Autonomy: States and Industrial Transformation* (Princeton NJ: Princeton University Press, 1995).

Granovetter, M., 'The strength of weak ties', *American Journal of Sociology*, 78(6) (1973) 1360–80.

Granovetter, M., 'Economic action and social structure: the problem of embeddedness', *American Journal of Sociology*, 91(November) (1985) 481–510.

Hannula, H., Radosevic, S. and von Tunzelmann, N. (eds), *Estonia, the New EU Economy: Building a Baltic Miracle?* (Aldershot: Ashgate, 2006).

Hewett, K., Roth, M.S. and Roth, K., 'Conditions influencing headquarters and foreign subsidiary roles in marketing activities and their effects on performance', *Journal of International Business Studies*, 34 (2003) 567–85.

Männik, K., Hannula, H. and Varblane, U., 'Country, industry and firm size effects on foreign subsidiary strategy: an example of five CEE countries', *University of Tartu, Faculty of Economics and Business Administration Working Paper Series*, no. 27 (Tartu, 2004).

McGowan, F., Radosevic, S. and von Tunzelmann, N. (eds), *The Emerging Industrial Structure of the Wider Europe* (London: Routledge, 2004).

Rugman, A.M. and Verbeke, A., 'Subsidiary specific advantages in multinational enterprises', *Strategic Management Journal*, 22(3) (2001) 237–50.

Statistical Office of the Republic of Slovenia 2004, Business Subjects, <http://www.stat.si/eng/tema_ekonomsko_poslovni.asp>.

UNIDO, Statistical Database 2004, <http://www.unido.org/regions.cfm?TY=R&RID=04>.

Yoruk, D.E., 'Patterns of industrial upgrading in the clothing industry in Poland and Romania', in McGowan et al. (2004), *op. cit.*, pp. 95–110.

Young, S. and Tavares, A.T., 'Centralization and autonomy: back to the future', *International Business Review*, 13 (2004) 215–38.

Zanfei, A., 'Transnational firms and the changing organisation of innovative activities', *Cambridge Journal of Economics*, 24 (2000) 515–42.

Appendix

Table A.15.1 Results of ordered regressions

Independent variables	Change in level of productivity in production		Change in level of technology of production equipment		Change in level of product quality		Change in share of exports from output	
	Model 1	Model 2	Model 1	Model 2	Model 1	Model 2	Model 1	Model 2
FACTTECH	−9.49E-02	3.944**	0.767	4.786**	0.553	4.385**	−0.838	0.628
FACTMARK	0.789	−0.517	1.160**	−0.896	1.273**	1.322	2.186**	3.352**
FACTMAN	−8.29E-02	1.544	−0.781	−0.932	−0.766	−2.655	6.126E-02	−0.466
FACTFIN	−0.457	0.445	−0.747	1.163	−1.782**	−1.858	−1.162*	8.243E-02
FACTTECH^2		−4.667**		−4.522**		−4.302**		−1.744
FACTMARK^2		1.390		2.260		9.353E-02		−1.274
FACTMAN^2		−1.996		−4.38E-02		1.738		0.452
FACTFIN^2		−1.803		−3.028		−0.132		−2.376
SLOVENIA	0.130	−0.186	−0.588	−0.617	−1.239**	−1.305**	−0.142	−0.174
POLAND	9.713E-02	8.667E-02	9.283E-02	0.132	4.859E-02	0.101	−0.858**	−0.878**
HUNGARY	0.255	0.381	−0.100	−1.43E-02	−0.780*	−0.733*	−1.012**	−0.940**
SLOVAKIA	−0.204	−0.223	−0.596	−0.639	−0.368	−0.396	8.402E-02	7.663E-02
ESTONIA	0[a]	0[a]	0[a]	0[a]	0[a]	0[a]	0[a]	0[a]
HIGH-TECH	−1.154**	−1.276**	−0.846*	−0.970**	−0.831*	−0.866*	0.317	0.262
MED-HIGH-TECH	−0.366	−0.434	−2.57E-02	−6.30E-02	−1.49E-02	−6.37E-02	0.323	0.295
MED-LOW-TECH	0.326	0.347	0.501*	0.536*	0.409	0.385	0.581**	0.592**
LOW-TECH	0[a]	0[a]	0[a]	0[a]	0[a]	0[a]	0[a]	0[a]
MINORITY FOREIGN OWNED	−0.634**	−0.387	−0.118	0.111	−0.194	−5.38E-02	−0.389	−0.274
MAJORITY	0[a]	0[a]	0[a]	0[a]	0[a]	0[a]	0[a]	0[a]

FOREIGN OWNED								
SME	−0.806***	−0.760***	−0.659**	−0.620**	−0.525**	−0.513**	−1.077***	−1.096***
LARGE	0[a]	0[a]	0[a]	0[a]	0[a]	0[a]	0[a]	0[a]
ESTBL. > 1990	−0.590*	−0.550*	−0.287	−0.282	−0.424	−0.470	0.259	0.255
ESTBL. 1991–1995	−0.130	−7.17E-02	6.702E-02	8.210E-02	2.179E-02	1.194E-02	0.260	0.264
ESTBL. > 1996	0[a]	0[a]	0[a]	0[a]	0[a]	0[a]	0[a]	0[a]
INTERMEDIATE GOODS	0.756**	0.688*	0.233	0.190	0.408	0.404	0.633*	0.681*
FINAL PRODUCTS	9.538**	0.575**	0.303	0.287	0.107	0.128	0.125	0.176
INTERMEDIATE GOODS AND FINAL PRODUCTS	0[a]	0[a]	0[a]	0[a]	0[a]	0[a]	0[a]	0[a]
Obs. (A)	426	426	426	426	426	426	426	426
Correct cases (B)	348	348	349	349	348	348	350	350
−2Log Likelihood	719.787	705.855	738.345	726.139	716.460	709.693	850.944	846.492
Final	(0.000)	(0.000)	(0.015)	(0.002)	(0.002)	(0.001)	(0.000)	(0.000)
Model fitting:	45.842	59.774	31.933	44.138	39.322	46.090	78.550	83.002
Chi-Square	(0.000)	(0.000)	(0.015)	(0.002)	(0.002)	(0.001)	(0.000)	(0.000)

Notes:
Link function: Logit.
[a] This parameter is set to zero because it is redundant.
*** Significant at 1% level; ** significant at 5% level; * significant at 10% level.

16
Effective Tax Rates as a Determinant of Foreign Direct Investment in Central and East European Countries: A Panel Analysis[1]

Christian Bellak and Markus Leibrecht

16.1 Introduction

Governments in Central and East European Countries (CEEC-5; see Table 16.1) intervene to influence the location choice of multinational enterprises (MNEs) by various measures. They provide incentive packages, fiscal and non-fiscal, and they try to shape various location factors in order to lower production costs for foreign firms. One location factor that figures prominently in actual policy-making as well as in the public debate is the corporate income tax rate. What is at issue therefore is, whether tax-rate cuts are an appropriate policy tool for attracting foreign direct investment (FDI)[2] and whether FDI responds significantly to changes of the corporate income tax burden in the CEEC-5.

A first look at the data reveals that a close relationship between FDI and corporate income taxation is indeed plausible. First, the data show a remarkable surge of European and US direct investment into the CEEC-5. A considerable variation over time and between host and home countries in the distribution of FDI is given (see Tables 16.1 and 16.2). As expected, larger countries receive the highest FDI inflows. Yet Poland, the country with the largest population, performs only slightly better than the Czech Republic. This manifests itself in a relatively low FDI stock per capita in Poland (not reported). Furthermore, Table 16.1 reveals that there is a surge in FDI inflows to all of the CEEC-5 since 1995.

Table 16.2 shows the origin of the FDI stock. The three most important home countries are Germany, The Netherlands and Austria. The large share of Austria in Slovenia as well as the large shares of Germany and the Netherlands in all countries but Slovenia are striking. The data also reveal

Table 16.1 Aggregate FDI flow into the CEEC-5 ($US m)

Year	Czech Republic (CZ)	Hungary (HU)	Poland (PL)	Slovakia (SK)	Slovenia (SI)
1990	72	311	89	93	4
1991	523	1462	291	81	65
1992	1003	1479	678	100	111
1993	654	2350	1715	179	113
1994	878	1144	1875	273	116
1995	2568	5103	3659	258	152
1996	1435	3300	4498	370	174
1997	1286	4167	4908	231	332
1998	3700	3828	6365	707	218
1999	6310	3312	7270	428	106
2000	4984	2764	9341	1925	137
2001	5639	3936	5713	1584	369
2002	8483	2845	4131	4123	1606
2003	2583	2470	4225	571	181
average 1990–1994	626	1349	930	145	82
average 1995–2003	4110	3525	5568	1133	364

Source: UNCTAD database.

Table 16.2 Origin of FDI in the CEEC-5 (bilateral stock in per cent of total stock)

Main home countries[3] Country (Year)	AUT	GER	FR	IT	NL	UK	US	Total
CZ (2001)	9.99	24.16	6.6	0.61	29.21	6.14	6.40	83.11
HU (2000)	12.21	25.81	6.52	2.73	22.52	1.06	8.21	79.06
PL (2001)	3.33	18.83	15.39	4.15	24.25	3.12	9.54	78.60
SK (2001)	18.45	22.35	2.65	9.18	15.83	6.83	6.20	81.49
SI (2001)	44.5	12.0	10.3	6.2	3.1	4.0	3.8	83.4

Source: OECD (2004); Bank of Slovenia (2004).

that most of the FDI stock is owned by European Investors. The observed surge in FDI inflow to the CEEC-5 was accompanied by a more or less pronounced drop in the overall statutory corporate income tax rates[4] in most of the CEEC-5.

Table 16.3 shows that all CEEC-5 but Slovenia reduced their rates, notably Slovakia and Poland. The average decrease of the rates is 10.9 percentage points. Note that Slovakia started to reduce its rate in 2000 whereas Poland experienced a more gradual fall. The slight increase in Hungary between 1998 and 2000 is due to an increase in the local business tax. In comparison, the drop in the rates in the seven main home countries was modest. The

Table 16.3 Overall statutory corporate tax rates, 1996–2004 (in per cent)

Year	CZ	HU	PL	SK	SI	AUT	FR	GER[5]	NL	UK	US	IT
1996	39.00	19.00	40.00	40.00	25.00	34.00	36.70	57.40	35.00	33.00	40.00	52.20
1997	35.00	19.00	36.00	40.00	25.00	34.00	36.70	57.40	35.00	31.00	40.00	53.20
1998	35.00	19.14	36.00	40.00	25.00	34.00	41.70	56.70	35.00	31.00	40.00	41.30
1999	35.00	19.40	34.00	40.00	25.00	34.00	40.00	52.30	35.00	31.00	40.00	41.30
2000	31.00	19.64	30.00	29.00	25.00	34.00	36.60	51.85	35.00	31.00	40.00	41.25
2001	31.00	19.64	28.00	29.00	25.00	34.00	35.30	38.67	35.00	30.00	40.00	40.25
2002	31.00	19.64	28.00	25.00	25.00	34.00	34.30	38.67	34.50	30.00	40.00	40.25
2003	31.00	19.64	27.00	25.00	25.00	34.00	34.30	39.58	34.50	30.00	40.00	38.25
2004	28.00	17.66	19.00	19.00	25.00	34.00	34.30	38.67	34.50	30.00	40.00	37.25

Source: Bellak et al. (2004).

largest reductions occurred in Germany and Italy, the countries with the highest rates in 1996. The average fall is about 5.9 percentage points.

The three tables presented above therefore suggest the possibility of competition for FDI *inter alia* via tax rate cuts. But is this relationship statistically and economically meaningful? The main purpose of this chapter is to investigate if there is indeed a significant causal relationship between the effective corporate tax burden and FDI-flows to the CEEC-5. This is done by estimating tax rate elasticities from panel data. We thereby focus upon FDI from the main home countries (i.e., Austria, Germany, France, Italy, The Netherlands, United Kingdom and the United States) to the CEEC-5 (i.e., Czech Republic, Hungary, Poland, Slovak Republic and Slovenia), because the latter are in the center of the ongoing public debate within the EU about an increase of (harmful) tax competition. The time span considered here ranges from 1996 to 2002.[6]

The remainder of the chapter is structured as follows. Section 16.2 gives some brief insights to the causes of FDI from a conceptual point of view. Section 16.3 includes a short review of existing empirical literature on corporate income taxes as a location factor. Section 16.4 discusses the variables and the methodology used in the estimation. In section 16.5 the estimation results are presented and discussed, section 16.6 summarizes.

16.2 Theoretical background

The causes of FDI into a particular country have been studied extensively in the literature. It is useful to separate these causes into two distinct questions (Frenkel et al., 2004). First, why does FDI emerge at all? Second, why does a *particular* country succeed in the competition for FDI?

Turning to the first question, broadly speaking two distinct motives why firms want to undertake FDI are given (Navaretti and Venables, 2004). The first motive is to supply a market directly through a subsidiary. This is termed horizontal or market-oriented FDI (HFDI). The second motive is to find low production-cost locations. This type of FDI is termed vertical or efficiency-oriented FDI (VFDI) and is primarily motivated by factor–cost differentials. Moreover, new FDI and expansion of existing FDI should be distinguished, since both decisions may depend on different determinants. The latter distinction is particularly relevant for taxation issues discussed below.

Despite this clear conceptual separation between HFDI and VFDI, it is difficult to separate FDI empirically. For example, a particular location may be chosen not only because of its low production costs but also for its proximity to large neighboring markets. This introduces a horizontal motive to the – at first sight – purely vertical one (Caves, 1996; Navaretti and Venables, 2004). Furthermore affiliates in the CEECs are often supplied with headquarter services from abroad (like controlling, R&D, advertising) and yet still are horizontal FDI. Discussion of these motives for FDI per se does not give a satisfying answer to the first question raised above. An answer is provided by the OLI paradigm (Dunning, 1988; Markusen, 1995). Based upon various theories (e.g. Trade Theory, Theory of the Firm and Theory of Industrial Organization) it proposes that FDI emerges if a firm has an Ownership (O) advantage (e.g. a patent) combined with a Location (L) advantage (e.g. low production costs; large market size) and an Internalization (I) advantage (e.g. economies of interdependent activities). If only an O-advantage is given, licensing results. If an O- and an I-advantage are given exports instead of FDI are used for servicing the foreign market. The predictions of the OLI-paradigm about the choice of the route of foreign market servicing are listed in Table 16.4.

The second question, why a particular country succeeds in the competition for FDI, is answered by identifying the most important host country determinants or L-factors which attract FDI, *conditional* upon a firm's decision to undertake FDI. Yet the OLI paradigm provides examples of potential determinants only. In particular, it does not suggest how to operationalize the L-advantages. The OLI paradigm neither does attribute weights to single location factors like taxation, nor does it assess their relative weights (e.g. taxes vs. relative unit labor costs). The selection of relevant L-advantages for an empirical analysis therefore remains a difficult issue, which can only be tackled by looking at the empirical evidence given so far.

Table 16.4 The choice for Foreign Direct Investment

Ownership-advantages	Internalization-advantages	Location-advantages	Lead to the following type of foreign market servicing ...	... resulting in the following location choice of production
Yes	Yes	Yes	FDI	Abroad
Yes	Yes	No	Exports	Domestic
Yes	No	No	Contractual resource transfers	Domestic
Why?	*How?*	*Where?*		

Source: Based on Dunning (1988).

16.3 The impact of taxation on FDI

It is difficult to come up with strong predictions about the consequences of tax rate cuts on FDI inflows in CEEC-5. This is due to a conceptual and an empirical argument. This subsection therefore takes a brief look at earlier evidence and discusses a number of conceptual points in the remainder.

There exist only a few empirical studies, which suggest a mixed picture. These studies suggest that taxes have only a relatively low impact on FDI to CEEC-5. In Bellak et al. (2004) we survey six papers which include taxes as a determinant of FDI and find a median tax-rate elasticity of −0.22. This implies that a 1 percentage point change in the tax rate will reduce FDI by 0.22 per cent. This value is well below the value of −3.3 found by DeMooij and Ederveen (2001, 2003) for FDI to mainly developed countries. The low semi-elasticity of −0.22 may be explained by the following facts, which are partly transition-specific:

- Tax-cutting strategies of governments may have little impact on FDI, since FDI may reflect strategic decisions by the management and are thus only partly cost-driven in the short run (compared to portfolio investment which reacts more directly to changes in profitability).
- As far as FDI-flows contribute to expansionary investment, it may react less than in the case of new investment, Greenfield investment in particular.
- Given the large number of location factors stated to be relevant for location decisions by firms themselves, taxes may well have a lower relative weight than other location factors.
- Also, the possibility for transfer pricing may turn the tax burden for MNEs *ceteris paribus* in a non-issue.
- But the validity of this relatively low value is also questionable from a conceptual point of view as most of the papers surveyed use the *statutory* corporate income tax rates as measures of the tax burden in the host countries instead of the (forward looking) bilateral effective average tax rates (*beatrs*), which are better suited for FDI. Using the statutory tax rate of the host country may therefore result in a sort of measurement error bias in the estimated tax-rate elasticities as the *beatrs* differ in level and variability from the statutory corporate income tax rates (see Bellak et al., 2004 for details).

It thus remains mainly an empirical question to determine the role of the tax burden for FDI for particular countries and particular time periods and thus raises interesting methodological issues. This study adds to the literature by considering the tax burden as a determinant of FDI in general, and more importantly by using *beatrs* instead of the statutory corporate income tax-rates as a measure of tax burden.

16.4 Variables, data and methodology

16.4.1 Dependent variable

Net-bilateral-FDI-outflow-to-GDP-ratio from home country (*i*) to host country (*j*) for the years 1996 to 2002 (*t*) is used as the dependent variable (*fdigdp*). FDI data are taken mainly from the OECD International Direct Investment Statistics Yearbook 1991–2002.[7] Because of various problems related to the measurement of FDI (e.g. valuation problems, negative FDI-flow values; cf. Falzoni, 2000) we do not use a logarithmic specification, but we normalize bilateral FDI flows by the host-country GDP in order to reduce existing heterogeneity between country-pairs in FDI flows.

16.4.2 Independent variables

As we are entirely concerned with the second question raised above (where to locate?) our independent variables have to be valid proxies for host country-related L-advantages. We base our choice of independent variables upon the findings of some recent and/or widely cited studies.[8] Specifically, we use as right-hand-side variables proxies for the L-advantage taxation, which is our variable of main interest and we control for other important location factors.

(a) Taxation (beatr)

In Bellak et al. (2004) we argue that from a conceptual and empirical point of view forward-looking effective tax rates should be used for assessing the role of corporate income taxation on FDI. More precisely, we argue that for location decisions of MNEs *beatrs* should be used. Table 16.5 summarizes this view. It shows the relation between FDI and taxation concerning the problem of location choice. From the OLI paradigm we conclude that the L-advantages determine the location choice (where?) in the case of FDI. Combining this FDI-related argument with the argument of the taxation literature, which states that for *discrete* choices the average tax rate is relevant, reveals that *beatrs* are the relevant L-factor to reflect the tax component of

Table 16.5 Parent company's location decision: OLI and taxation

Decision problem	Determinants according to the OLI paradigm	FDI-decision	Relevant effective tax rate
1. FDI?	O, L, I		
2. Where?	L_j (e.g. tax burden)	Choice of particular host country$_j$	Average
3. Conditional upon location: How much?	–	Scale of FDI	Marginal

Source: Based on Devereux and Griffith (2002).

the location decision of MNEs. Hence, we use *beatrs* as a measure of the tax burden. The rates are calculated using the methodology developed by Devereux and Griffith (1999). For further details concerning assumptions and the calculation, see the Appendix and Bellak et al. (2004). We expect a negative sign of the estimated coefficient.

(b) Market size (popmio)

In theory host-country market size increases FDI since a larger host market increases the likelihood that MNEs will be able to recoup the costs of their foreign investment (Navaretti and Venables, 2004). We therefore expect a positive sign of the estimated coefficient. Market size is proxied by the population (in millions) of the host country.

(c) Privatization (share)

The share of the private sector in the host country's GDP as published annually by the European Bank for Reconstruction and Development (EBRD) is intended to capture the privatization process in the CEEC-5 (EBRD various years; Holland and Pain, 1998). An increasing private market share implies privatization of former state-owned property and hence the possibility for foreign firms to engage in FDI. Moreover, privatization may act as a signal of commitment to private ownership (Holland and Pain, 1998) and a larger private sector market share per se implies more possibilities to engage in FDI. We therefore expect a positive sign of the estimated coefficient. Figure 16.1 depicts the close interrelationship between FDI and privatization revenues on a descriptive basis.

(d) Distance (dist)

Distance is an important determinant of FDI (Brainard, 1997). It is especially relevant for production FDI where economies of scale on the plant level of the affiliate have to be weighed against the costs of exporting. This measure has been frequently used in gravity-type models as well as in specifications in empirical studies explaining FDI. The expected sign of the estimated coefficient is ambiguous a priori. While large distance may encourage FDI due to an I-advantage it also may discourage it due to the lack of market know-how, higher communication and information costs and differences in culture and institutions (Buch et al., 2004 and 2005; Buch and Lipponer, 2004).

(e) Home-country population (pophmio)

In order to account for size differences of the home countries, we include population of the home countries (in millions) as a control variable. Since the outward investment potential of larger countries is higher than that of smaller countries, home-country size is expected to be positively related to inward FDI in the host country.[9]

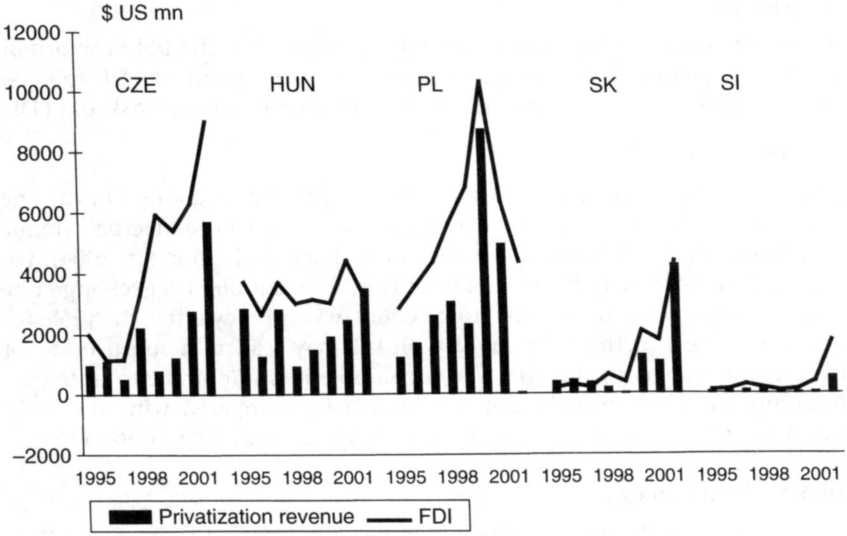

Figure 16.1 Privatization revenues and FDI
Source: EBRD Transition Reports, WIIW database.

(f) Tariffs (tar)

Tar is the ratio of tariffs on imports over imports of goods and services. From a theoretical point of view the sign of the coefficient of this variable is a priori ambiguous depending upon the underlying motive for FDI. If the observed FDI is mainly HFDI then the market imperfection theory of FDI suggests a positive sign. High trade costs may encourage HFDI because servicing more distant markets via exports is more expensive, not least because of transport costs. In this case HFDI occurs due to an internalization advantage (tariff-jumping FDI). On the other hand if FDI is mainly VFDI then theory suggests a negative sign (Frenkel et al., 2004; Navaretti and Venables, 2004). In the case of VFDI high trade costs can be seen as a location-disadvantage, which deters FDI. At first sight, for the CEECs we emphasize the tariff-jumping hypothesis, which refers to final-goods production to supply the host-country market. Yet this reasoning, which suggests a positive relationship between trade costs and FDI, is not relevant here, since trade costs have been lowered substantially in absolute terms during the last years, especially vis-à-vis the EU. Rather, we expect a negative relationship between trade costs and FDI flows, i.e. the lower the tariffs on imports, the higher the incentive to undertake FDI with a high input share supplied mostly by the parent company from abroad.

(g) Risk (risk)

In countries in transition, where not only economic, but also political turmoil may arise, *political risk* may play a role as a determinant of FDI, too. We expect a negative relationship (a positive coefficient) between risk and FDI.

(h) Inflation (infl)

The sign of the coefficient of this variable is a priori ambiguous. On the one hand, it may have a negative impact upon FDI due to the macroeconomic instability which high inflation rates imply (Buch and Lipponer, 2004). On the other hand, it may have positive effects, for example, via exchange rate changes: an appreciation of the home country's currency alters the price for acquiring assets in the host country and it may lead to a substitution of exports for foreign production. For our sample it is important to note that inflation has been brought down substantially compared with the early transition period. Hence it may no longer impact (negatively) upon FDI.

16.4.3 Methodology

Our data set includes a panel of seven home countries (*i*), five host countries (*j*) and seven years (*t*). The estimated model is generally specified as follows:

$$fdigdp_{ijt} = X'_{ijt}B_1 + W'_{jt}B_2 + Z'_{it}B_3 + \gamma_t + \alpha_i + \beta_j + e_{ijt} \qquad (16.1)$$

$fdigdp_{ijt}$ is the FDI to GDP ratio and $\mathbf{X}_{ijt}$, $\mathbf{W}_{jt}$ and $\mathbf{Z}_{it}$ are (1 x k_l, l = 1,2,3) vectors of right-hand side variables described above. All of the right-hand side variables except *beatr, dist, popmio* and *pophmio* are measured in levels. The *beatr* is used in log-form to allow for possible non-linearities. More specifically, the log-form implies that the higher the *beatr* the higher the absolute change in *beatr* for a given change in the dependent variable needs to be. *Popmio, pophmio* and *dist* are used in log-form as this reduces the range of values of the variables and makes estimation procedures more robust. Owing to the inclusion of relatively large countries like Poland (among the host countries) and the US (among the home countries), respectively, or relatively distant home countries (US), a log-transformation is needed. γ_t,α_i, and β_j are fixed time, host- and home-country effects. This specification follows the gravity-model literature (e.g. Mátyás, 1997; Frenkel et al., 2004). e_{ijt} is the remainder error term.

The model is estimated by pooled OLS. In a first round of specification search severe outliers are detected by using common descriptive statistics (studentized residuals, Cook's Distance) and added variable plots. Ten data points are detected as severe outliers (GER-SK (2002), GER-SK (2000), GER-HU (1998), GER-CZ (2002), US-SK (2001), US-HU (2000), UK-HU (1999), FR-SK (2002), FR-PL (2000), AUT-SK (2001)). These data points make up about four per cent of our sample and are dropped from the analysis.

In a second round we estimate various models starting from one which includes all of the variables described above. Since tests (Breusch–Pagan-test for heteroskedasticity and Arellano–Bond-test for serial correlation; see Table 16.6) show the presence of non-spherical residuals, we use Newey–West robust covariance estimates (lag of 2) throughout the analysis. We first test the fixed effects for significance via robust Wald tests. This shows that time as well as host-country fixed effects can be removed from the specification. Home-country fixed effects are jointly highly significant.

Next we analyse the sign and significance of our substantive variables in X'_{ijt}, W'_{jt}, and Z'_{it}. Insignificant variables are dropped stepwise. As insignificance may be due to multicollinearity we check for its presence via (robust) Wald tests, cross-correlations, variance inflation factors and condition numbers. High multicollinearity does not seem to be a problem (see Table 16.6 below). Four variables (*lnpophmio, risk, tar* and *infl*) are dropped one by one, with little impact upon sign and magnitude of the remaining coefficients. We include time and host country fixed effects again after dropping *lnpophmio, risk, tar* and *infl* in our specification, but they remain insignificant.

Table 16.6 Correlation matrix

	lnbeatr	**lnpopmio**	**lnpophmio**	**lndist**	**share**	**infl**	**tar**	**risk**
lnbeatr	1.00							
lnpopmio	0.39	1.00						
lnpophmio	0.21	−0.00	1.00					
lndist	0.11	0.09	0.76	1.00				
share	0.24	0.26	−0.01	−0.05	1.00			
infl	−0.18	0.13	−0.00	0.05	−0.30	1.00		
tar	0.47	0.61	0.00	0.06	−0.28	0.30	1.00	
risk	−0.51	−0.20	0.00	0.02	−0.24	0.16	−0.24	1.00

Table 16.7 Summary statistics: dependent variable

Variable	**Obs**		**Mean**	**Std. Dev.**	**Min.**	**Max.**
fdigdp	235	Overall	0.356	0.460	−0.689	2.486
		Between		0.344	0.004	1.332
		Within		0.318	−0.787	2.044
fdigdp by host country						
		CZ	0.450			
		PL	0.353			
		HU	0.519			
		SI	0.180			
		SK	0.293			
gdp		Overall	63010.50	55858.89	15882.10	207128.20
		Between		53988.70	19463.52	164549.10
		Within		15471.67	19556.15	108264.90

Table 16.8 Summary statistics: independent variables

Variable	Obs		Mean	Std. Dev.	Min.	Max.
lnbeatr	235	Overall	3.421	0.262	2.936	3.839
		Between		0.250	2.938	3.728
		Within		0.085	3.164	3.623
share	235	Overall	72	7.612	55	80
		Between		6.599	61.428	77.857
		Within		3.807	64.142	79.5
lndist	235	Overall	6.899	1.024	4.036	8.900
		Between		1.054	4.036	8.900
		Within		0	6.899	6.899
lnpopmio	235	Overall	2.135	0.981	0.684	3.654
		Between		0.980	0.687	3.651
		Within		0.0038	2.127	2.143
lnpophmio	235	Overall	3.844	1.064	2.074	5.663
		Between		1.076	2.079	5.616
		Within		0.0127	3.809	3.897
tar		Overall	5.663	8.992	0	46.230
		Between		6.779	0.005	19.360
		Within		5.944	−6.440	33.864
infl		Overall	8.240	4.614	1.290	21.200
		Between		2.629	5.386	13.273
		Within		3.835	0.238	17.913
risk		Overall	14.194	3.283	7.600	19.170
		Between		3.088	9.597	17.170
		Within		1.216	10.154	17.194
beatr		Overall	31.650	7.866	18.850	46.520
		Between		7.389	18.887	41.931
		Within		2.943	22.235	38.021

16.5 Results

Our final specification includes *lnbeatr, share, lnpopmio, lndist* and a full set of home-country dummies as regressors.

Table 16.9 shows that controlling for major determinants of FDI flows tax rate reductions had a statistically significant impact on *fdigdp* in the past. From a more substantive point of view the coefficient on *lnbeatr* of −0.330 implies that a one per cent decrease in the *beatr* increases *fdigdp* by 0.0033 percentage points. Evaluated at the overall averaged *fdigdp* (about 0.356 per cent; see Table 16.7) and the overall averaged host-country GDP (about EUR 63010.5 mn, see Table 16.7), a one per cent reduction of the tax rate *ceteris paribus* would lead to an increase of FDI inflows of about EUR 2.10 mn on average. From the coefficient on *lnbeatr* a tax-rate elasticity (semi-elasticity) of about −2.93 (evaluated at the overall averaged mean of *fdigdp* and *beatr*) is derived. Given the overall averaged mean *beatr* of about 31.65 per cent (see Table 16.8) the

Table 16.9 Estimation results

Dep. Var.: *fdigdp*	**Pooled OLS with Newey–West robust standard errors (lag 2)**
lnbeatr	−0.330**
	(−2.99)
share	0.0082**
	(2.38)
lndist	−0.281***
	(−3.55)
lnpopmio	0.117***
	(3.88)
constant	3.004***
	(3.30)
Number of Observations	235
R-squared	38.48
F-Test (10; 224)	12.85***
Arellano–Bond-Test AR(1)	2.92***
Arellano–Bond-Test AR(2)	2.02**
Arellano–Bond-Test AR(3)	0.87
Breusch–Pagan-Test	87.07***

Note: t-values in parenthesis.
*/**/*** denote significance at the 90/95/99% confidence level.
Home-country fixed effects included and significant.
Host-country and time-fixed effects are not significant.
risk, tar, infl and *lnpophmio* are not significant.

semi-elasticity implies that a decrease of the *beatr* to 30.65 per cent would increase FDI inflows by EUR 6.7 mn. The semi-elasticity of −2.93 is substantially larger than the median semi-elasticity of −0.22 surveyed by Bellak et al. (2004) and it is much more in line with the results derived by DeMooij and Ederveen (2003).

The coefficient of *share* implies that a one percentage point increase in *share* would *ceteris paribus* increase of *fdigdp* by 0.0081 percentage points. This results in a semi-elasticity of +2.27. Hence, a one percentage point increase in the private market share results in an increase of FDI flows by 2.27 per cent. Given an overall mean share of 72 per cent (see Table 16.8) an increase of share by 1 percentage point will increase FDI flows by EUR 5.10 mn. The semi-elasticities for *popmio* and *dist* are +1.03 and −2.50, respectively (all semi-elasticities are evaluated at the overall averaged mean of *fdigdp* and *beatr*).

Concerning other explanatory variables the insignificance of *pophmio* may be explained with the inclusion of home-country fixed effects. The insignificance of *tar* can be explained with the minor importance of tariffs (overall mean of 5.6 per cent; see Table 16.8) almost throughout our sample. Political risk may play no role, because our host countries are not that different in this respect and, because the host country with the lowest risk level (Slovenia) also has the lowest *fdigdp* value (see Table 16.7). The insignificance of *infl* is in line with other studies (e.g. Frenkel et al., 2004). Moreover inflation has been brought down considerably in the host countries considered here. Hence the insignificance of inflation seems to be a plausible result.

Finally, standardized coefficients (not shown) suggest that *dist* is the most important determinant of FDI in our specification. From a policy perspective perhaps more interesting is that taxation (*beatr*: beta = −0.189) exerts a slightly stronger impact upon FDI than the privatization process (*share*: beta = +0.135).

16.6 Summary

During the past few years lowering the corporate tax rate was seen as a key policy instrument for attracting FDI of foreign MNEs into the CEEC-5. Tax rates are an attractive policy tool for policy-makers, because they can easily be changed and are thought to affect the behavior of economic agents immediately. This chapter investigates whether there is indeed a significant empirical relationship between corporate taxation and FDI. Our empirical study, which is based on a panel data set, includes *beatrs* instead of the statutory tax rate and as controls mainly variables which intend to capture the differences in L-advantages of the CEEC-5. Contrary to earlier evidence which suggests a minor role of taxes for FDI in the CEECs, our regression analysis of FDI flows into five East European host countries from the seven most important home countries confirms the importance of the tax rate as a determinant. The deduced tax-rate elasticity is about −2.93. This result is, *inter alia* a consequence of replacing the statutory tax rate in the estimation by a more appropriate measure, namely the *beatr*.

Hence, from an individual country perspective, tax-lowering strategies have been successful in attracting FDI in the past. Yet, we doubt that the results of our analysis provide a good guidance for future policy strategies (i.e. further tax-rate cuts) in the CEECs, as *created assets* such as high quality public infrastructure certainly will become more decisive location factors, particularly if these countries are to profit more from headquarter services, R&D activities etc. of foreign MNEs.

Notes

1 This study has been prepared under FWF contract Nr. 1008, Sonderforschungsbereich 'International Tax Coordination', <http://www.sfb-itc.at/>.

2 From now on, it is convenient to use the term FDI for the location choice of MNEs, it being understood that FDI is an operationalization for the 'activity of MNEs in the host country'.

3 AUT = Austria, GER = Germany, FR = France, IT = Italy, NL = The Netherlands, UK = United Kingdom, US = United States.

4 'Overall' means that local business taxes are included.
5 The overall tax rate for non distributed profits is shown.
6 The analysis starts with the year 1996, due to the deep recession in the CEECs up to 1995, and ends with the year 2002, due to restrictions in data availability.
7 A detailed data description can be found in the Appendix.
8 In particular we base our choice upon the following papers: Holland and Pain (1998); Carstensen and Toubal (2004); Frenkel et al. (2004); Desai et al. (2004); Navaretti and Venables (2004).
9 In per capita terms, however, small countries typically invest more abroad, e.g. Switzerland, The Netherlands.

References

Bank of Slovenia, *Foreign Direct Investment Report 1994–2003* (Ljubljana 2004).

Bellak C., Leibrecht, M. and Römisch, R., 'New evidence on the tax burden of MNC activities in Central- and East-European new member states', *SFB International Tax Coordination Working Paper* No. 2 (Vienna: Vienna University of Economics, 2004).

Brainard, S.L., 'An empirical assessment of the proximity-concentration trade-off between multinational sales and trade', *American Economic Review*, 87(4) (1997) 520–44.

Buch, C., and Lipponer, A., 'Clustering or competition? The foreign investment behaviour of German banks', *Deutsche Bundesbank Discussion Paper Series* 1: Studies of the Economic Research Centre No. 06 (2004).

Buch, C., Kleinert, J. and Toubal, F., 'The distance puzzle: on the interpretation of the distance coefficient in gravity equations', *Economics Letters*, 83(3) (2004) 293–98.

Buch, C., Kleinert, J., Lipponer, A. and Toubal, F., 'Determinants and effects of foreign direct investment: evidence from German firm-level data', *Economic Policy*, January (2005) 51–110.

Carstensen, K. and Toubal, F., 'Foreign direct investment in Central and Eastern European countries: a dynamic panel analysis', *Journal of Comparative Economics*, 3 (2004) 3–22.

Caves, R., *Multinational Firms and Economic Analysis* (Cambridge: Cambridge University Press, 1996).

DeMooij, R.A. and Ederveen, S., 'Taxation and foreign direct investment: a synthesis of empirical research', *Paper presented at the OCFEB Conference* (The Hague: October 2001).

DeMooij, R.A. and Ederveen, S., 'Taxation and foreign direct investment: a synthesis of empirical research', *International Tax and Public Finance*, 10 (2003) 673–93.

Desai, M.A., Foley, F.C. and Hines, J.R., 'Foreign direct investment in a world of multiple taxes', *Journal of Public Economics*, 88 (2004) 2727–44.

Devereux, M.P. and Griffith, R., 'The taxation of discrete investment choices', *IFS Working Paper Series* No. W98/16 (1999).

Devereux, M.P. and Griffith, R., 'The impact of corporate taxation on the location of capital: a review', *Swedish Economic Policy Review*, 9(1) (2002) 79–105.

Dunning, J.H., *Explaining International Production* (London: Unwin Hyman, 1988).

European Bank for Reconstruction and Development (various years), *Transition Report* (London: EBRD).

Eurostat, *European Union Foreign Direct Investment Yearbook 2001* (Luxembourg: Eurostat, 2002).

Falzoni, A., 'Statistics on foreign direct investment and multinational corporations: a survey', *Research Network on Foreign Direct Investment and the Multinational Corporation: New Theories and Evidence* (Milan: Bocconi University, 2000).

Frenkel M., Funke, K. and Stadtmann, G., 'A panel analysis of bilateral FDI flows to emerging economies', *Economic Systems*, 28 (2004) 281–300.

Holland, D. and Pain, N., 'The diffusion of innovations in Central and Eastern Europe: a study of the determinants and impact of foreign direct investment', *National Institute of Economic and Social Research Discussion Paper* No. 137 (London: NIESR, 1998).

Markusen, J.R., 'The boundaries of multinational enterprises and the theory of international trade', *Journal of Economic Perspectives*, 9(2) (1995) 169–89.

Mátyás, L., 'Proper specification of the Gravity Model', *World Economy*, 20(3) (1997) 363–8.

Navaretti, G.B. and Venables, A.J., *Multinational Firms in the World Economy* (Princeton, N.J.: Princeton University Press, 2004).

OECD, *Taxing Profits in a Global Economy: Domestic and International Issues* (Paris: Organization for Economic Cooperation and Development, 1991).

OECD, 'Recent Privatization Trends in OECD Countries', *Financial Market Trends* (Paris: Organization for Economic Cooperation and Development, 2002) pp. 43–55.

OECD, *International Direct Investment Statistics Yearbook 1991 – 2002* (Paris: Organization for Economic Cooperation and Development, 2004).

Databases

Austrian Institute of Economic Research database
European Commission AMECO database
Eurostat New Cronos database
European Innovation Scoreboard 2004 database
OECD Foreign Direct Investment Statistics database
UNCTAD Foreign Direct Investment database
Vienna Institute of International Economic Studies database
World Development Indicators database

Appendix: Detailed description of data and data sources

fdigdp

FDI reflects the bilateral net-FDI outflows from the home countries (I) to the host countries (*j*) for the years (*t*) 1996 to 2002. FDI flow data were first converted into a common currency (EUR million) using the average bilateral exchange rate in *t*. FDI data are taken mainly from the OECD International Direct Investment Statistics Yearbook 1991–2002 and the OECD Foreign Direct investment database. Missing values for the US are substituted by information provided by the Bureau of Economic Analysis (BEA). For the UK missing values are amended by data provided by Eurostat *European Union Foreign Direct Investment Yearbook 2001* vis-à-vis Slovakia and the Direct Investment Report of the Bank of Slovenia vis-à-vis Slovenia. GDP data are taken from the New Cronos database.

beatr

The bilateral average effective tax rate is calculated using the Devereux–Griffiths (1999) methodology, based on the following assumptions and parameters:

- Three different assets (machinery, building and inventory in the manufacturing sector).
- Seven ways of financing a cross border investment of 1 with a pre-tax financial return of 20: (i) retained earnings subsidiary; (ii) new equity subsidiary and retained earnings parent; (iii) debt subsidiary and retained earnings parent; (iv) new equity subsidiary and new equity parent; (v) debt subsidiary and debt parent; (vi) new equity subsidiary and debt parent; (vii) debt subsidiary and new equity parent.
- economic depreciation rates of the various assets: 3.61 per cent for buildings, 12.25 per cent for machinery, 0 for inventory.
- nominal interest rate of 7.625 per cent.
- common inflation rate of 2.5 per cent.
- constant nominal exchange rate.
- a weighted average structure of assets (buildings/machinery/inventory) of 55/35/10.
- a weighted average structure across the various types of financing (retained earnings/equity/debt): 55/10/35 for parent and 1/3/ 1/3/ 1/3 for subsidiary.

Our assumptions about the asset structure differ from those of other studies, which mainly follow OECD (1991), because data on inventories in the CEEC-5 show that they are far less important than they have been within the OECD as reported in 1991. Instead, we assign a higher weight to investment in buildings. Note also that we do not include any tax incentives in our measure since the choice of relevant incentives in each home and host country would be arbitrary. *beatr* is measured in per cent.

share

This variable is taken from various issues of the Transition Report published by the European Bank for Reconstruction and Development. It is measured in per cent.

dist

Distance is defined as the geographical distance between the capital cities of the home and the host country in kilometers. Data are taken from various internet sources.

pophmio

Home-country population is defined as total population (in millions) of the seven home countries. Data are taken from the WDI database and from Eurostat.

popmio

Host-country population is defined as the total population (in millions) of the five host countries. Data are taken from Eurostat.

risk

Political risk data are taken from various issues of 'Euromoney'. 25 is the maximum value (lowest possible risk level) and zero the minimum value.

infl

As a proxy for inflation the GDP-deflator of each host country is used, taken from the AMECO database.

tar

tar is defined as the ratio of 'tariffs on imports' (from Eurostat, Main National Accounts, Position D212: 'taxes and duties on imports excluding VAT', position S13: General Government) over 'imports of goods and services', taken from the European Commission's AMECO database.

17
Multinationals and National Systems of Innovation: Strategy and Policy Issues

Robert Pearce and Marina Papanastassiou

17.1 Introduction

One of the most significant journeys in our understanding of international business has been that from an essentially centralized view of innovation in MNEs toward one that encompasses an increasing range of decentralized inputs and strategic postures. This change in perspective can then be seen as decisively embodied within comparable changes in the way in which the *effects* of international business on individual host countries have been analysed. Here we can see a refocusing from an FDI-based interpretation of flows of separate firm attributes (increasingly technology and other intangible assets rather than finance capital per se) toward a more MNE-strategy oriented evaluation of how firms position their operations in a specific location within wider globalized programs (Pearce, 2001, forthcoming). The aim of this chapter then is to generate a methodology for the assessment of the ways in which MNEs' globalized strategies for innovation involve themselves with the attempts of national economies to generate and operationalize innovation competences as a source of growth and international competitiveness.

An extremely valuable formulation on which to build an understanding of these changes and their current implications remains Vernon's (1966) original product cycle model (PCM). In the next section, therefore, we show how external and internal forces altered the environment for MNE innovation from the home-country focus of the PCM towards the need for more decentralized approaches to innovation. From this, Section 17.3 derives the elements of a stylized global innovation strategy (GIS). This postulates MNEs approaching a decentralized innovation strategy by drawing distinctive capabilities and attributes from several different countries into an integrated program that enhances their *global* competitiveness, and thus raises vital questions about how the individual *national* economies are rewarded. The core aim of the study is then to provide some insights on these questions. Section 17.4 therefore investigates the implications of the ways in which components of MNEs' GIS interact with different elements of countries' National Systems of Innovation (NSI). Policy interpretation of this is provided in the concluding section.

17.2 From the product cycle to global innovation strategies

It is not necessary to recapitulate here the full detail of Vernon's (1966, 1979) PCM. However, several specific points that provide platforms for the development of our subsequent analysis can be briefly brought out. Thus the key characteristic of the first stage of the PCM is the *centralization* of the innovation process in the home country of a firm that is not yet in any way concerned with foreign markets.[1] This centralization derives, in Vernon's analysis, from two factors. First, that without value-adding activities in other countries there would be no effective means of detecting or accessing externally customer-driven ideas for new goods (in Vernon's essentially market-driven view of innovation) or emerging technologies. Secondly, Vernon also elucidates the benefits centralization offers by providing for close day-to-day contact between key top-level functional participants in innovation (scientists, market researchers, engineers, entrepreneurial strategic management). We will see that even within the dispersed innovation networks that have emerged in MNEs the continued benefits of such functional interaction often persist where individualized locally-responsive completion of product development is pursued.

At the start of the second stage of the PCM overseas demand for the good emerges in an ad hoc fashion, beyond the current strategic horizons of the firm.[2] Thus, at some time after the initial innovation, at least the more prosperous and cosmopolitan citizens of other countries reach income levels that allow them to begin to aspire to ownership of the good and to take the initiative to secure its import. After a period of more systematic exporting this stage culminates in firms initiating local production in some of the good's more significant foreign markets. Though this decision can involve routine economic influences, such as trade barriers or input costs, it can also have a clear market-seeking strategic purpose. Part of this is that local production can allow for response to local tastes and supply conditions through adaptation of the product and process.

By the third stage of the PCM the product and its technology are highly standardized and familiar so that, with the entry of many rival suppliers, the market has become very price competitive. This then leads to a further strategic motivation in the form of efficiency-seeking, where firms carefully pursue the most cost-effective production locations. This ability now to separate where a good is produced from where it is sold was, in practice, facilitated by increasing freedom of trade. Here one of the key recent changes in the global economy allows for the internationalization of one of firms' two core strategic imperatives, in terms of cost-effective supply of their well-established mature goods. Our analysis then suggests that two other changes (one external to the MNE and one internal to the firm and, indeed, endemic to the original PCM process itself) also led to international restructuring of firms' other core priority, in the form of generation of new products and the sourcing of new competitive attributes. Thus the PCM's process of home-country innovation followed by a relaxed and gradual diffusion of new goods into foreign markets needed to be replaced by more proactive and internationalized approaches to technology generation and product innovation.

The external change that we see as conditioning the need for new approaches to innovation in MNEs is that, through most of the last third of the twentieth century, the spread of sustained economic growth led to an increased number of countries at quite similar and high levels of income. Taken with increased numbers of firms competing internationally and the shortening of product life cycles[3] this meant that, even if the essence of a new innovation emerged in one location, firms would need to get the new good into all key markets with the greatest possible speed.

Internal to the firm, once it has been through the PCM, the assumption that it is not a MNE has been removed and this alters its organizational structure in a way that, in effect, provides it with the mechanism to deal with the new competitive need provoked by the external change. Thus the second stage of the original PCM saw the establishment of local-market-focused subsidiaries in other high-income markets. These can provide the vehicle for getting new goods into key markets very promptly. Indeed it would be an expectation of internal competition in MNEs that ambitious subsidiaries in major markets would *demand* early access to important new products emerging elsewhere. If these subsidiaries had, as suggested by our interpretation of the second stage of the PCM, already developed in-house capacity to individualize established products to meet specific distinctive characteristics of their local market and production conditions, they could then become, in effect, part of new innovation processes. Thus, rather than adapting well-established goods in an *ex post* fashion, they could acquire early access to a development process whose essence (new technology and market ideas) is emerging elsewhere in the MNE and complete its own version of the innovation that responds to aspects of its own competitive context. This would, in some ways, collapse the first two stages of the original PCM into one internationally-differentiated innovation process.[4]

From the viewpoints previously outlined we can distil three aims for the global innovation strategy that we describe in the next section and which become central to the subsequent analysis. First, the MNE needs to secure the key ground-breaking inputs into a major new innovation in the most effective manner possible. Importantly the emphasis of this effectiveness does not relate primarily to costs, though it must be accepted that adequate budgeting for these expenditures becomes a delicate balancing act in firms facing the myopic short-termism of immediate profitability and support for the share price (with a problematic concomitant reluctance to allow for the need to reinforce future sources of profitability and, ultimately, competitive survival). Instead, the primary aim of this effectiveness is to maximize the possibilities for securing access to important new scientific knowledge (adding science-driven potentials to the likely origins of major innovations) and customer-originated ideas for radical extensions of an industry's competitive scope. Since we would see increasing technological- and market-heterogeneity as a central feature of the evolving global economy the implication of this first aim of the GIS is a need to monitor, and secure first-mover access to, potential sources of the primary inputs that drive major innovations, through internationalized approaches to precompetitive R&D and market research.

If the MNE does put together the fundamental components of a major new innovation (the 'new product concept' that we elaborate in the next section) then the second aim of the GIS will be to get it into all key markets as quickly as possible. Thus the intensity of globalized competition, and the ability of rival firms to quickly generate their own variants of the new good, means that the innovating MNE needs to maximize returns during its period of monopoly supply and therefore must target all relevant markets as an integral part of the innovation process. The third aim then adds an important dimension to this by suggesting that the new good should enter all segments of the global marketplace not only quickly, but responsively. In many industries it is accepted that differences in economic conditions, regulations and tastes, mean that extensive variation of products yields competitive advantage in different segments of the global economy. Responding to such differences *ex ante*, within the innovation process of new goods, can then represent a significant step forward compared to earlier *ex post* adaptation of mature products.

17.3 A global innovation strategy

In this section we outline an idealized approach to a GIS available to a contemporary MNE[5] (Pearce and Papanastassiou, 1996; Pearce, 1997; Papanastassiou and Pearce, 1999). We see this as implemented through two stages, each of which is articulated around a different type of R&D laboratory.[6] Thus the first, precompetitive, stage of the GIS is positioned (Figure 17.1) around a network of Internationally Interdependent Laboratories (IIL),[7] and seeks to contribute to achieving the first aim (or benefit) of globalizing innovation activity. Thus MNEs with a perception of a range of separate scientific disciplines that could provide new results that contribute to a

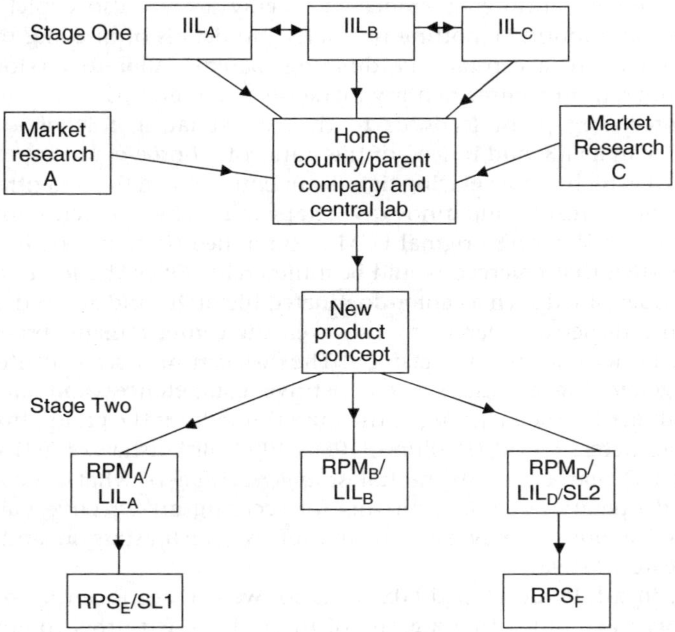

Figure 17.1 Global innovation strategy
Source: Authors.

technology-driven component of innovation, acknowledge technological heterogeneity (i.e. that different countries' science-bases provide frontier-quality research capabilities in different areas of specialization) and set up a portfolio of IILs to tap into the distinctive potentials of different NSIs. Though the set of IILs can be seen as seeking to be a balanced portfolio, in the sense that each aims to take independent responsibility for covering research in a potentially important area of science (that reflects a strong distinctive capacity of its host NSI), the MNE also needs them to generate at least ad hoc networking (interdependency) in terms of exchange of results, questions and opinions. Thus the perception of potentials for future commercial innovation emerging from basic research may well derive from insights representing synergistic combinations of results from different projects in different IILs, and the further progress of these potentials may then benefit from the building in of complementary research reflecting the specialized disciplines of additional parts of an IIL network.[8] Thus Figure 17.1 includes lines and arrows connecting IILs, reflecting the benefits of knowledge flows that can enrich and condition the work of individual labs in terms of their contribution to the ultimate strengthening of the MNE group's core scientific competences. Crucially arrows also depict a more persistent and routine reporting of results and details of ongoing programs from each IIL to a central co-ordinating 'parent' laboratory (along with similar flows from complementary market-research groups).

The preceding point leads us to the co-ordinating role of a 'parent' laboratory in a GIS, and indeed of the status of a home-country parent HQ in an MNE which is now placing significant emphasis on the decentralization of much of its creative and innovative strategic activity. The centralization of innovation in Vernon's original PCM also implied that, at least in its early life, the MNE that emerged would be a hierarchy. Once the forces outlined earlier undermined such a center-dominated hierarchy and allowed dynamic forces into dispersed operations, however, the center remains pre-eminent but in a rather reformulated fashion. Thus we can now see a 'center' as no longer generating all the key competitive competences and taking full responsibility for the planning of their use throughout the group. Instead, in this more heterarchical (Hedlund, 1986, 1993; Birkinshaw, 1994) context, it is the HQ that aspires to the fullest *understanding* of what is evolving in dispersed operations and, from this, of securing an effective balance in networks of both supply and creative units (Papanastassiou and Pearce, 1999; Pearce, 1999a).

Then, in a fully-developed GIS scenario, we can see a central or parent laboratory as co-ordinating a group of IIL projects, with their potential for overlaps and synergies, and then collating the outputs of these separate units. One responsibility of this central lab may be to nurture and facilitate interdependence between IILs by encouraging the sharing of results and by inculcating a non-defensive culture in which it is routinely acceptable for one IIL to see its work taken up by another.[9] Crucially it may also be such a central lab that is most naturally committed to discerning applied research potentials in the basic research output. Thus the commercial potentials of precompetitive work may often not be perceived from individual isolated basic research projects, but only when various results are brought together and evaluated dialectically in terms of their resonance (consistencies, inconsistencies, overlaps and interdependencies). The parent lab may then allocate back into the appropriate parts of the precompetitive IIL network responsibilities for necessary applied research that is needed to complete the technological basis for possible commercial innovations.

In the GIS the first stage is completed when the center (bringing together market research perceptions as well as scientific results) is able to fully define the essential elements of a new product concept (NPC). This NPC does two things. First, it fully defines the essential nature of the major new service through which the innovation will extend the competitive scope of the firm and industry. Secondly, it puts into place (from basic and applied research of stage one) all the technologies necessary to secure effective supply of a prototype of the good. However, to secure the second and third aims of a GIS this new good needs to be got into each segment of the global market place quickly and in forms that fully respond to particular localized tastes and conditions. Thus the remaining responsibility of the 'center' in a GIS is to allocate product development responsibilities to particular subsidiaries, and to provide them with the core details of the NPC (i.e. to articulate and organize stage two of a GIS).

As depicted in Figure 17.1, the second stage of a GIS is implemented through product mandate (PM) subsidiaries and associated locally integrated laboratories (LIL). Thus a PM gains, from its MNE parent, full responsibility for developing a product (here by filling in the locally-responsive details of a NPC), its initial production, marketing and further competitive evolution. To do this a PM builds a strong functional scope, including R&D, market research and assertive marketing, creative engineering and an impetus from an ambitious management that is committed to the unit's individualized competitive status in the MNE. A LIL is a central element in the PM's creativity, and essentially mediates new science emerging from applied research to the other functions (i.e. those with which it works in a closely *integrated* fashion) involved in the innovation process. Thus a LIL will not itself carry out further pure scientific investigation, but instead learn a core new technology, interpret it to associated functions (marketing, engineering) and thereby play a key role in its practical manifestation in new goods, services and production techniques.

In a fully-developed GIS an MNE will authorize a separate PM/LIL nexus in *each* of several regional markets to generate a distinctive variant of the NPC that responds to all the idiosyncratic tastes and needs of its regional customer base and optimizes the production technology in its use of the input environment (availability and price of factors) of the host country. A subsidiary operating in this manner can be designated as having a *regional* product mandate (RPM).

With the completion of separate product developments in its group of RPMs the GIS provides its contribution to *one* of the MNE's two core strategic priorities, by providing major additions to the product scope that are based on new knowledge competences. In a way presaged within the PCM these new products will eventually lose the competitive edge of originality and uniqueness, so that they may fall within the purview of an MNE's second strategic priority, i.e. the need to supply maturing products in more cost-competitive ways. At this point some supply of these goods may be relocated to cost-based rationalized product subsidiaries (RPS) elsewhere in the group. The role of an RPS is to take on production of well-established goods of the MNE in a cost-effective manner. Thus Figure 17.1 shows an RPS in country E taking on production of a good originally developed in RPM_a and one in country F similarly inheriting supply responsibility for a good created in RPM_d.

This efficiency-seeking behavior takes us back into the worlds of hierarchy and technology transfer, in the sense that the roles of RPS_e and RPS_f depend on using technologies finalized by RPM_a and RPM_d. Such technology transfer we see as mediated by another type of R&D unit, the support laboratory (SL). Two variants of the SL role are depicted. Thus RPS_e includes an SL1-type unit, whose aim is to learn the technology from RPM_a and secure its effective assimilation and activation in the RPS's supply. Alternatively SL2 activity is depicted as a supplementary aim of the LIL in RPM_d, seeking to organize the outward transfer of those parts of the RPM/LIL's technology that will be used in RPS_f. In effect the SL2 teaches RPS_f to apply and use its new technology.

The emergence of this efficiency-seeking priority emphasizes the vulnerability of any innovation-based unit to product-life-cycle forces. In the case of PM/LIL operations the question of the ability to sustain dynamic creativity is enhanced in a GIS scenario where it is dependent on access to group technology and new product ideas (the NPC). Thus the persistence of PM/LIL units often depends (externally) on the parent MNE group's ability to generate new innovation potentials, and (internally) on the host-country's ability to support those aspects of its NSI that enable the facility to continue to attract elements of the group's creative programs. The presumed ideal of a PM would in fact be to escape from a GIS and internalize more local inputs (science, market research, etc.) into a fully localized creation of a unique new good with world market scope (i.e. become a *world* product mandate, WPM).

In this characterization of the GIS MNEs achieve major increments in their *global* competitiveness by combining different inputs from many *national* systems of innovation. The implications of this for the technological progress and competitiveness of these individual national economies is the subject of the next section. To facilitate this we adopt a simple linear NSI comprising basic research, applied research, product development and adaptation of products and processes. Though this is a one-dimensional perspective on a complex and interactive system[10] it does delineate the types of work that can interact with facets of the GIS.

17.4 MNEs' participation in NSIs

This section uses the concept of a GIS to evaluate how MNEs participate in individual NSIs. For each stage of an NSI we indicate the institutions (subsidiaries and/or laboratories) through which MNEs operate, what they interject into the NSI through such units, and the ways in which they co-opt and utilize the relevant outputs.

17.4.1 Basic research

The institutional arrangement through which MNEs' GIS enters the basic research component of an NSI is the IIL. This provides inputs into the NSI in two ways: funding and new dimensions of technology. As we have defined it an IIL is a wholly-owned and fully-controlled research facility of an MNE operating within another country's NSI. The aim is to generate new research results whose value is likely to be perceived in a manner that is synergistic with other results (or supportive of ongoing research agendas) elsewhere in the group. A MNE's financial commitment to an IIL is likely to be manifest, first, in expanded scientific infrastructure, through the building of new labs or the extensive refurbishment and re-equipment of existing ones.

The second funding component of an IIL will take the form of salaries to locally-trained scientists. Thus the distinctive research capacity that emerges in an IIL is expected to derive from the recruitment of a balanced team of local scientists, who possess specialist capabilities that reflect their education, training and previous research work within the country's science-base. This immediately raises the crucial point that these are potentially scientists (experienced and successful researchers and, perhaps, their emerging 'star' protégées) with a very high opportunity cost to those local institutions that fail to retain their services. MNE funding through salaries may, therefore, be merely crowding-out some of the higher-quality possibilities in local labs.[11] This must be acknowledged as an important factor, albeit one that is hard to evaluate practically since it invokes a particularly complex version of the counterfactual situation. Two positive possibilities of the IIL can be suggested, though.

First, the work the scientists do in the IIL may be more productive (in terms of strengthening local technology scope and then, perhaps, feeding through to better performance of the wider local economy) than they could have achieved in a host-country institution. How this might eventuate is a key element in the analysis here. Secondly, the top local scientists recruited by IILs might *not* have perceived the continued development of their careers to be in the research units of the firms and universities of their country of origin. They may initially have been candidates for the international migration that is increasingly common in high-quality human capital, with the improved funding (their salaries and project support) and/or the renewed and extended research stimulus of IIL work instead now retaining their local commitment. This is likely to have positive externalities, or spillovers, into the local scientific community beyond their direct benefits to the (admittedly 'foreign') lab they now work for. This may include charismatic stimulus (through public lectures, university visits, contributions to broader controversies and intellectual debates, etc.) to younger scientists outside their own institution, or inputs to the formulation of wider scientific policies and programs (serving on government advisory boards, funding bodies, committees of enquiry, etc.).

The second input into an NSI deriving from the operation of an IIL can be seen as technology itself, in the sense of new research options emerging from access to an additional, but essentially complementary, body of scientific knowledge and competence (i.e. that of the parent MNE). Thus *where* an IIL is located is determined by the ability of the host-country's technological heritage and current research capacity to support the type of investigation required. But *what* that research is determined by the MNE's own technology trajectory (Papanastassiou and Pearce, 1999). This trajectory can be seen to comprise the MNE's current stock of core technology and its embodiment in a product range, and a broadly understood view of the directions in which these are expected to evolve through basic research (IILs) and product innovation (LILs). Thus the perceived needs of extending and enriching the technological trajectory will determine the range of scientific disciplines to be researched in IILs. The essentially evolutionary nature of this will also indicate that particular IIL projects, while clearly seeking to benefit distinctively from the specific strengths of the host-country inputs, are likely also to be defined in the light of existing MNE technologies (those seen as potential bases for valuable progress) and indeed, to some degree, will use these technologies among the building blocks from which research programs are generated.

An idealized interpretation of the effects of IILs on the basic research component of an NSI is that they can both deepen and widen the scope of the work undertaken. Access to improved funding, and to a body of complementary technologies, can reinforce the scope for the NSI to further pursue those lines of basic investigation that are dictated by its own technological heritage and established specialisms. Here the IIL supports the processes of agglomeration that deepen the focus on particular areas of science for which the NSI already has an established reputation of world research leadership. But IILs' agendas may, to some degree, also work against these agglomerative forces, without necessarily weakening the ability of distinctive basic research to benefit the rest of an NSI (and economy).

Thus research issues and current technologies of MNEs may ask somewhat different questions of the NSI's strengths than would have been articulated by purely local scientific and commercial interests. If the symbiotic process between the two scientific communities (MNE and NSI) works effectively the IIL research agenda will then differ from that which is purely locally driven, but be no less logical as an evolution of the research programs. Compared to what would have happened in their absence, IILs may widen the basic research agenda of the host country, but in ways that remain coherent and cohesive with the balanced and logical progress of the NSI. Ideally IILs carry out projects that would not otherwise have been undertaken but which, nevertheless, make distinctive use of the defining strengths of the local science-base and research community.

Within the concept of a purely *national* innovation system the justification for support (financial and institutional) for *pre*competitive basic research is that ultimately some of the output *will* fuel commercial progress and supply part of the basis for the competitive evolution of the economy. However, while the networked position of MNEs' GIS participation may bring in resources that enrich and stimulate the basic research component of an NSI, these interdependencies may also influence the future application of their results in ways that diminish the contribution to the rest of the NSI. The international linkages and aims of GIS programs may mean that the innate tendency is to see the output of IILs more in terms of potential 'lateral' flows to other elements of the group than 'horizontal' to another phase of the NSI.[12]

There is, of course, a very real potential for MNE basic research output to move into applied research in the same NSI (indeed, almost inevitably, in the same unit). Thus the perception of possible eventual commercial use for particular basic research results may emerge in the IIL that carried out the project, and this may secure for it the permission to accede to the appropriate applied research. A strong IIL can internalize the crucial basic/applied research transition.

But the scope for leakage out of the NSI is, as observed, also very strong. Basic research results of an IIL may be transferred to another part of the MNE science community in two forms. In the first the possible commercial applications may, again, have been perceived in the basic researching IIL, but permission for further applied research may this time be denied. Here a group-level decision determines that another IIL, in another country, is better equipped to perform the type of applied research that appears necessary. The second case is the GIS scenario where research results flow to, and are assimilated by, a central/parent laboratory.

17.4.2 Applied research

MNEs' applied research in a particular NSI still takes place mainly within an IIL-type facility and still targets the solution of an essentially scientific problem. From the point of view of the GIS, however, the investigation is now less purely speculative, in the sense that the need to answer a more specifically-defined scientific question *is* perceived in terms of an emerging commercial possibility. The applied research problems posed here are now likely to derive extensively from a new body of knowledge, and thus be positioned within a complementary range of questions, that resulted from antecedent basic research in the MNE and the perception of possible competitive uses for it.

Additional funding clearly remains a routinely significant input of MNEs to the applied research component of an NSI. However, it is the ability now to ask very specific and potentially highly-rewarding questions, that are defined within a complementary body of supportive new basic research, that becomes relatively more important. This can happen in two ways. First, as part of the wider group-defined GIS program. Secondly, in a more self-contained (horizontal) fashion, where the IIL is able to carry through to an applied research context the investigation of issues deriving from its own basic work.

In ways that closely resemble those indicated earlier for basic research the MNE inputs can both enrich generally, and refocus the content of, the applied research component of an NSI. However, the latter element may have particularly significant consequences here. While MNEs' involvement may well increase the amount and scientific quality of the applied research done in an NSI it will mean that an increased proportion of output is likely to be diverted away from support of the local economy towards the global technological and innovation programs of the sponsoring companies.

Of course it remains possible that an MNE's important applied research results could feed forward into product development operations in the same NSI. Two factors endemic to the global aims and options of MNEs can mitigate against this, however. First, even if an IIL has resolved all the scientific questions relating to an innovation (or, at least, has access to all the needed technology) it may still be that the MNE's operations in the same country do not provide the ideal context for the fulfillment of the commercial innovation process. Thus the other functional inputs that complement science in the process of innovation may be perceived as better equipped to take forward the commercial activation of the applied work in another location. Even where an innovation is self-contained around a small body of new science from one precompetitive research location the parent group's evaluation of its range of options in other functional inputs may provoke the outward leakage of the new knowledge to the benefit of other NSIs. Secondly, the applied research done in an IIL may in fact be far from self-contained, and it is then innate to its role in a GIS that the results will flow from the NSI to feed into a broader program that is co-ordinated elsewhere. Though the results may ultimately be a significant component in building the technological base of a NPC this cannot influence the country's participation in the eventual commercialization process.

17.4.3 Product development

MNEs' involvement in the product development facet of an NSI is operationalized through two institutions; a product mandate (PM) subsidiary and an associated locally integrated laboratory (LIL). The PM, we recall, takes full responsibility for the innovation, initial production, marketing and subsequent competitive evolution of a product. The LIL is a central element in the PM's creativity, and essentially mediates the new science emerging from applied research to the other functions involved in the innovation process. The framework envisages two scenarios through which new technology enters an MNE's operations in the product development stage of an NSI.

First, the PM's innovation process may be predominantly driven by the results of applied research secured through the internalized (horizontal) transfer of scientific output from precompetitive IIL-type work in the same country. This independent and self-contained subsidiary-level approach to innovation will also need, and be partly defined by, top-quality marketing inputs, since it will derive locally the broad new product concept (NPC) as well as then fill in the precise details of its commercialized form. Although the scientists recruited for the LIL will embody less distinctive capacities than the basic researchers of IILs, they will still need the talent to comprehend, articulate and apply new results and often, therefore, represent a significant opportunity cost in terms of their non-availability to indigenous enterprise. As suggested, the PM's market researchers are certainly also likely to be among the most creative available and their recruitment, again, represents a significant opportunity cost to local firms.

Against this (and subject to the usual counterfactual uncertainties) this scenario has powerful potential to enrich the product innovation phase of an NSI. Notably it means that where an MNE's involvement in the previous stages has generated distinctive outputs in precompetitive science these are co-opted to secure equally distinctive outcomes in market competitiveness. Also it means that a product innovation secured in a PM can have greater competitive impact than would a comparable one in a local enterprise, since its markets may be determined by the MNE's wider global perspectives. Thus, we can suggest, if a new good is derived essentially within one NSI it is likely to represent a unique addition to the group's product range and potentially have access to all the world's markets. We thus designate the subsidiary activating this self-contained scenario as a *world* product mandate (WPM).

The alternative GIS scenario determining the role of the RPM/LIL nexus in innovation involves it in the completion of a process begun elsewhere in the MNE group, by picking up an outline NPC and its associated science and generating from it a fully market-responsive and efficiently-produced good.

Where a RPM wins the responsibility to fill in the competitive detail of an NPC it is likely to do so by asserting the capacity of its host NSI to supply the necessary creative inputs (scientists, marketing, engineering), rather than any potentials for low-cost production possessed by the wider local economy.[13] To do this, we have noted, it will be seeking to recruit top quality local personnel with potentially high opportunity cost to indigenous firms' innovation activity.[14] Against this, potential external benefits or spillovers may also be discerned in an enhanced learning scope available to local personnel that are recruited for the RPM/LIL's operations. Thus the process of assimilating the technology and ideas underpinning the NPC will involve these personnel in interaction with the high-level creative activity of the MNE group, which may inculcate valuable new attitudes and perceptions on the organizational procedures of efficient innovation. This will immediately strengthen the NSI by deepening the LIL's in-house ability to retain and enhance its position in its group's innovation network. Beyond this the increased experience of these personnel in the formulation and operationalization of innovation processes (additional to an enhanced competence in their specialized functional area) may be of immense value if they decide to move back into indigenous enterprise.[15]

Overall there are realistic reasons for an expectation that MNEs' use of the RPM/LIL nexus in a globalized approach to innovation can interject knowledge and resources into individual NSIs in a manner that can result in enhanced competitiveness of the local economy. The sources of this competitiveness in technology and creative personnel impute to these MNE operations a role in generating a country's dynamic comparative advantage. Within the host economy the immediate benefits of this competitiveness may take the form of new and higher-quality products available to consumers and an upgraded employment structure encompassing higher-value jobs. However, the innate involvement of RPM/LIL operations in MNEs' global strategies also implies the benefit of improved *international* competitiveness. Thus the key manifestation of success for the MNE takes the form of significant exports from the host country, as the PM fulfills its responsibility to supply new goods to, at least, parts of the company's global markets. The origins of this in successful innovation characterize this as technology gap trade.

17.4.4 Adaptation

While the previous stage of our NSI involved the generation of the original competitive use of an initially disembodied new source of technology, here the final stage addresses the frequent need to adapt technologies that are already successfully embodied in established competitive goods in order to sharpen their value in the specific context of a particular national economy. Our understanding of the global competition strategies of MNEs suggests that production of an existing good may commence in a new country for two possible reasons.

First, a market-seeking truncated miniature replica (TMR) subsidiary pursues the objective of increasing the returns from supply of the good to local consumers.[16] Traditionally a dominant factor provoking this was avoidance of restraints on trade, leading to the interpretation (Kojima, 1978) of such a relocation of production as sub-optimal, trade-destroying, behavior. In more recent perspectives of competitive strategy this localization of supply may, instead, acknowledge the distinctiveness of certain important markets and the advantage to then be obtained from individualized responsiveness to their idiosyncratic needs. Thus the TMR, working with an SL1 lab and a strong marketing group, will adapt the goods (and perhaps, as a result, also the production technology) to enhance their competitiveness in the perception of local consumers. The core benefit to the host economy is then manifest in greater consumer satisfaction. To the extent that this then results in greater demand, successful TMR/SL1 activity can also expand employment and tax revenues.

The second motivation (included in the GIS of Figure 17.1) for starting production of an established good in a new location is the efficiency-seeking one of securing an additional low-cost source of supply to an increasingly price-competitive international market. Thus an RPS is expected to realize economies of scale and utilize the host-country's most abundant inputs, in order to assert itself as a specialized low-cost location in an MNE's supply network for successful mature products. The importance, in an MNE's supply network rationalization process, of locating manufacture of mature goods in countries where the abundant inputs fit the existing production techniques underpins a trade-orientation that helps activate these countries' sources of static comparative advantage (Pearce, 2001). Thus it is construed (Kojima, 1978) as a welfare-enhancing trade-creating mode of behavior. In some cases MNEs may feel the need to optimize this process by adapting the production technique to further sharpen its match with available local inputs. This may be done locally (in-house) through an SL1 or externally through advice from an SL2 that is already familiar with the process technology.

17.5 Conclusions

The key theme of this analysis is that contemporary MNEs address globalization as a strengthening of the context in which they can leverage *difference* between economic areas[17] in terms of tastes, production capacities and technological and research scope. Paralleling this is the view that as countries' development proceeds the strength and distinctiveness of their NSIs increasingly define the extent and form of their international competitiveness. The scenarios reviewed here indicate the generalized way in which the interjection of MNEs' global programs for technological and competitive enrichment can strengthen individual NSIs, but also that the GIS's tendency to do this on a selective basis can alter the balance of an NSI (between stages) and its content (what is done in a particular stage). It is the need to understand the qualitative detail of MNEs' participation, rather than quantitative extent of attraction, that is the basic policy recommendation here. This can take the form of two, complementary, warnings.

The first is a warning against a short-termist over-emphasis on attracting PM/LIL operations as a key element in a country's product development activities. Though this can strengthen the immediate scope for innovation-based trade success it often does so in a way that diminishes the depths of the roots of such competitiveness in the host economy's wider capabilities. Thus, basing localized innovation on technology and product concepts already generated by the MNE can shortcut aspects of the development sequence, but also makes it extremely dependent and with a diminished reflection of distinctive localized science and competences. Sacrificing 'backward' roots in the NSI's precompetitive activity increases the dependency of the country's competitive development (innovation) on both the ability of MNEs' wider GIS to generate new product concepts, and the ability of the economy to supply the types of skills needed to *play a role* within a mainly externally-driven creative process. Lack of real roots could make MNEs' product development activity in a country almost as potentially 'footloose' as cost-based supply operations.

The second warning is the corollary of the above, suggesting that governments should normally also welcome MNE participation in basic and applied research activity. Three factors in our analysis could lead governments to display reluctance in attracting IIL-type operations. First, that the results of successful IIL research may leak from the NSI (to fuel MNEs' operations elsewhere). Secondly, that precompetitive work is per se expensive, unpredictable and high-risk. Thirdly, and opportunistically, that the product development stage can now depend on MNEs' global operations as sources for key scientific inputs.[18] As observed above, allowing the emergence of an imbalance in an NSI (towards product development) has innate short-termist risks by diminishing potential sources of creative sustainability. With explicit regard to IILs we have suggested that, even where the next *use* of powerful research results may be elsewhere in the MNE's network, these units can still strengthen an NSI in significant ways.

First, it is reasonable to treat new basic/applied results as a public good. Thus, even when utilized outside the NSI by an MNE, such results do not become lost or forgotten, and can still represent a useable part of the technology stock of the country's science-base. Secondly, the NSI may be strengthened through its human capital (scientists). At an extreme, MNEs' IIL operations may retain the local participation of top-quality scientists, who might otherwise migrate. More generally, MNE involvement with precompetitive science may benefit the motivation and performance of local researchers by (in addition to salary improvements) setting them more interesting challenges. This can stem from placing their work at the creative interface of two science communities, that of the MNE (its GIS) and that of the host NSI.

Overall, then, the plea is for a policy based on a careful and detailed understanding of heterogeneity and distinctiveness; both in the interdependencies between elements of an NSI and in the range of technological needs and strategic motivations in MNEs.

Notes

1 This is implied in Vernon's exposition of the first stage, but becomes more clear when the second stage analysis focuses on explaining how the sources of competitiveness generated in the home country innovation process provide the firm with the capability to enter overseas markets for the first time (i.e. become MNEs).

2 In the practicalities of Vernon's modeling the product will have been innovated in the US and eventually have come within the income and taste compass of (mainly) European consumers.

3 See, for example, Grandstrand and Sjolander (1992).

4 As early as the late 1970s Giddy (1978) noted that the product cycle was becoming 'highly compressed' into a series of near simultaneous innovations in several major markets.

5 We do not argue that this represents an approach that is pervasive in its full articulation among current MNEs. Rather it is intended to serve, for analytical

purposes, as a summary of various institutional forms (e.g. laboratories) and organizational approaches and linkages that we *do* believe have been validated in the wider literature and in our own research results.

6 The main focus of our analysis here is built around the science and technology inputs as pursued through different types of R&D units. This is not to play down the importance of market research in both seeking out new product ideas and needs from customers as early inputs into innovation (stage one) or in later helping to develop the optimal variants of new products for the tastes and conditions of particular market areas (stage two). However, it is the R&D networks that we see as being consciously articulated by MNEs as a response to *ex ante* perceptions of *different* scientific potentials and technological competences of particular NSIs. Similarly it is the policy commitment to scientific and technological progress that generates the distinctive strengths of NSIs as sources of potential economic development pursued by governments.

7 For elaboration of the typology of MNE laboratories used here see Papanastassiou and Pearce (1999, pp. 149–59), Pearce (1999b). It is originally derived from Hood and Young (1982) and Haug et al. (1983).

8 This reflects one manifestation of the importance of interdependent individualism (Papanastassiou and Pearce, 1998, 1999) in the innovation programs of MNEs (also relevant to PM subsidiaries and their LILs). Here one element (laboratory or subsidiary) seeks to build up a core of individualized competitive competences, but is also willing to operate interdependently by seeking supplementary advice from other parts of the group when needed, and similarly to provide support to other parts of the group from its own specialized competences.

9 Thus part of the central laboratory's responsibility can be seen as the generation of a sense of the operation of procedural justice (Kim and Mauborgne, 1991, 1993; Taggart, 1997, 1999) within the collaborative R&D network.

10 Of course the detail of fully-developed analyses of NSIs (Lundvall, 1992; Nelson, 1993) focuses on the social institutions, competitive organizations and collaborative arrangements whose interdependencies comprise the activity at the different stages. However, we can then present our analysis as, in essence, focusing on the activity and effects of one subset of these institutions and systems, in the form of those involved in GISs of MNEs.

11 Though an element of crowding-out may also apply to the physical capital (infrastructure), it is likely to be more precise for human capital (scientists). Thus for infrastructure (laboratory capacity) net expansion is clearly feasible. The host-country stock of the quality of researchers sought by MNEs is, however, fixed at a particular point in time.

12 For discussion of project mobility in MNEs' science and innovation programs see Pearce and Singh (1992, pp. 73–5, 144–5). The associated ideas of 'reverse transfer' of technology in MNEs has been analysed by Yamin (1999) and Håkanson and Nobel (2000, 2001).

13 In fact this does not imply *inefficient* production of the new goods because the innovation process can seek to engineer production technologies that are oriented to use of the available input mix and competences.

14 The demands on local marketing personnel will be less here, however, since they do not create the truly path-breaking aspects of a NPC but only focus on its locally-responsive refinement.

15 Indeed a certain frustration with the somewhat dependent position of an MNE's PM/LIL activity could encourage such personnel to pursue more individualized

creative scope in a local enterprise once they have acquired the confidence to do so.

16 A truncated miniature replica is expected to supply most of the product range of the group (as a 'miniature replica' of the parent) to its local host-country market, but will be functionally 'truncated' through the absence of creative scope (no role in innovation) and strategic decision-making.

17 Discussed for convenience during the exposition as national economies, but also applicable to wider (integrated) areas or subregions (clusters).

18 In fact if all, or most, developed industrial economies adopted the same opportunistic approach and lowered support for basic/applied research the ultimate result would be a slowing of technology-driven economic progress world-wide. It could then be MNEs that perceive this first and most clearly. A response could then be an attempt to reinvigorate precompetitive investigation through commitment to IIL networks.

References

Birkinshaw, J.M., 'Approaching heterarchy: a review of the literature on multinational strategy and structure', *Advances in International Comparative Management*, 9 (1994) 111–44.

Giddy, I.H., 'The demise of the product cycle model in international business theory', *Columbia Journal of World Business*, 13(1) (1978) 90–7.

Grandstrand, O. and Sjolander, S., 'Internationalisation and diversification of multi-technology corporations', in Grandstrand, O., Håkanson, L. and Sjolander, S. (eds), *Technology Management and International Business* (Chichester: Wiley, 1992) pp. 181–207.

Håkanson, L. and Nobel, R., 'Technology characteristics and reverse technology transfer', *Management International Review*, 40 (2000) 29–48.

Håkanson, L. and Nobel, R., 'Organisational characteristics and reverse technology transfer', *Management International Review*, 41 (2001) 395–420.

Haug, P., Hood, N. and Young, S., 'R&D intensity in the affiliates of US-owned electronics companies manufacturing in Scotland', *Regional Studies*, 17 (1983) 383–92.

Hedlund, G., 'The hypermodern MNC: a heterarchy?' *Human Resource Management*, 25 (1986) 9–35.

Hedlund, G., 'Assumptions of hierarchy and heterarchy, with applications to the management of the multinational corporation', in Ghoshal, S. and Westney, E. (eds), *Organisation Theory and the Multinational Corporation* (London: Macmillan, 1993) pp. 211–36.

Hood, N. and Young, S., 'US multinational R&D: corporate strategies and policy implications for the UK', *Multinational Business*, 2 (1982) 10–23.

Kim, C. and Mauborgne, R.A., 'Implementing global strategies: the role of procedural justice', *Strategic Management Journal*, 12 (1991) 125–43.

Kim, C. and Mauborgne, R.A., 'Procedural justice, attitudes and subsidiary top management compliance with multinationals' corporate strategic decisions', *Academy of Management Journal*, 36(3) (1993) 502–26.

Kojima, K., *Direct Foreign Investment: A Japanese Model of Multinational Business Operations* (London: Croom Helm, 1978).

Lundvall, B.A. (ed.), *National Systems of Innovation: Towards a Theory of Innovation and Interactive Learning* (London: Pinter, 1992).

Nelson, R.R. (ed.), *National Systems of Innovation: A Comparative Study* (Oxford: Oxford University Press, 1993).

Papanastassiou, M. and Pearce, R., 'Individualism and interdependence in the technological development of MNEs: the strategic positioning of R&D in overseas subsidiaries', in Birkinshaw, J. and Hood, N. (eds), *Multinational Corporate Evolution and Subsidiary Development* (London: Macmillan, 1998) pp. 50–75.

Papanastassiou, M. and Pearce, R., *Multinationals, Technology and National Competitiveness* (Cheltenham: Edward Elgar, 1999).

Pearce, R., *Global Competition and Technology* (Basingstoke: Macmillan, 1997).

Pearce, R., 'The evolution of technology in multinational enterprises: the role of creative subsidiaries', *International Business Review*, 8(2) (1999a) 125–48.

Pearce, R., 'Decentralised R&D and strategic competitiveness: globalised approaches to generation and use of technology in multinational enterprises', *Research Policy*, 28(2–3) (1999b) 157–78.

Pearce, R., 'Multinationals and industrialisation: the bases of "inward investment" policy', *International Journal of the Economics of Business*, 8(1) (2001) 51–73.

Pearce, R., 'Globalisation and development: an international business strategy approach', *Transnational Corporations* (forthcoming).

Pearce, R. and Papanastassiou, M., *The Technological Competitiveness of Japanese Multinationals: The European Dimension* (Ann Arbor: University of Michigan Press, 1996).

Pearce, R. and Singh, S., *Globalising Research and Development* (Basingstoke: Macmillan, 1992).

Taggart, J.H., 'Autonomy and procedural justice: a framework for evaluating subsidiary strategy', *Journal of International Business Studies*, 28(1) (1997) 51–76.

Taggart, J.H., 'MNC subsidiary performance, risk and corporate expectation', *International Business Review*, 8(2) (1999) 233–55.

Vernon, R., 'International investment and international trade in the product cycle', *Quarterly Journal of Economics*, 88 (1966) 190–207.

Vernon, R., 'The product cycle hypothesis in a new international environment', *Oxford Bulletin of Economics and Statistics*, 4 (1979) 255–67.

Yamin, M., 'An evolutionary analysis of subsidiary innovation and 'reverse' technology transfer in multinational companies', in Burton, F., Chapman, M. and Cross, A. (eds), *International Business Organisation* (Basingstoke: Macmillan, 1999) pp. 67–82.

18

Conclusions and Policy Issues

Ana Teresa Tavares and Aurora Teixeira

18.1 Conclusions

Multinationals, innovation and regional development are, and have been for a long time, topical and debatable issues. Despite the vast bodies of literature focusing on each of such aspects, and the fact that they are undoubtedly linked, there is a need to contribute to a more integrated understanding of how these issues relate and interact, as their very interaction underpins to a considerable extent the prosperity of countries and regions. This volume aims exactly at shedding light onto the complex and dynamic, yet specific, relationships that can be established between multinational firms, key actors in today's global economy, industrial clusters (as agglomeration and proximity are deemed important to promote linkages and innovation), and the innovative activity itself. Underlying this project is also the need to develop and specify the policy implications arising from these phenomena, especially in terms of devising concrete scenarios and making practical proposals for positive policy intervention, so that the developmental impact of multinationals, clustering processes and innovative activities are maximized.

It is widely accepted that multinationals impact on their host economies at multiple levels, leading to various kinds of (positive or negative) externalities, generally referred to as 'spillovers' (Blömstrom, 1986; Blömstrom and Kokko, 1998; for a survey see Tavares and Young, 2005). One of the aspects that provides more scope to stimulate spillovers is the generation of linkages between domestic and foreign-owned firms. This is exactly one of the key themes explored in the book, together with innovative activities and clustering, as will be emphasized later.

General evidence gathered in the present volume seems to support a positive view of multinationals' impact on host economies, notably through innovation. For instance, Girma, Gong and Görg (Chapter 4) find that FDI originating from OECD countries and domestic-market seeking FDI are significant channels of innovation spillovers, and in-house R&D and employees' skill upgrading are important determinants of firms' innovation activity. The positive impact of multinationals through innovative activities is also highlighted by Damijan, Jaklič and Rojec (Chapter 3). The authors conclude that in the case of Slovenia, MNEs' own R&D expenditures and previous innovation activity (internal sources) are consistently confirmed as significant determinants of innovation activity; however, this innovative process is much more efficient when accompanied by diffusion of knowledge from outside sources. Thus, they found that external knowledge spillovers, either domestic or international, are important innovation incentives. Specifically, R&D subsidies, both domestic and from international sources, and intra-sectoral innovation spillovers complement internal sources and increase considerably the ability of Slovenian firms to innovate. Inward FDI also significantly increases firms' ability to innovate. Compared to their domestic counterparts, foreign-owned firms even show a lower average level of R&D expenditures, suggesting that innovation activity must be driven by other factors such as knowledge and technology spillovers. Interestingly Iammarino, Sanna-Randaccio and Savona (Chapter 5), analysing the case of Italy, a country with a relatively weak magnitude of inward investment (relative to the size of the economy), conclude that innovative firms, and particularly MNEs, seem to be more aware of the problems encountered as compared to non-innovative firms. They note an interesting positive structural association between firms' perception of obstacles to innovation and their propensity to innovate.

Another main theme dealt with here refers to linkages between foreign-owned and domestic firms. Key conclusions of this volume in this regard are as follows. It is shown that the linkage effect of MNEs on the host country is more likely to be favorable when multinationals use intermediate goods intensively, there are large costs of communication between the headquarters and the production plant, and the home and host countries are not too different in terms of the variety of intermediate goods produced. However, if these conditions are reversed, multinationals may create 'enclave economies' within host countries (Rodríguez-Clare, 1996).

Delving further into the topic of linkages established between multinationals and their local environment, Giroud (Chapter 11) discusses potential substantial impact that MNEs have on Malaysia's economic development, both directly (through their activities) and indirectly, looking specifically at inter-firm technology transfer through backward linkages. She found that for the period in analysis suppliers in Malaysia improved substantially aspects such as delivery, quality, costs and service focus, in line with requirements by foreign-owned firms. Such evidence seems to indicate the rise of specific skills/capabilities in the country. A similar outcome, this time at the level of the firm, was encountered by Varum (Chapter 14). Her findings support the occurrence of positive spillovers derived from international buyer–supplier relations, demonstrating that such relations motivated and forced domestic suppliers to develop their practices and capabilities. She emphasizes that most of these relationships occur within global production networks, creating new opportunities for international knowledge diffusion that suppliers should strive to exploit.

Moreover, in the case of Brazil, the findings and analyses of Figueiredo (Chapter 12) show that outward-looking industrialization has stimulated firms located in that country to improve their export performance. The technological capability of most MNE subsidiaries and local firms, far from being confined to very basic levels for long periods, has consistently been upgraded to carry out diverse types of innovative activity.

Despite all the positive impacts of multinationals, Dachs and Ebersberger (Chapter 2) note some difficulties foreign MNEs have when trying to tap into local (Austrian) knowledge. They justify such inability by the fact that MNEs may already possess knowledge superior to that accessible in that host country, and because the transfer of such knowledge from the affiliate to the parent company is not cost-free. Dachs and Ebersberger nevertheless emphasize the small differences in the Austrian case between foreign-owned and domestically-owned companies, and the surprisingly high percentage of foreign affiliates tapping into local knowledge, which they consider as 'good news for policy'.

Ferreira, Tavares and Hesterly (Chapter 6) also recognize that the local subsidiary does not always undertake high value-added activities, or more sophisticated, than those of indigenous firms. In such cases they state, spillovers may be limited, as they probably are when the subsidiary has too large a technological gap vis-à-vis local firms. In the case of Malaysia, Giroud stresses the little impact of FDI on innovative capabilities, which may probably be attributed to a similar reason. A parallel outcome is conveyed in Costa, Videira and Veloso (Chapter 13) who concluded that, although the Autoeuropa investment created significant value to the Portuguese economy, not only did it forge an operation that had much higher level of internal value creation than any existing assembly line in Portugal at the time, but it also grew to represent now 75 per cent of the income generated in the auto sector, while creating an important demand for almost all other sectors that supply this industry. The economic lever of the investment was largely determined by greater value added in the unit itself. Its indirect effect on the rest of the economy is equivalent to other assembly lines in terms of quality or intrinsic value. Such results suggest that the arrival of the Ford–VW Group in the early 1990s provided an important scale factor in the assembly lines sector and in the purchases from its local supply chain, but it did not imply a substantial difference in value and diversity of first-line supply offer in the national industrial fabric.

One way to overcome difficulties in developing linkages might be through clustering, in particular when led by 'flagship' firms. Ferreira et al. (Chapter 6) recognize that the scenario of having a MNE as the cluster's flagship may be interesting from a host economy perspective – especially for those without a very consolidated/sophisticated industrial base, a foreign subsidiary may be a source of novel technologies and business models. Thus, if the foreign MNE operates in an industry not present in the host economy, or if such subsidiary is a step above local counterparts in the same or related industries, clearly the host economy may have scope to take advantage of, for instance, relevant demonstration and learning effects. Nonetheless, having a MNE as a flagship bears an additional risk, as MNEs may be more footloose (less rooted) than domestic firms. Foreign-owned subsidiaries tend to have less local strings and carry out divestments and/or relocations more easily, with potential dramatic effects on the local economic and social environment. In turn, one could also counter-argue that, because MNEs have by definition their international network and are used to operate flexibly they have usually greater survival probability than less flexible non-internationalized firms. This argument is harder to defend when the domestic-owned firm would be itself a MNE, in which case it could arguably have a similar potential advantage of a foreign-owned firm. Hence, the idea that there are alternative pathways to industrial development, based on MNEs and/or local firms, is again present. A key conclusion that is restated here is the potential that MNEs have to bring distinctive resources and capabilities to a host economy, but that these possible benefits may not ensue if there are no linkages with the local industrial fabric, notably when the indigenous sector is not able to become a potential partner of a given subsidiary.

Following up on this argument that the host economy's characteristics matter, this less researched direction – host country's environment influence on MNE affiliates – is investigated by Dachs and Ebersberger (Chapter 2). They express the idea that spillovers from the host country to MNE affiliates are extremely relevant to science and technology policy. If innovative MNE affiliates mainly exchange knowledge within the group to which they belong, and have only little or no exchange with the host country, one may conclude that MNEs are not dependent on local knowledge in the innovation process ('home base exploiting'). In this case MNE research, development and innovation activities are therefore extremely mobile, and can be moved elsewhere to benefit from cost advantages without losing spillovers from the host-country knowledge base. On the contrary, if one finds strong links with domestic enterprises and universities ('home base augmenting'), it may be an indication that the innovative activities of MNEs are less 'footlose' than the critics of globalization assume. Results obtained by Dachs and Ebersberger tend to confirm the behavior predicted in the IB literature – foreign-owned companies in Austria prefer to rely on MNE internal sources rather than on external knowledge absorbed by either formal co-operation or informal channels. This case reflects the 'home base exploiting' rather than the 'home base augmenting' view with respect to the intensity with which foreign-owned companies draw on knowledge bases in their home countries. Nonetheless, they conclude that a high proportion of foreign affiliates often enter into co-operative arrangements.

Another relevant element for this discussion relies on the recognition that MNE subsidiaries are heterogeneous, and, therefore, their potential impact on the host economy is also likely to be asymmetric according to the strategic roles performed by the locally based subsidiaries, as well as the evolution of such subsidiaries' status within the MNE group. This issue is related to the important aspect of embeddedness, for some subsidiaries may be perfect 'cathedrals in the desert', or, on the contrary, develop important linkages with the host environment. One of the critical variables when analysing subsidiaries' local embeddeness is the level of autonomy enjoyed by the subsidiary (vis-à-vis its MNE group). Regarding autonomy, the usual hypothesis is that subsidiaries less integrated in their groups tend to generate greater network benefits than highly integrated ones. It has been argued that autonomous subsidiaries tend to be more locally embedded, and that local embeddedness may create uniqueness that underlies getting more status, thus autonomy, from the group. However, as Männik and von Tunzelmann (Chapter 15) note, there seems to be an interesting U-shaped relationship whereby too much autonomy may mean that the subsidiary is 'stand alone', not sharing much with the group, and that such behavior severely limits technological exchange and other forms of information/knowledge-based interactions. Having neither too much autonomy from nor too much dependence on the parent company positively affects the increase in productivity, the level of technology and product quality. Some intermediate position of technological autonomy satisfies the parent company for achieving the maximum subsidiary performance in the host economy (the host countries studied are CEECs). The authors also highlight that subsidiaries use both the local potential and international sources for technology upgrading. From the perspective of international knowledge flows, subsidiaries locating in technologically underdeveloped transition countries tend to maintain deep corporate links and this does imply the presence of an additional knowledge inflow. From another perspective, the subsidiary exploits the local knowledge sources, which in turn are expected to develop local innovation potential.

The importance of the local context, and in particular the dynamic of the local business environment, has also been discussed in IB research, especially that focused on subsidiaries (Birkinshaw and Hood, 2000; Holm et al., 2002). This strand of research often looks at the strength/relevance of the local context versus corporate control as determinants of subsidiary evolution. Several chapters (e.g., Chapter 9 by Pedersen, and Chapter 10 by Andersson and Persson) concluded that embeddedness in a set of relationships is conducive to organizational value creation and achievement of economic goals. An important implication of network embeddedness is that organizations may develop higher degree of trust towards each other, something that can limit expectations of opportunistic behavior and as such reduce governance cost (an issue also strongly emphasized by Ferreira et al. in Chapter 6). These interactions and lower transaction costs facilitate business in general and exploitation of opportunities, as they ease exchange and contracting out work to other firms that have complementary resources and skills; in turn, inter-firm linkages and co-operation tend to generate spillovers, leading to a self-reinforcing process and to a virtuous dynamic.

A relevant facet of local embeddedness consists in the fact that MNEs are fraught by the 'costs of foreignness', i.e. they are at a disadvantage vis-à-vis their domestic counterparts in such aspects as establishing networking ties with local firms. However, these costs may be offset by reputation and other 'ownership advantages' (Dunning, 1977) that MNEs may hold that make local network stakeholders want to relate to the foreign subsidiary. Examining the factors of subsidiary development Pedersen (Chapter 9) concludes that although the most important driver of subsidiary development is the subsidiary's own set of entrepreneurial initiatives for upgrading its resources and competences, the dynamism of the business environment emerges also as an important source for the upgrading of subsidiary competences. The author finds that subsidiaries with high level of competences tend to be relatively narrow in their scope, while subsidiaries with broader scope have lower average level of competence for the activities they carry out. In this vein, high levels of competence will only be associated with high levels of interdependence when the competence is useful and important for other units in the same multinational group. Pedersen's results indicate that headquarters mainly has discretion over the scope of subsidiary activities and the extent to which subsidiaries are integrated in the MNE network, having less influence on the level of subsidiary competence. In that sense the parent company drivers may seem to have a double character, enhancing one dimension of subsidiary development (scope of activities), but at the cost of another dimension (level of interdependence).

Focusing on MNEs' re-investments, but similarly to Pedersen, Andersson and Persson (Chapter 10) bring to the fore the importance of local embeddedness. They show that subsidiaries that have deep and intense relationships with local business partners tend to become important to other sister units within the MNE. A subsidiary's external embeddedness not only develops knowledge of use for other internal units but also is of importance for the subsidiary's business performance on the host market. Such social capital or 'relational resource' in Tsai and Ghoshal's (1998) sense, positively influences resource exchange and combination, which in turn positively impacts on innovativeness, an aspect also stressed by Ferreira et al. (Chapter 6). As highlighted, the strategies/roles of foreign subsidiaries vary considerably across locations and across clusters, and are associated to the dynamism of the cluster and to the overall level of foreign ownership: subsidiaries in leading edge clusters are more embedded than in other sectors (Birkinshaw and Hood, 1998).

The issue of clustering and its relation with MNEs was also extensively discussed in the present volume. In the case of Ireland, Gleeson et al. (Chapter 7) stress that local firms and MNEs exhibit a similar trend in spatial concentration, growing more clustered from 1985 until the early 1990s before tending toward greater spatial dispersion up to 2001. Throughout the period, the level of spatial concentration among MNEs remained far greater than that of local firms. This trend from greater to less pronounced spatial concentration since the late 1990s in part reflects the spatial policy direction implemented at the time, which focused on developing the regional areas of Ireland while at the same time reducing congestion in the main cities of Dublin and Cork. However, as the authors argue, this effect can only partly be due to policy as the real emphasis in policies toward spatial dispersal came late in the 1990s when the economy reached full employment, while it is evident that the process of spatial dispersal had begun in the early 1990s. This suggests that there were market forces already driving enterprises out of the more concentrated locations so that while policy reinforced those pressures, it cannot be said to have led them.

Following up this interest in policy measures and the impact of such policies (and other determinants) on industrial development and clustering processes, notably by analysing whether policy-driven high-tech cluster development is a feasible strategy by 'peripheral' countries, Peters and Young (Chapter 8) investigate the cases of Ireland and Scotland. These two economies have both invested heavily in biotechnology over an extended period; yet, Peters and Young conclude that these economies are far from possessing anything like a fully developed cluster or an innovative milieu. It has proved difficult in Scotland to attract talent from the Cambridge/London cluster, let alone internationally; while leading figures have exited Scotland for improved prospects abroad. According to the authors, Ireland is better placed in this regard because of its better-integrated global diaspora. The example of Ireland highlights the challenges for public policy in interfering on the structure of the biotech sector, and seeking to nurture a flagship firm. There are attractions to this approach that can be integrated closely with a cluster model of development, but the risks of picking winners are nevertheless high. In Scotland, while the number of players suggests a buoyant, innovative industry, most are small start-ups. The Scottish approach is termed a 'nursery-model', where an environment of early-stage development revolving around university science is encouraged. The high costs and risks of progressing along the value chain suggest high failure rates for such firms and a high propensity for mergers and acquisitions. Policy-makers are faced with high funding demands for only a handful of embryonic dominant players that may emerge from the pack and have the potential to go to the market.

The Malaysian government, targeting the development of specific industrial agglomerations, has also adopted a cluster-based approach. One of the main objectives in terms of clustering was related to international linkages generated among the electronics and electrical appliances, and textiles industries. With lower levels of inward FDI, in part due to the emergence of China, the Malaysian government faces the challenge of nurturing its own industries and enhancing their competitiveness in the international markets (Giroud, Chapter 11).

The role of public policy is considerably highlighted by a number of chapters in the present volume. A hot topic nowadays is whether tax incentives are significant influences on FDI attraction. Focusing on the CEECs, and contrary to earlier evidence suggesting a minor impact of taxes on FDI in those countries, Bellak and Leibrecht's (Chapter 16) econometric analysis of FDI flows into five East European host countries from their seven most important home countries shows the significance of the tax rate as an inward investment determinant.

Moreover, Girma et al. (Chapter 4) recommend that notwithstanding the benefits of being the 'export-processing zone of the world', Chinese policy-makers should strive harder to attract foreign investors that are likely to have stronger linkages with the domestic economy. They recognized nevertheless that the scope for ensuring the flow of the 'right kind' of FDI – high-tech domestic market oriented – can be rather limited, as placing performance requirements on foreign invested firms is against the rules of the WTO, which China joined in 2001. Thus, in their regard, getting the firm level fundamentals right – the right organizational and incentive structures to develop adequate technical capacity – seems to be the most potent policy option conducive to fostering SOEs' innovative capacity.

Last but not least, Pearce and Papanastassiou (Chapter 17) claim that, overall, the plea is for a policy based on a careful and detailed understanding of heterogeneity and distinctiveness (referred to earlier, in terms of the scope of MNE subsidiaries); both in the interdependencies between elements of a National System of Innovation, and in the range of technological needs and strategic motivations in MNEs. According to them, and to several other contributions in this book, there is the need to understand the qualitative detail of MNEs' participation, rather than merely focusing on the quantitative extent of inward investment attraction. Although not neglecting the potential effects of important in-flows of foreign capital, the heart of the matter is that qualitative dimensions are what makes the most substantial difference as regards the overall impact of MNEs' activities in host economies, and what may stimulate often needed structural change in countries and regions eager to grow and develop rapidly. For this reason, qualitative dimensions were repeatedly emphasized throughout this volume. Because the prosperity of host (and home) economies should not be envisaged as a fait accompli, or a static reality, we chose also to focus on, highlight and debate as much as possible relevant policy issues and, when appropriate, issuing specific recommendations for policy-makers.

18.2 Policy recommendations

All contributions included in the present volume unanimously convey that public policy *matters a lot* in what concerns multinationals' activities and their relationship with innovation and local development.

Reflecting the increasing interest and favorable attitude towards FDI and multinationals, a growing number of countries adopted policies specifically focused on attracting inward investment, and the success of such attraction became very important politically for host country governments. Nevertheless, most countries started by adopting piecemeal approaches to woo foreign investors, mainly consisting of offering traditional fiscal and financial incentives, like tax exemptions and grants according to the amount of investment and number of jobs created (to be developed in section 18.2.1 below). Some countries referred to in this book (such as Ireland and Singapore) were pioneers in more systemic FDI policies, complementing them with measures at other levels (e.g. macroeconomic policies, measures to improve the supply of qualified human capital, provision of infrastructure, and so forth). Yet, the majority of host economies has been, until recently, giving importance only to the *quantity* (amount) of inward investment attracted, neglecting to a large extent the characteristics related to the *quality* of the investment. Another problem in most host economies had to do with policy management and implementation, given that often the responsibility for accompanying policy implementation was not clearly institutionalized.

Recent trends in FDI policy world-wide (documented in several chapters of the book) show that nowadays governments have in general a more discerning approach, benefiting from their own experience and trial and error with policy design and implementation, and from the example of the most successful competitors (e.g. it is not uncommon to see former members of staff of IDA Ireland, considered a paradigm of a successful inward investment agency, helping to shape FDI policy and institutions in other countries, from Costa Rica to Eastern Europe). Hence, a considerable number of countries currently have inward investment attraction agencies/departments, supplemented by a variety of developments in the legal system and a host of proactive measures to enhance their ability to seduce potential foreign investors.

Another positive development is the increasing awareness of the relevance of the qualitative dimension associated to inward investment. Each investment has its own characteristics, and they can vary considerably from project to project. Hence, projects with equal amounts of investment, and the same number of jobs created (the basic criteria mentioned above) may have very distinct types and magnitude of impact. Even projects by apparently similar companies, for instance, from the same sector, may represent significantly different kinds of operation in a specific host economy (e.g. from a simple assembly line to a center of excellence for the group). The increasing consciousness of these asymmetries has motivated policy-makers in several countries to try to discern better the potential quality and characteristics of each investment, in order to calibrate more adequately the type and amount of incentives offered to each project. It is nevertheless a hard task, given that multinationals are more and more demanding in terms of incentives and overall support for their operations, as they capitalize on the head-to-head 'race', or scramble, that countries engage in order to attract FDI (Oxelheim and Ghauri, 2003). This makes the bargain between governments and companies a tough undertaking, and underscores the need for more 'clever' policies rather than escalating grants or more and more tax exemptions. Moreover, the public opinion (and also several academic studies) increasingly question the effectiveness and economic efficiency of the generous incentives provided.

The recent controversy on the effects of globalization, and on the opportunity cost of FDI incentives (motivated mainly by frequent divestments/ closure or re-location of subsidiaries) generated even more debate of the public opinion on these issues, leading to institutions such as local firms' and SME associations to demand greater support, feeling that they were less helped by governments than large multinationals. This was also supplemented by a growing awareness of the real difficulty with FDI: what is hard is to 'embed' foreign companies in the host economy/local context, much more than to attract their operations per se.

Given these developments, governments seem to be increasingly actively involved in stimulating embeddedness, through networking, linkages, agglomeration/clustering and innovative activities (see section 18.2.2 below). The option, of promoting embeddedness, as stated before, is however much more difficult as it requires softer, more integrated policies, often not even FDI-related. These themes will be developed subsequently in this chapter.

There are various motives justifying why it is difficult for MNE subsidiaries to be embedded in the local context: for instance, 1) difficulty in finding suppliers that have enough scale and ability to respond to the strict standards of quality and delivery time that often multinationals require; 2) not unusually the technological gap is too wide between indigenous companies (frequently SMEs) and foreign subsidiaries, this impeding effective spillovers; 3) multinationals, especially those with global operations, tend to have a vast array of choices available as regards suppliers, partners, etc., hence often they find better alternatives globally. All these forces go against local linkages. However, an economy may decide to do something to improve the potential for local linkages, such as: 1) targeting inward investment projects that fit the local context, i.e. for which the host economy has resources and capabilities that already match the needs and requirements of foreign investors; 2) developing these resources and capabilities (e.g. like Ireland did when targeting electronics companies, deciding to stimulate human capital formation – engineers in electronics and information technologies – specifically geared to fit the needs of such investors), or stimulating initiatives like supplier consortia, and purchasing centrals; 3) stimulating cluster formation, hoping that industrial agglomeration will generate more opportunities for information exchange, linkages and synergies among the actors in the cluster. All these themes received great focus throughout the chapters of this book.

Also related to the controversy about the impact of globalization, the skepticism about the effectiveness and efficiency of FDI incentives (questioning their opportunity cost), and the discussion among several actors on what should be done in the future to enhance prosperity, is the current debate between distinct models of development and policy choices. This is about the important strategic options countries must select, as resources are limited. For instance, a country has to decide what model of endogenous development it wants – based on MNEs, on clusters of indigenous SMEs, or in a mix of the two – i.e. indigenous entrepreneurship or foreign owned flagships? All policies have an opportunity cost and FDI policy constitutes no exception. Cluster promotion, namely focused on high-tech related industries, has been recently regarded as a 'miraculous' policy solution in order to foster growth, development and prosperity (we will further develop and discuss this theme in section 18.2.3 below).

As the various contributions in the present volume highlight, these are critically important themes from the point of view of host-country development, and there are no easy answers, no standard recipes for all cases. Hence, even if based on some common principles, FDI (and accompanying policies) need to be tailored, based on an in-depth knowledge of the real circumstances and resources/capabilities associated to each host economy. The next sections will treat some of the key issues summarized above with greater depth, incorporating the specific suggestions and recommendations from the chapters included in this book.

18.2.1 'Traditional' versus 'soft' FDI incentives

As Bellak and Leibrecht (Chapter 16) emphasize, governments in general intervene to influence the location choice of multinational enterprises through distinct measures, namely providing incentive packages, fiscal and non-fiscal, and trying to shape various location factors in order to lower production costs for foreign firms. One location factor that figures prominently in actual policy-making as well as in the public debate is the corporate income tax rate, but many others exist, such as financial grants (provided indiscriminately/non-tied, or tied to specific activities, such as R&D and training of employees), provision of infrastructure and other facilities, and so forth.

Traditionally, tax-lowering strategies have been widely used and relatively successful in attracting FDI in the past. Bellak and Leibrecht's analysis of FDI flows into five East European host countries confirms the importance of the tax rate as an important FDI determinant. Despite such evidence, the bulk of the chapters emphasize that 'softer' policies need to be adopted by governments if they wish to promote local embeddeness by encouraging supply linkages and integrating high quality public infrastructure with the availability of more intangible assets (e.g. human capital and R&D). The chapters by Giroud (Chapter 11), by Dachs and Ebersberger (Chapter 2), and by Girma et al. (Chapter 4), for instance, are very specific about the need for such variants of policy. Costa et al. (Chapter 13) also emphasize very strongly the fact that development impact is highly contingent on the depth of the linkages generated with local firms.

18.2.2 Beyond attraction: promoting embeddedness

Several countries have been proactive both in terms of FDI attraction, and as regards the promotion of linkages. As Giroud (Chapter 11) notes, government support in the case of Malaysia was 'really worth it'. She emphasizes that without incentives the Malaysian electronics industry would not have developed. However, the picture is different regarding policies toward the development of supply linkages. Whereas in the case of Singapore the government has been successful in developing its own domestic industrial base, and in attracting inward investment into higher-value-added activities, Malaysian FDI policies are no longer sufficient to make it attractive. According to Giroud, it is imperative that further attention be paid to the deepening of inter-firm relationships.

Also related to the topic of local embeddedness, Dachs and Ebersberger's (Chapter 2) results show that the R&D activities of foreign companies with operations in Austria are less mobile than some may have feared, simply because they rely to a considerable extent on the knowledge and expertise of their home country. They additionally highlight that domestic organizations benefit from the presence of foreign affiliates, since co-operative arrangements are nevertheless frequent, usually implying the mutual exchange of knowledge, and intense contacts that end up increasing the likelihood of spillovers. These authors advise governments to try to stimulate and increase the frequency and quality of foreign subsidiaries' contacts with other organizations in the national innovation system. This may be achieved by providing incentives to enterprises to link up with universities, research centers or other firms in the country (an idea conveyed in several of the chapters included in this volume). A number of OECD countries have already taken action to strengthen such links, particularly between the public science system and industry (OECD, 2004). One of the important side effects of such initiatives may be to embed foreign-owned companies more strongly. Damijan et al. (Chapter 3) call for government's stimulation of innovation activity by encouraging non-innovative firms to launch such activity, increasing R&D subsidies, attracting inward FDI and, in general, creating and upgrading the competitive environment in each sector.

Girma et al. are rather specific, suggesting to the Chinese government a 'two-pronged' policy challenge: to concentrate on attracting high-tech domestic market-oriented FDI, and at the same time to ensure that state-owned enterprises have the right organizational and incentive structures to develop adequate technical capacity. These authors argue that Chinese policy-makers should strive harder to attract foreign investors that are likely to have stronger linkages with the domestic economy.

Some chapters (e.g. Chapters 13 and 14) in this book are focused on buyers–suppliers' linkages, highlighting the importance of such linkages (and their eventual multiplier effect, as Costa et al. argue) both quantitatively (through material linkages) and qualitatively (through demonstration effects and transfer of best practices, for instance in terms of the adoption of quality standards, as Varum notes in Chapter 14). She also emphasizes that there is a need to ensure that domestic firms actually change their routines and do not adopt practices merely formally, so as to get subsidies or quality certification. Varum also highlights the need to adopt after-care policies and post-subsidization monitoring in order to guarantee an effective change of routines and mentalities on the part of local suppliers.

In a complementary perspective, Pearce and Papanastassiou (Chapter 17) put forward 'warnings' against a short-termist over-emphasis on attracting product mandate/locally integrated laboratories' operations as a key element in a country's product development activities. Though this can strengthen the immediate scope for innovation-based trade success it often does so in a way that diminishes the depth of the roots of such competitiveness in the host economy's wider capabilities. According to these authors, sacrificing 'backward' roots in the National System of Innovation's precompetitive activity increases the dependency of the country's competitive development (innovation) on both the ability of MNEs' wider global innovation strategy to generate new product concepts, and the ability of the economy to supply the types of skills needed to play a role within a mainly externally-driven creative process. Lack of real roots could make MNEs' product development activity in a country almost as potentially 'footloose' as cost-based supply operations. Governments should thus promote MNE participation in basic and applied research activity. More generally, MNE involvement with precompetitive science may benefit the motivation and performance of local researchers by (in addition to salary improvements) setting them more interesting challenges.

The integrated modes of knowledge creation within contemporary multinational firms suggest that the quite widespread host location obsession with MNE 'footlooseness' and crowding-out potential has become obsolete. According to Iammarino et al. (Chapter 5), to disregard MNEs' evolutionary and dynamic scope may lead to myopic policy perspectives, which fail to address the possibility for mutual knowledge enrichment for both MNEs and territorial systems, therefore missing fundamental opportunities for local growth. However, how to attract asset-seeking and knowledge-producing foreign investment, or how to promote innovation-conducive environments (i.e. fostering the local capacity to generate, absorb and diffuse knowledge), is something still rather unclear, and further research is still needed to provide a sounder base for public intervention.

18.2.3 Cluster promotion: a panacea?

Intimately related to the issue of embededdness in the local environment, regional agglomerations or clusters are tackled with reasonable depth in some chapters of the book (e.g. Chapter 6, 7 and 8). The relevance of this topic is justified by the failure of several cluster initiatives that have been attempted in many countries (where governments, for example, threw subsidies without a systems view of the cluster and of the networks at stake). It is also justified because regions increasingly seek to figure out ways to promote endogenous growth and development though inward investment, namely, stimulating innovation. As Peters and Young detail, both Ireland and Scotland have invested heavily in promoting life-science clusters over an extended period of time. Although their respective government strategies have been quite different historically, reflecting their country capabilities and attractive potential, they are now converging as Irish policy shifts toward the development of its science-base. The example of Ireland highlights the challenges for public policy in taking a position on the structure of the biotech sector, and seeking to nurture a flagship firm. There are interesting features related to this approach that can be integrated closely with a cluster model of development, but the risks of picking winners are high.

It is important to note, as Ferreira et al. (Chapter 6) rightly state, that for a flagship-led cluster to emerge, the first requirement is the very existence of the flagship. Policy-makers should thus look at the industrial landscape seeking to identify who are the flagships with more 'fertility potential'. Government should thus first take stock of the companies offering more opportunities for spin-offs, and adopt proactive/selective policies to stimulate what appear to be the 'right' ones. This does not mean the former obsession of some countries with 'picking winners' and just giving them subsidies, but rather more sophisticated and targeted measures – for instance, as referred to above, promoting innovation through funding selected projects; promoting ties with universities, either via matchmaking (decreasing transaction costs), supporting joint projects with universities and other research institutions, and so forth. In the case where the flagship is foreign-owned, the government needs to support proactive and very selective FDI attraction policies to the point of identifying a specific company and trying to persuade it to invest (like Ireland and others did, through fine-grained targeting up to the firm/subsidiary level). The 'right' type of firm to chase, as Ferreira et al. affirm, would vary according to the initial conditions of the country, the expected upgrade, extant resources and capabilities of workers. In short, it would presuppose a serious evaluation of the match between the economy's locational determinants (Dunning, 1977) and the firm's characteristics. In the case of potential re-investments a set of measures can be enacted: promote linkages, material and intangible/knowledge-based (as highlighted by several chapters in this volume) – hopefully with spin-offs; trying to upgrade extant innovative activities of the flagship – hence stimulating a 'learning environment'; and making sure labor laws stimulate the design of contracts that do not prevent the use of previously accumulated knowledge. Restrictive employment contracts can seriously hinder entrepreneurship and they are to be avoided where a country wants to stimulate industrial dynamism through new firms and spin-offs.

Further recommendable 'soft' measures relate to stimulating human capital formation and on-the-job training to help employees learn as much as possible before spinning off, more equipped to understand markets and technologies, thus enhancing new firms' survival prospects. Other survival-enhancing policies could be to provide selected business services as small, new ventures have in general few or no resources to establish such services on their own.

Any selective measure to stimulate networking (suppliers' consortia, purchasing centrals, information exchange, increased use of information technology, joint collaborative projects), would potentially bear fruit. So would, in principle, innovation – targeted policies, with an underlying long-term vision of what is needed for the economy's industrial strategy. For instance, intellectual property measures such as assisting SMEs in applying for patents may be an example of a 'soft' measure that would eventually enhance the (measurable) innovative potential of such companies. Also helping companies to acquire university/R&D collaboration and consultants' services, e.g. for strategy definition, and for technology development and commercialization, could help such companies considerably to access resources and competencies that alone they are not able to tap into.

Some of the contributions to this volume emphasize that special attention should be given to policy implementation and evaluation, implying also the courage to correct mistakes. This has not often been done, even when the adopted panaceas proved ineffective. As argued in Chapter 6, one thing is certain: any policy is as good as its implementation; policy experimentation should occur, and the old tactic of subsidy-throwing avoided; policy should be selective, carefully planned and evaluated, always with a long-term, systemic, perspective.

18.2.4 Policy is important, but not sufficient per se

Even if the policies adopted by a host economy are well designed and implemented, there are always other factors that act in parallel toward the final outcome. Hence, policy measures may be useful, though they are seldom sufficient. The local environment has to be able to absorb those policies, and the market and institutional dynamics need to act in complementary ways, sometimes even starting processes that can be enhanced by policy measures, or conversely magnifying the impact of adopted policies (see, for instance, the considerations made by Gleeson et al. in Chapter 7 regarding this aspect). Hence, there is usually (or there must be) a logic of complementarity between policies and the local milieu; and also complementarity between FDI-related policies and other variants of policy (e.g. macroeconomic policies, horizontal policies in terms of education, and so forth). This complementarity indicates that there are no a priori recipes for the best policy, as what constitutes the most appropriate avenue differs among contexts.

For instance, it is clear that linkage promotion policies are highly context specific and need to be adapted to the specific circumstances prevailing in each host economy (UNCTAD, 2001). Thus, as noted by Giroud in Chapter 11, the role of host government is three-fold: promote the competitive development of indigenous firms; promote information exchange and raise awareness of foreign firms' requirements to local firms and of local firms' existence to foreign firms; and address issues specific to foreign firms. As Pedersen (Chapter 9) demonstrates, the main direct effect of policy-makers goes through the creation of a dynamic local business environment, which has a strong positive influence on the level of competence. Recognizing that the direct effect of policy-makers on the scope of activities and level of local-foreign firms' interdependence is very limited, Pedersen's results emphasize that the local business environment is almost as important as own subsidiary initiatives in upgrading the level of competences of a given subsidiary. Damijan et al. (Chapter 3) also reinforce the importance of ensuring that there is an appropriate competitive environment in each sector. Thus, broad systemic market conditions are extremely important to enable the impact of multinationals' activities to materialize to the full.

Also important are the characteristics of the local industrial fabric; for instance, Girma et al. (Chapter 4) are very emphatic about the need to get the firm-level fundamentals right. Without this, linkage and innovation-enhancing policies cannot operate miracles. This can be related to the importance of the absorptive capacity of the local actors, that has been highlighted in most chapters in this volume.

Although a high degree of external embeddedness is important both for a subsidiary's market performance and its position within the MNE, embeddedness as such does not trigger re-investments. As Andersson and Persson (Chapter 10) point out, for policy-makers in a country or region, influencing domestic companies to participate in knowledge development projects together with foreign-owned sub-units can potentially pay off, if these foreign-owned units also become important for the development processes within their MNEs. These authors highlight the relevance of local linkages for positive subsidiary evolution, an issue that is also featured in Männik and von Tunzelmann's Chapter 15, who also note the importance of linkages with the local environment for subsidiary autonomy and thus for the subsidiary's evolutionary prospects.

This book has tried to provide evidence on a wealth of country and industry contexts, aiming not only at being informative and diagnosing potential problems and challenges, but also trying to contribute with practical solutions and policy proposals that could help host countries to capitalize on the potential impact from FDI, from local-foreign firms' linkages, as well as understanding better the agglomeration/clustering and innovation-related phenomena underlying economic growth and prosperity. We do hope this has been a pleasant journey.

References

Birkinshaw, J. and Hood, N., 'Characteristics of foreign subsidiaries in industry clusters', *Journal of International Business Studies*, 31(1) (2000) 141–54.

Birkinshaw, J. and Hood, N., 'Multinational subsidiary evolution: capability and charter change in foreign-owned subsidiary companies', *Strategic Management Journal*, 23(4) (1998) 773–95.

Blomström, M., 'Foreign investment and productive efficiency: the case of Mexico', *Journal of Industrial Economics*, 35 (September) (1986) 97–110.

Blomström, M. and Kokko, A., 'Multinational corporations and spillovers', *Journal of Economic Surveys*, 12(3) (1998) 247–77.

Dunning, J., 'Trade, location of economic activity and the MNE: a search for an eclectic approach', in Ohlin, B., Hesselborn, P. and Wijkman, P. (eds), *The International Allocation of Economic Activity* (London: Macmillan, 1977) pp. 395–418.

Holm, U., Malmberg, A. and Sölvell, Ö., 'MNEs impact on local clusters: the case of foreign owned subsidiaries in Sweden', *IIB Working Paper* 02/11 (Stockholm: Stockholm School of Economics, Institute of International Business, 2002).

Nelson, R., 'The simple economics of basic scientific research', *Journal of Political Economy*, 67 (1959) 297–306.

OECD, *Science, Technology and Industry Outlook* (Paris: Organization for Economic Cooperation and Development, 2004).

Oxelheim, L. and Ghauri, P. (eds), *European Union and the Race for Foreign Direct Investment in Europe* (Oxford: Elsevier, 2003).

Rodríguez-Clare, A., 'Multinationals, linkages and economic development', *American Economic Review*, 86(4) (1996) 852–73.

Tavares, A.T. and Young, S., 'FDI and multinationals: patterns, impact and policies', *International Journal of the Economics of Business*, 12(1) (2005) 13–6.

Tsai, W. and Ghoshal, S., 'Social capital and value creation: the role of intrafirm networks', *Academy of Management Journal*, 41 (1998) 464–76.

UNCTAD, *World Investment Report 2001: Promoting Linkages* (New York and Geneva: United Nations, 2001).

Index